IET TRANSPORTATION SERIES 5

Sliding Mode Control of Vehicle Dynamics

Other related titles:

Volume 1 **Clean mobility and intelligent transport systems** M. Fiorini and J.-C. Lin (Editors)
Volume 2 **Energy systems for electric and hybrid vehicles** K.T. Chau (Editor)
Volume 3 **Low carbon mobility for future cities: Principles and applications** Hussein Dia (Editor)
Volume 7 **Evaluation of intelligent road transportation systems: Methods and results** M. Lu (Editor)
Volume 38 **The electric car** M.H. Westbrook
Volume 45 **Propulsion systems for hybrid vehicles** J. Miller
Volume 79 **Vehicle-to-grid: Linking electric vehicles to the smart grid** J. Lu and J. Hossain (Editors)

Sliding Mode Control of Vehicle Dynamics

Edited by
Antonella Ferrara

The Institution of Engineering and Technology

Published by The Institution of Engineering and Technology, London, United Kingdom

The Institution of Engineering and Technology is registered as a Charity in England & Wales (no. 211014) and Scotland (no. SC038698).

First published 2017

The Institution of Engineering and Technology
Michael Faraday House
Six Hills Way, Stevenage
Herts SG1 2AY, United Kingdom

www.theiet.org

British Library Cataloguing in Publication Data
A catalogue record for this product is available from the British Library

ISBN 978-1-78561-209-1 (hardback)
ISBN 978-1-78561-210-7 (PDF)

Typeset in India by MPS Limited

Contents

About the editor **ix**
Preface **xi**

1 Introduction to sliding mode control **1**
Jorge Dávila, Leonid Fridman, and Antonella Ferrara

1.1 Historical background 1
1.2 Definitions of solutions 3
1.2.1 Filippov definition 3
1.2.2 Equivalent control 5
1.3 Sliding manifold design 6
1.4 First-order sliding-modes 10
1.4.1 Conventional sliding-modes 10
1.4.2 Integral sliding mode 12
1.4.3 Integral sliding modes for linear systems 14
1.5 Second-order sliding-modes 17
1.5.1 The concept of high order sliding-mode 17
1.5.2 Twisting second-order sliding-modes algorithm 19
1.5.3 Super-twisting second-order sliding-modes algorithm 21
1.5.4 Suboptimal second-order sliding-mode algorithm 24
1.6 Robust exact arbitrary order differentiator 26
1.7 Conclusions 29
Acknowledgments 29
References 29

2 Longitudinal vehicle dynamics control via sliding modes generation **33**
Enrico Regolin, Gian Paolo Incremona, and Antonella Ferrara

2.1 Introduction 33
2.2 TC and ABS control systems 35
2.2.1 Traction control systems 35
2.2.2 Antilock braking systems 36
2.2.3 Overview of sliding mode applications in wheel-slip control 37
2.3 Modeling of vehicle longitudinal dynamics 38
2.3.1 Longitudinal tire model 39
2.3.2 Vehicle model 44

2.4 Sliding mode observer for tire–road friction and velocity 46
2.4.1 Sliding mode tire–road friction observer 46
2.4.2 Observer evaluation 49
2.5 Sliding mode control algorithms for wheel-slip control 53
2.5.1 Integral sliding mode control 54
2.5.2 Higher order sliding mode control 55
2.5.3 Suboptimal second order sliding mode control 56
2.5.4 Integral suboptimal second order sliding mode control 57
2.5.5 Super twisting sliding mode control 57
2.5.6 Simulations 58
2.6 Sliding mode low level actuation in conventional ABS 62
2.6.1 HIL braking system configuration 63
2.6.2 Low level control algorithm 67
2.6.3 Low level control results 69
2.7 Conclusions 71
Acknowledgments 71
References 71

3 Sliding mode control of traction and braking in two-wheeled vehicles 77
Mara Tanelli, Matteo Corno, and Antonella Ferrara

3.1 Introduction 77
3.2 State of the art 78
3.2.1 Braking control 81
3.2.2 Traction control 85
3.3 Control-oriented modeling of the longitudinal dynamics of two-wheeled vehicles 88
3.3.1 The complete motorcycle traction dynamics 89
3.3.2 The complete motorcycle braking dynamics 96
3.4 The SOSM control approach 102
3.4.1 Preliminaries 103
3.5 Traction and braking control design 104
3.5.1 Traction controller design 104
3.5.2 Braking controller design 107
3.6 Analysis of the closed-loop performance 108
3.6.1 Traction control 108
3.6.2 Braking control 111
3.7 Concluding remarks 114
References 115

4 Lateral vehicle dynamics control via sliding modes generation 121
Enrico Regolin, Dzmitry Savitski, Valentin Ivanov, Klaus Augsburg, and Antonella Ferrara

4.1 Vehicle modeling and problem formulation 121
4.1.1 Dynamic equations 122

4.1.2 Cornering forces 123
4.1.3 Yaw moment generation 126
4.1.4 Vehicle steering properties 128
4.1.5 Control problem formulation 131
4.2 Yaw rate control survey 132
4.2.1 Vehicle stability control: yaw rate and side slip angle 132
4.2.2 Brake-based stability control: example 136
4.2.3 Torque-based stability control: example 138
4.3 Vehicle stability control via sliding mode control 139
4.3.1 Overall control structure 140
4.3.2 Reference model 141
4.3.3 Feedforward control 142
4.3.4 Second-order sliding mode control design 143
4.3.5 Integral sliding mode control design 145
4.4 Controllers assessment 147
4.4.1 Vehicle specifications and model validation 147
4.4.2 Simulation results 148
4.5 Conclusions 154
Acknowledgment 155
References 155

5 Stability control of heavy vehicles **159**
Hocine Imine and Leonid Fridman

5.1 Introduction 159
5.1.1 Main risks of accidents related to HVs 160
5.1.2 State of the art review 163
5.1.3 Main contributions of the chapter 166
5.2 Vehicle model description 166
5.3 Stability control approach 171
5.3.1 Sliding mode observer design 171
5.3.2 Steering control 174
5.4 Validation results 179
5.4.1 Rollover avoidance results 181
5.4.2 Lane keeping results 183
5.5 Conclusion 187
Acknowledgement 187
References 188

6 Sliding mode approach in semi-active suspension control **191**
Dzmitry Savitski, Dmitrij Schleinin, Valentin Ivanov, and Klaus Augsburg

6.1 Introduction 191
6.2 Semi-active suspension system 193
6.3 Active suspension system 196

6.4 Methods on suspension control 197
6.4.1 Vehicle modelling 197
6.4.2 Single degree of freedom model 198
6.4.3 Quarter car model 201
6.4.4 Half car model 204
6.4.5 Control methods overview 205
6.5 Semi-active suspension control using sliding mode approach 210
6.5.1 Problem statement 210
6.5.2 Vehicle dynamics model 211
6.5.3 Actuator model 212
6.5.4 Overall control specifications 215
6.5.5 High-level controller 215
6.5.6 Allocation of the control demand 218
6.5.7 Low-level controller 219
6.5.8 Results and discussions 220
6.6 Conclusions 225
Acknowledgements 225
References 226

7 Observer-based parameter identification for vehicle dynamics assessment 229
Robert Tafner, Markus Reichhartinger, and Martin Horn

7.1 Introduction 230
7.1.1 Modeling of selected vehicle dynamics 234
7.1.2 Chassis roll motion 238
7.1.3 Model parameters and sensor setup 241
7.2 State-of-the-art review 242
7.3 Contributions 247
7.3.1 Robust lateral dynamics parameter estimator 248
7.3.2 Robust roll dynamics parameter estimator 254
7.3.3 Simulation-based concept evaluation 263
7.3.4 Experimental results 274
7.4 Conclusions 281
Acknowledgment 282
References 282

Index 289

About the editor

Antonella Ferrara is Full Professor of Automatic Control at the University of Pavia. Her research activities are focused on sliding mode control, automotive control, and robotics; she has authored or co-authored more than 300 papers on these topics. She was Associate Editor of several scientific international journals. At present, she is Associate Editor of the *IEEE CSS Control Systems Magazine and of the International Journal of Robust and Nonlinear Control,* as well as member of the program committee of numerous international conferences. She served as Coordinator of the Pavia Research Unit in the European project PROTECTOR on ADAS and automatic control of vehicles, and is Principal Investigator in the European project ITEAM on vehicle dynamics control. She is Senior Member of the IEEE Control Systems Society, and member of several related Technical Committees. She was chair of the IEEE CSS Women in Control Standing Committee.

Preface

Starting to give lectures on vehicle dynamics control in the post-bachelor Master Course on 'Design and Development of Vehicle Dynamics Engineering', organized at The University of Pavia since 2016, I realized that a book on advanced automatic control of vehicle dynamics, specifically of variable structure or sliding mode type, was missing. So, the idea of proposing a book which could be of interest to both undergraduate and post-graduate students, as well as to researchers and practitioners came to my mind. The book that has come out is something in between a monography and a conventional collection of chapters. Different authors have contributed to the book writing, but each of them was committed to respect a common format for the chapters. Each chapter contains the problem formulation and a general overview of the physical aspects of the considered automotive control problem. Then, each chapter provides a rather general survey of the relevant literature on the chapter topic. The last part of each chapter, finally, reports the authors' contributions on the specific topic, describing solutions and results included in authored publications.

The common factor of the last part of the book chapters is the use of sliding mode control as a methodology to solve the considered control problems. This methodology results in being particularly appropriate to be used in the automotive field by virtue of its intrinsic robustness versus disturbances and modelling inaccuracies, which is accompanied by an appreciable ease of implementation.

By respecting the prescribed format, all the authors have contributed to the logical didactical build-up of the book. The chapters are of mathematical nature, dealing with the theoretical aspects in a clean and rigorous way, but they also include simulation and experimental results to better illustrate the theory. When possible, also some coverage of the advanced technology used to implement the control algorithms is provided.

The present book covers the control of longitudinal, lateral and vertical dynamics of four-wheeled vehicles. Conventional (combustion driven) vehicles and fully electric vehicles are considered, also making reference to advance control approaches such as torque vectoring. In addition, a chapter is devoted to motorcycles dynamics control, while another to heavy vehicles control. The book also includes a chapter devoted to present and discuss advanced estimation and observation methods for wheeled vehicles. Also these methods are based on the theory of sliding mode control.

For the readers' convenience, the book begins with a chapter which is an introduction to sliding mode control. It reviews all the concepts and algorithms used throughout the book to deal with the considered vehicle dynamics control problems.

This introductory section provides a brief summary of the main concepts and findings developed by the community of variable structure and sliding mode control researchers in the last two decades. This part can be used as a stand-alone reading, since it is sufficiently general to be of interest also to the readers who are not interested in the automotive application.

I think that this book provides a sufficiently comprehensive picture of the recent and present research in sliding mode control of vehicle dynamics and I am grateful to all the authors for accepting of being involved in the small venture that writing a book always is, especially by many hands. I hope that the readers can benefit from our joint effort.

Antonella Ferrara

Chapter 1
Introduction to sliding mode control

Jorge Dávila[1], *Leonid Fridman*[2], *and Antonella Ferrara*[3]

This chapter summarizes the basic concepts used in the design of sliding mode controllers, from the definition of conventional sliding set, and the main concept of sliding motion, to the design of the advanced robust exact high-order sliding modes differentiator. These pages describe the basis on which the methodologies presented along the book are designed, with this aim a short historical background is presented in Section 1.1, the main concepts related to the sliding mode theory are presented in Section 1.2. Section 1.3 describes the design process applied to the manifold, which restricts the movement of the state trajectories during the sliding motion. The basis of the design of standard sliding mode controllers are presented in Section 1.4. The main second-order sliding-mode algorithms are presented in Section 1.5 and the robust-exact high-order sliding-mode differentiator is described in Section 1.6.

1.1 Historical background

The history of Variable Structure Systems dates from the early fifties as has been described in [1]. By 1980, the main part of classical Sliding Mode Control (SMC) theory had been finished and later reported in Prof. Utkin's monograph in Russian in 1981 (English version [2]). In this monograph (see also the accomplished result by Drakunov et al. in [3]), the two-step procedure for SMC design was clearly stated:

1. Sliding surface design;
2. Discontinuous (relay or unit) controllers ensuring the sliding motion.

The main advantages provided by the Sliding Mode Controllers were the following:

(a) theoretically exact compensation (insensitivity) with respect to bounded matched uncertainties;

[1]Instituto Politécnico Nacional (IPN) – ESIME Ticomán, Mexico City, Mexico
[2]Universidad Nacional Autónoma de México – Facultad de Ingeniería, Mexico City, Mexico
[3]Dipartimento di Ingegneria Industriale e dell'Informazione, University of Pavia, Pavia, Italy

(b) reduced order dynamics when the sliding surface has been reached, that is when the sliding motion has started;
(c) finite-time convergence to the sliding surface.

However, the following disadvantages appear: chattering; insensitivity only with respect to matched perturbations (perturbations that affects the same coordinates than the control); the sliding variables converge in finite-time, however the state variables only converge asymptotically; non-ideal closed-loop performance in presence of parasitic dynamics, discretization and noises.

By the early 80s, the control community had understood that the main disadvantage of SMC is the "chattering" effect [2]. It has been shown that this effect is mainly caused by unmodeled cascade dynamics that increase the system's relative degree and perturb the ideal sliding mode existing in the system [4].

As one of the ideas to overcome the chattering problem in the sliding mode controllers, the Second Order Sliding Mode (SOSM) concept was introduced by Prof. Levant (Levantovskii). The first and simplest SOSM algorithm is the so-called "twisting algorithm" (TA), designed for a relative degree two system. With the use of the TA, the *sliding surface design is no longer needed for the one degree of freedom mechanical systems*. Moreover, the twisting algorithm collapses *the dynamics of such systems to the origin*. Even when the use of the twisting algorithm in a system with relative degree one allows to compensate exactly a (Lipschitz continuous) disturbance by means of an absolutely continuous control, this approach requires the derivative of the sliding variable. At the end of the 80's, the main issue was the question about if it was really possible to generate a continuous control without the use of derivatives.

The super-twisting algorithm presented in the early nineties in [5], provided a continuous control signal, generated without the use of the derivative of the sliding variable, and able of maintain a SOSM for first order systems with Lipschitz bounded uncertainties/disturbances. This last property of the super-twisting algorithm allowed to construct a robust exact sliding mode differentiator [6] and gave further impetus to the development of the mathematical theory and applications of SOSM algorithms.

Second Order Sliding Modes attracted the full attention of the international control community after a presentation at the Third IEEE Workshop on Variable Structure and Lyapunov Theory [7] and the publication of the first tutorial paper on High Order Sliding Modes (HOSMs) [8]. Since 1996, the number of publications on SOSM theory and its applications have grown exponentially, generally through the efforts of Prof. Bartolini and his team (see, for instance, [9,10] with a similar idea). The results of Prof. Bartolini's group are based on another SOSM controller that does not (explicitly) need derivatives, the Sub-Optimal Algorithm. It is important to remark that this SOSM algorithm has been used to solve numerous relevant control problems, see for instance [11] or [12] and the references therein.

Second order sliding mode controllers have been successfully applied to solve several control problems, however they are restricted to the class of second order dynamics. In 2001, the first arbitrary order SM controller was introduced in [13]. Such controller allowed to solve the finite-time output stabilization and exact matched perturbation/uncertainties compensation problem for an output with an arbitrary relative degree. For further information see [14].

1.2 Definitions of solutions

Systems driven by SMC can be typically described by nonlinear discontinuous differential equations. Since the classical results on the solution of differential equations do not hold in case of SMC, two different approaches will be analyzed: the *Filippov's method* and the *equivalent control approach* [2,15,16].

Consider the following Single-Input Single-Output (SISO) system of order n:

$$\dot{x}(t) = f(x, u, t) \tag{1.1}$$

with $x \in \mathbb{R}^n$ being the state vector and $u \in \mathbb{R}$ the control signal. Let assume that $f(\cdot)$ is differentiable with respect to x and absolutely continuous with respect to time. Let $\sigma(x)$, called the sliding variable, be the output variable and assume that it is differentiable.

Let define the sliding manifold, or surface, as the following restriction in the state space:

$$S = \{x \, : \, \sigma(x) = 0\} \tag{1.2}$$

An ideal sliding mode take place on the manifold (1.2) if the state $x(t)$ evolves in time satisfying $\sigma(x(t)) = 0$ for all $t > t_r$, for some finite $t_r \in \mathbb{R}^+$.

During the sliding motion, the system dynamics "collapses" and is restricted to move along the manifold (1.2), this restricted movement implies the reduction of the dynamical order.

If the control $u = u(x)$ in equation (1.1) is discontinuous, then the resulting closed-loop state-equation takes the following form:

$$\dot{x}(t) = f^c(x) \tag{1.3}$$

where $f^c : \mathbb{R}^n \to \mathbb{R}^n$ is discontinuous with respect to the state vector.

1.2.1 *Filippov definition*

The classical theory of differential equations is not applicable due the discontinuous nature of $f^c(x)$. However, the solution concept, proposed by Filippov [15], for equations with discontinuous right-hand sides can be applied to construct a solution that is obtained from approaching the point of discontinuity from different directions.

To prove that (1.3) determines an unique evolution of the system when is restricted to the sliding manifold S, let us highlight that the closed-loop system on $\sigma(x) = 0$ coincides with the solution of the following equation:

$$\dot{x}(t) = \alpha f^+ + (1 - \alpha) f^- = f^0 \tag{1.4}$$

where f^+ and f^- are the limits of $f^c(x)$ as the point of discontinuity is approached from opposite sides of the tangent to the manifold S, $0 \leq \alpha \leq 1$ is a scalar chosen such that the resulting vector field f^0 is tangential to S, see Figure 1.1.

This new representation allows us to represent equation (1.4) as a differential equation whose right-hand side is defined by the convex set:

$$F(x) = \{\alpha f^+ + (1 - \alpha) f^- \, : \, \text{for all } \alpha \in [0 \quad 1]\} \tag{1.5}$$

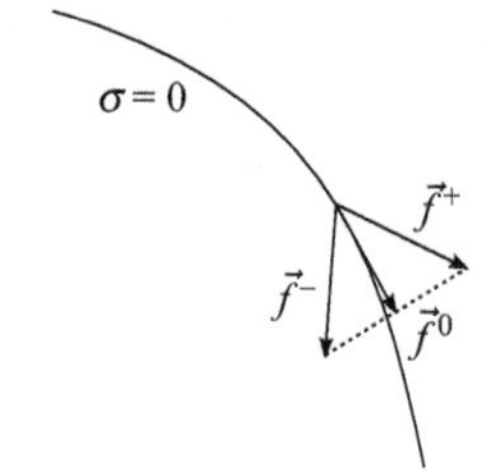

Figure 1.1 Schematic representation of Filippov's definition

Thus

$$\dot{x} \in F(x) \tag{1.6}$$

In sliding mode, the equalities $\sigma(x) = \dot{\sigma}(x) = 0$ are satisfied. Thus, it follows that the next equality should also be satisfied:

$$G_\sigma \cdot f^0 = 0 \tag{1.7}$$

where G_σ is the gradient of σ. By solving with respect to α, one obtain an unique value:

$$\alpha = \frac{G_\sigma \cdot f^-}{G_\sigma \cdot (f^- - f^+)}. \tag{1.8}$$

Therefore

$$\dot{x}(t) = \frac{G_\sigma \cdot f^- f^+ - G_\sigma \cdot f^+ f^-}{G_\sigma \cdot (f^- - f^+)} \tag{1.9}$$

Equation (1.9) holds only if $G_\sigma \cdot (f^- - f^+) > 0$, and $G_\sigma \cdot f^+ \leq 0$ and $G_\sigma \cdot f^- \geq 0$.

Let us define that a differential inclusion $\dot{x} \in F(x)$, $x \in \mathbb{R}^n$ is called a Filippov differential inclusion if the set $F(x)$ is nonempty, closed, locally bounded and upper semi-continuous.

Definition 1.1. *A differential equation $\dot{x} = f(x)$ with a locally bounded Lebesgue-measurable right-hand side has as solution in the Filippov sense, if it is replaced by a special Filippov differential inclusion $\dot{x} = F(x)$ where,*

$$F(x) = \bigcap_{\delta>0} \bigcap_{\mu N=0} \bar{co} f(O_\delta(x) \setminus N)$$

μ is the Lebesgue measure, $O_\delta(x)$ is a δ-neighborhood of x, and $\bar{co}\, M$ is the convex closure of the set M.

According to the Filippov's method, the solution of a differential equation with nonlinear discontinuous right-hand side exists and is uniquely defined on $\sigma(x) = 0$ [15]. Therefore this method represents a possible way to determine the evolution of the controlled system in sliding mode.

1.2.2 Equivalent control

Another way to undertake the analysis is by the equivalent control method, attributed to Utkin [2], which is defined as the control action necessary to maintain an ideal sliding motion on S. The idea is to exploit the equality $\dot{\sigma} = \sigma = 0$ that can be written as:

$$\dot{\sigma} = \frac{\partial \sigma}{\partial x}\frac{dx}{dt} = \frac{\partial \sigma}{\partial x} f(x, u, t) = 0 \tag{1.10}$$

Last is an algebraic equation in x, u and t, thus from this equation is possible to obtain the equivalent control signal $u_{eq}(t)$, which is the continuous control function necessary to maintain the equality:

$$\frac{\partial \sigma}{\partial x} f(x, u_{eq}, t) = 0 \tag{1.11}$$

Notice that, once the "equivalent control" signal is obtained, it can be substituted in (1.1) to obtain a continuous differential equation:

$$\dot{x} = f(x, u_{eq}, t) \tag{1.12}$$

Therefore, for initial conditions $\sigma(x(0)) = 0$ and in view of (1.10), the solution of the ordinary differential equation (1.12) represents the state trajectories in the tangential plane to the manifold S. It is remarkable that this solution coincides with the Filippov's method only for affine systems. To illustrate the obtention of this method, let us consider the following SISO system:

$$\dot{x} = f(x, t) + B(x, t)u \tag{1.13}$$

where $x \in \mathbb{R}^n$ is the state, $u \in \mathbb{R}$ is the control variable, while the output is the sliding variable $\sigma(x)$. Moreover assume that the sliding mode is enforced at the time instant $t_r \geq t_0$, so that one has $\sigma(x) = \dot{\sigma}(x) = 0$, for any $t \geq t_r$. According to the equivalent control method in [2], one can write:

$$\sigma(x) = \left[\frac{\partial \sigma}{\partial x}\dot{x}\right] = \left[\frac{\partial \sigma}{\partial x}\right]\left[f(x, t) + B(x, t)u_{\mathrm{eq}}\right] = 0 \tag{1.14}$$

where u_{eq} is the equivalent control [2]. Substituting u_{eq} in (1.13) one has

$$\dot{x} = f(x, t) + B(x, t)u_{\mathrm{eq}} \tag{1.15}$$

which is a differential equation with continuous functions describing the evolution of the system starting from $x(t_r)$ and $\sigma(t_r) = 0$.

Assume that $\frac{\partial \sigma}{\partial x}B(x, t)$ is not singular for any t and x, i.e.,

$$\det\left(\frac{\partial \sigma}{\partial x}B(x, t)\right) \neq 0$$

Thus

$$u_{\mathrm{eq}} = -\left\{\frac{\partial \sigma}{\partial x}B(x, t)\right\}^{-1}\frac{\partial \sigma}{\partial x}f(x, t). \tag{1.16}$$

Then, starting at t_0 with $\sigma(t_0) = 0$, the dynamics of the controlled system can be expressed as

$$\dot{x} = \left\{ I_n - B(x,t) \left[\frac{\partial \sigma}{\partial x} B(x,t) \right]^{-1} \frac{\partial \sigma}{\partial x} \right\} f(x,t), \tag{1.17}$$

where the right hand side is continuous and represents the corresponding equivalent evolution of the system when it is controlled by the discontinuous SMC law. Note that, in case of a linear system, $\sigma(x) = Cx$, (1.17) becomes:

$$\dot{x} = \left\{ I_n - B(x,t) \left[CB(x,t) \right]^{-1} C \right\} f(x,t) \tag{1.18}$$

with $\partial\sigma/\partial x = C$.

1.3 Sliding manifold design

SMC has been one of the most popular robust control techniques through the last 30 years. Sliding-mode controllers use high (theoretically infinite) frequency switchings to restrict the system dynamics to a reduced order one, which represents the desired behavior. The reduced order dynamics is resulting from the restriction of the natural free movement of the system to a manifold, the dynamics in the manifold is usually designed to be stable. Thus, the restricted movement on the manifold, when exists, can be seen as a sliding motion along a desired hyper-surface towards the origin, that is why this phase of movement is called sliding motion.

Let us show the features of a sliding-mode controller through a scalar system with first order dynamics. The system is given by

$$\dot{x} = f(x,t) + u \tag{1.19}$$

where $x \in \mathbb{R}$ is the state, $u \in \mathbb{R}$ is the control input and $f(x,t) : \mathbb{R} \times \mathbb{R}_+ \to \mathbb{R}$ is an uniformly bounded function of the state and time, i.e., $|f(x,t)| \leq \Gamma^+$ for a positive scalar Γ^+.

The control objective is bring to zero in a finite-time the state of the system in spite of the dynamics $f(x,t)$. To reach this goal, let introduce the control as a discontinuous function of the state, the so-called conventional SMC:

$$u = -\alpha \, \mathrm{sign}(\sigma(x)) \tag{1.20}$$

where $\sigma : \mathbb{R} \to \mathbb{R}$ is called the sliding variable, and the constant α should satisfy $\alpha > \Gamma^+$. As was mentioned before, one of the main characteristics of SMC consists in provide the possibility of bringing to zero the sliding variable in a finite-time. In this particular example, let $\sigma = x$, and for illustration purposes let $f(x,t) = 2\sin(0.5t)$. The only information required to design the controller is the value of Γ^+, or an estimation of this value, which is the upperbound of the uncertain function $f(x,t)$. Then, the gain of the controller is set equal to $\alpha = 2.01$. Setting the initial condition as $x(0) = 3$, from Figure 1.2 is possible to conclude that the sliding-mode controller guarantees the convergence of σ to zero, therefore it guarantees the existence of the sliding motion after a finite-time transient, in this case this settling time is equal to 5.17 s.

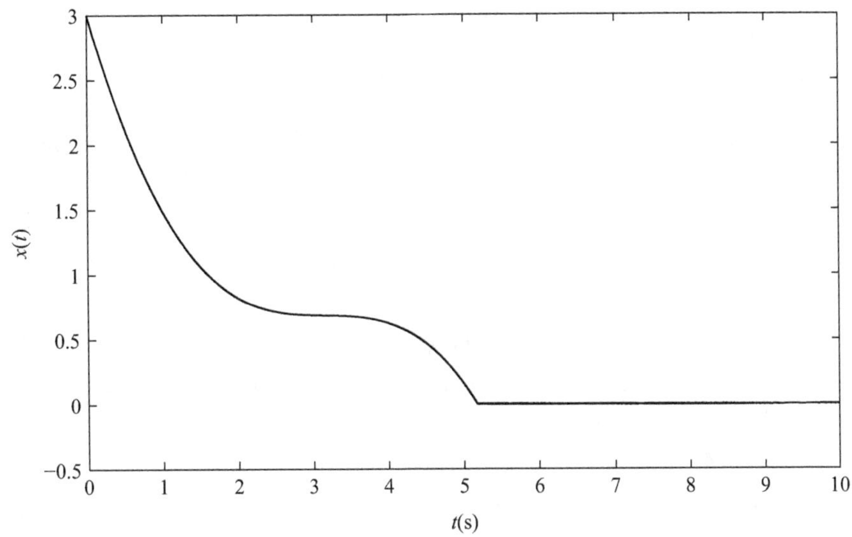

Figure 1.2 State trajectory for the system $\dot{x} = 2\sin(0.5t) + u$

The above given example illustrate well the main features of the sliding mode controllers, which are the finite-time convergence and the robustness, notice that the state of the system is equal to zero in spite of the unknown value of $f(x,t)$. However, given the selection of the sliding variable, $\sigma = x$, the state of the system is directly set to zero after the finite-time transient.

It is important to remark that in this example, the sliding motion is not appreciable given that the sliding variable is equal to x, therefore when the reaching phase ends, the equality $x = 0$ is guaranteed, then the sliding set comprises only one point.

Let us show the design of the σ variable using a more complex system, consider a mechanical system defined by the equation:

$$M(q)\ddot{q} + C(q,\dot{q})\dot{q} + G(q) = \tau + v \tag{1.21}$$

where $q \in \mathbb{R}^p$ is the vector of configuration variables, $M(q)$ is the matrix of masses and inertias, $C(q,\dot{q})$ comprises Coriolis' and centrifugal forces, $G(q)$ is the vector of conservative forces, v is the vector of external disturbances or uncertainties and τ is the vector of control forces and/or torques. Without loss of generality, let us consider the scalar case, where $p = 1$.

Introducing the state variables $x_1 = q$ and $x_2 = \dot{q}$, and assuming that $u = \tau$ is the control input, the system is written in the state space form as

$$\begin{bmatrix} \dot{x}_1 \\ \dot{x}_2 \end{bmatrix} = \begin{bmatrix} x_2 \\ M^{-1}(x_1)\left(-C(x_1,x_2)x_2 - G(x_1) + u + v\right) \end{bmatrix} \tag{1.22}$$

Let us propose the controller as the discontinuous function:

$$u = -\alpha\,\mathrm{sign}(\sigma) \tag{1.23}$$

Recall that two elements comply to the controller, the sliding variable $\sigma(x)$ and the discontinuous control which makes possible the convergence of σ to zero in finite time. The design of these elements is described in the following sentences.

First, the desired behavior is written in the form of a restriction using the sliding variable σ. In particular, let the desired behavior be chosen as a linear decreasing line with attenuation factor equals to $\lambda > 0$, the sliding variable takes the form:

$$\sigma = x_2 + \lambda x_1 \tag{1.24}$$

Second, using the derivative of σ, compute the necessary gain that guarantees the satisfaction of the condition $\sigma\dot{\sigma} < 0$.The derivative of σ is given by

$$\dot{\sigma} = M^{-1}(x_1)\left(-C(x_1, x_2)x_2 - G(x_1) + u\right) + \lambda x_2 \tag{1.25}$$

Due to the characteristic of any mechanical system, the states remain bounded, then there exists a scalar constant Γ^+ such that

$$\Gamma^+ \geq ||M^{-1}(x_1)\left(-C(x_1, x_2)x_2 - G(x_1)\right) + \lambda x_2||_\infty \tag{1.26}$$

Then, by selecting the gain α such that $\alpha > \Gamma^+$, the controller (1.23) guarantees that for all $\sigma \neq 0$:

$$\sigma\dot{\sigma} \leq \sigma\left(-\alpha \text{sign}(\sigma) + \Gamma^+\right) \leq -\left(\alpha|\sigma| - \Gamma^+\sigma\right) < 0 \tag{1.27}$$

Therefore

$$-\left(\alpha - \Gamma^+\right)|\sigma| < 0 \tag{1.28}$$

In order to numerically illustrate the SMC design, let us consider an inverted pendulum system, which is obtained by setting $q = \theta$, $M(q) = ml^2$, $G(q) = -mgl\sin\theta$, $C(q, \dot{q}) = b$, $\tau = T$ and $v = f(t)$, thus (1.21) takes the following form:

$$ml^2\ddot{\theta} = mgl\sin\theta - b\dot{\theta} + T + f(t) \tag{1.29}$$

The SMC had been tested using the numerical values $l = 1$ m, $m = 1$ kg, $b = 0.1$ N $(\text{ms})^{-1}$, $g = 9.8$ m s^{-2}, $f(t) = 1.1 + 0.5\cos(8t)$, with the initial conditions $\dot{q}(0) = 0$, $q(0) = 1.5707$, and setting the gains of the controller as $\alpha = 11$ and $\lambda = 0.5$. In Figure 1.3 the convergence of the states to the origin is shown, remarkable is that two behaviors are easily distinguishable there, these behaviors being separated by a discontinuity in $t = 1.03$ s. These two behaviors are called reaching phase, that is the period of time, comprised between $t = 0$ s and $t = 1.03$ s, in which the movement is directed towards the surface, and the sliding phase, that appears when the trajectories reach the sliding surface and the sliding movement begins.

The described movements are easily distinguishable in the phase plane and are shown in Figure 1.4. It is clear that, in this example, the sliding movement is characterized for a linear behavior, that was just the selected dynamics introduced in the sliding variable σ. It is also remarkable that the desired behavior and the exponential convergence to zero of the states are obtained in the presence of the uncertain function $f(x, t)$ that comprises a constant value and a bounded oscillation. The robustness provided by the sliding mode controller is the result of a high-frequency signal that is applied to the system, see Figure 1.5, from this figure it is clear that in the reaching phase the control signal has a constant value, after this period the control signal

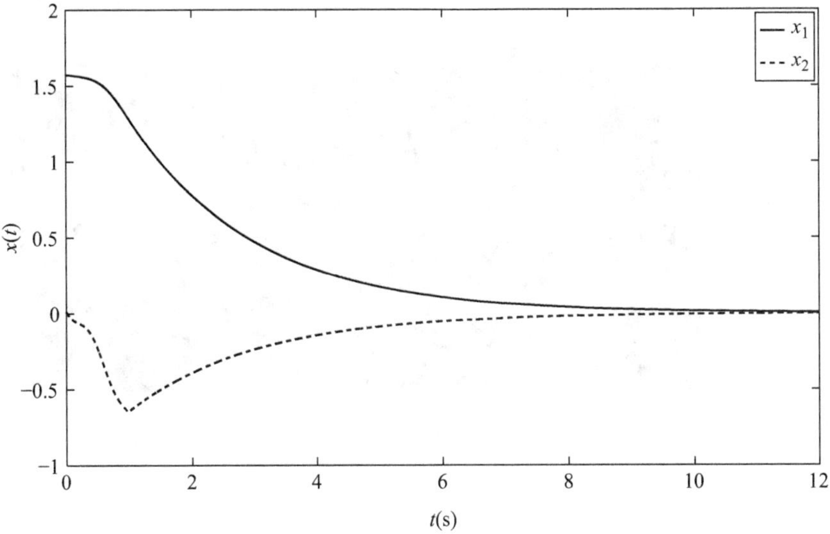

Figure 1.3 States of the mechanical system

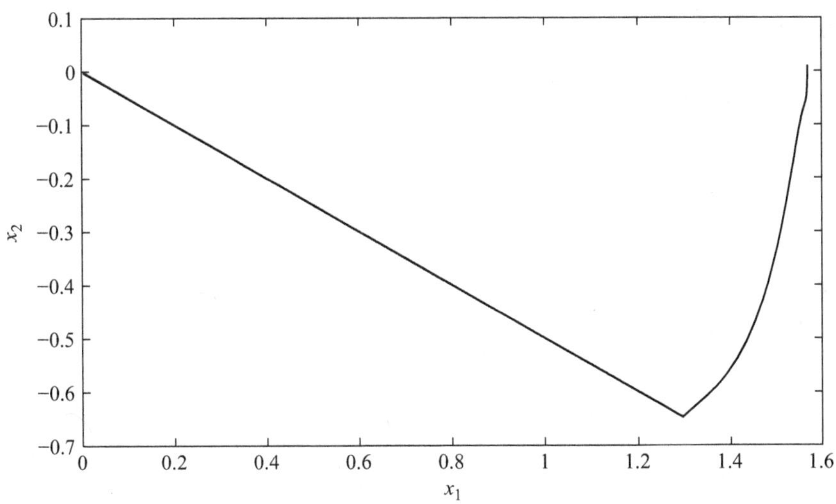

Figure 1.4 Phase plane of the controlled mechanical system

becomes a high frequency signal, this is a characteristic of the sliding motion, during this phase the system is not sensitive to bounded disturbances.

Now, the general methodology for the design of sliding mode controllers for linear systems is described in the following section.

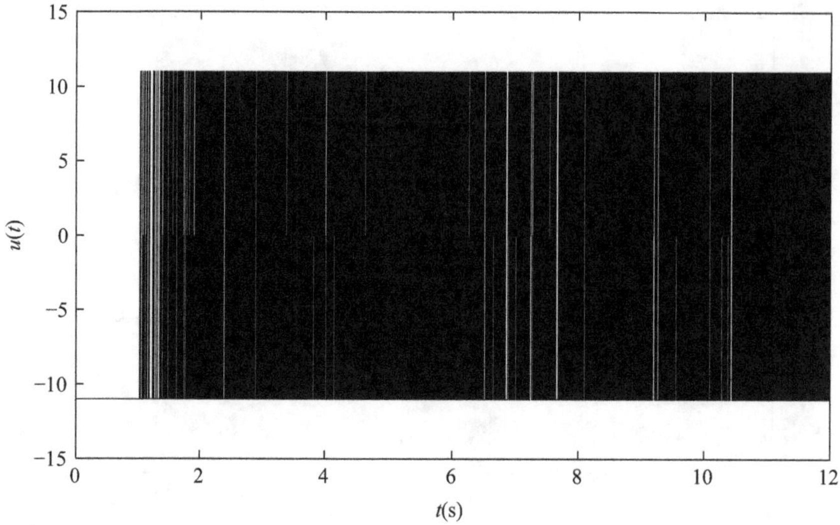

Figure 1.5 Control signal for the controlled mechanical system

1.4 First-order sliding-modes

1.4.1 Conventional sliding-modes

Consider the MIMO linear system:

$$\dot{x} = Ax + Bu + Df \tag{1.30}$$

where $x \in \mathbb{R}^n$, $u \in \mathbb{R}^m$ and $f \in \mathbb{R}^q$ are the states, control inputs and disturbances, respectively. It is assumed that B is a full rank matrix, the number of disturbances are less than or equal to the number of outputs, i.e., $q \leq m$, the pair (A, B) is controllable and there exists a vector function $\gamma(t) : \mathbb{R}_+ \to \mathbb{R}^m$ such that $Df = B\gamma(t)$, i.e., $Df \in span(B)$.

The control law takes the following form:

$$u = Kx + u_{nl} \tag{1.31}$$

where $K \in \mathbb{R}^{m\times n}$ is selected such that the matrix $\tilde{A} = A + BK$ is Hurwitz, the close loop system becomes:

$$\dot{x} = \tilde{A}x + B(\gamma(t) + u_{nl}) \tag{1.32}$$

Notice that, given that $\tilde{A}$ is Hurwitz, then the system (1.32) is stable but affected by the unknown input $\gamma(t)$. In order to design the nonlinear part of the controller u_{nl} to guarantee exponential stability of the origin, the state transformation $\bar{x} = Tx$ should be applied to the system (1.32). This transformation decouples the states that are directly affected by the unknown inputs from the rest of states. The transformation matrix T is chosen as:

$$T = \begin{bmatrix} B^{\perp} \\ B^{+} \end{bmatrix} \tag{1.33}$$

where B^{+} is the left pseudo-inverse of the matrix B and $B^{\perp}$ is the orthogonal complement of B^{+}, i.e., $B^{\perp} \in \mathbb{R}^{(n-m)\times n}$, $rank(B^{\perp}) = n - m$, $B^{+}(B^{\perp})^{T} = 0$. Under this transformation, system (1.32) becomes:

$$\begin{bmatrix} \dot{\bar{x}}_1 \\ \dot{\bar{x}}_2 \end{bmatrix} = \begin{bmatrix} \bar{A}_{11} & \bar{A}_{12} \\ \bar{A}_{21} & \bar{A}_{22} \end{bmatrix} \begin{bmatrix} \bar{x}_1 \\ \bar{x}_2 \end{bmatrix} + \begin{bmatrix} 0 \\ I_m \end{bmatrix} (u_{nl} + \gamma(t)) \tag{1.34}$$

Using the above described transformation, the nonlinear term that is just the SMC term is given by

$$u_{nl} = -\alpha \text{sign}(\sigma) \tag{1.35}$$

where α is a positive constant which satisfies $\alpha \geq ||\gamma(t)||_{\infty}$ and

$$\sigma = \begin{bmatrix} -L & I_m \end{bmatrix} Tx \tag{1.36}$$

with $L \in \mathbb{R}^{m\times(n-m)}$ selected such as

$$\bar{A}_{11} + \bar{A}_{12}L \tag{1.37}$$

is a Hurwitz matrix. The vector form of the nonlinear function $\text{sign}(\sigma)$ is defined as

$$\text{sign}(\sigma) = \begin{bmatrix} \text{sign}(\sigma_1) \\ \text{sign}(\sigma_2) \\ \vdots \\ \text{sign}(\sigma_m) \end{bmatrix} \tag{1.38}$$

where $\text{sign}(\sigma_i)$ is the signum of the ith row of σ. This SMC guarantees the convergence of σ to zero in finite-time, i.e., in finite time $\sigma = -Lx_1 + x_2 = 0$ and therefore $x_2 = Lx_1$. Then, in view of the design of L, given in (1.37), the sliding dynamics $\dot{x}_1 = (\bar{A}_{11} + \bar{A}_{12}L)x_1$ is exponentially stable. Therefore this controller guarantees the finite-time convergence of σ to zero, and the subsequent exponential convergence to zero of the sliding dynamics.

With the purpose of illustrating the applicability of the methodology, let us consider the linear system described by the state space form:

$$\dot{x} = \begin{bmatrix} 1 & 2 & 3 & 1 & 1 \\ 2 & 3 & 4 & 1 & 0 \\ 0 & 1 & 1 & 0 & 0 \\ 0 & 0 & 0 & 1 & 2 \\ 1 & 1 & 0 & 0 & 0 \end{bmatrix} x + \begin{bmatrix} 1 & 0 \\ 0 & 0 \\ 1 & 1 \\ 0 & 0 \\ 0 & 0 \end{bmatrix} u + \begin{bmatrix} 0 \\ 0 \\ 1 \\ 0 \\ 0 \end{bmatrix} w(t) \tag{1.39}$$

for simulation purposes let us consider that $w(t) = 0.5 + 0.5\sin(t) + w_2(t)$ where $w_2(t)$ is a sawtooth signal with amplitude 0.5 and frequency 0.5 Hz starting at $t = 5$ s. The controller, according to (1.31) and (1.47), is designed such that the closed loop

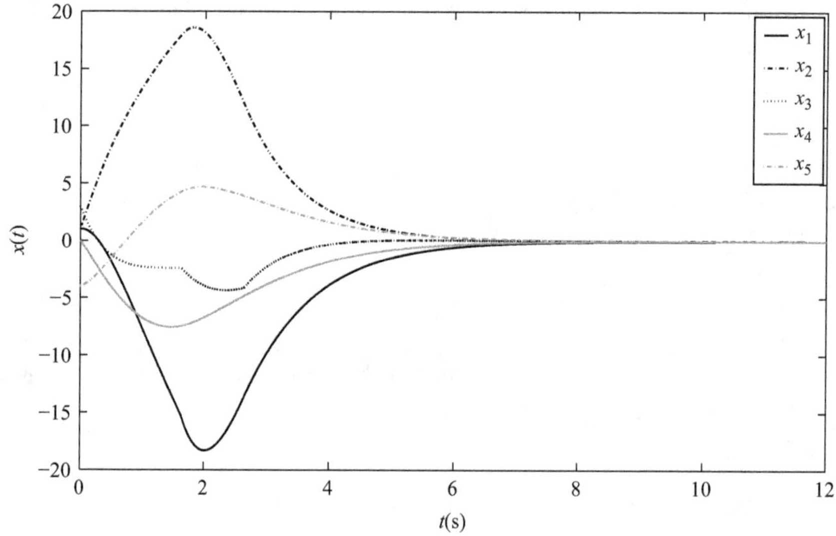

Figure 1.6 Linear system state driven by SMC

poles are set to -1, -0.5, -0.7, -0.6, -0.8 and the sliding dynamics possesses the closed loop poles -1, -1.1, -1.2, this goal is reached with the following set of gains:

$$K = \begin{bmatrix} -6.2106 & -8.2171 & -7.0155 & -4.5578 & -5.7567 \\ 6.6406 & 6.9316 & 3.6261 & 5.0132 & 7.6815 \end{bmatrix}$$

$$L = \begin{bmatrix} 1 & -2.2 & -3.2 \\ -0.4750 & 3.0473 & 4.7974 \end{bmatrix}$$

$$\alpha = 5 \tag{1.40}$$

The state trajectories are presented in Figure 1.6, where it is evident that the exponential convergence of the states to zero after the time $t = 2.63$ s, time in which the sliding variable σ_2 reaches zero, see Figure 1.7. Notice that the sliding variables reach the surface at different times.

The control signal that provides the robust stabilization of the system is presented in the upper image in Figure 1.8. The disturbance affecting the system is shown in the lower image in Figure 1.8, notice that the disturbance is always affecting the system.

1.4.2 Integral sliding mode

Recently, Integral Sliding Mode (ISM) methods have been studied to generate ideal sliding modes of the controlled system on the so-called *integral sliding manifold* starting from the initial time instant t_0. This approach allows the elimination of the so-called reaching phase. The other benefit of ISM is to protect the nominal controllers, which are already predesigned for the systems without uncertainties, adding

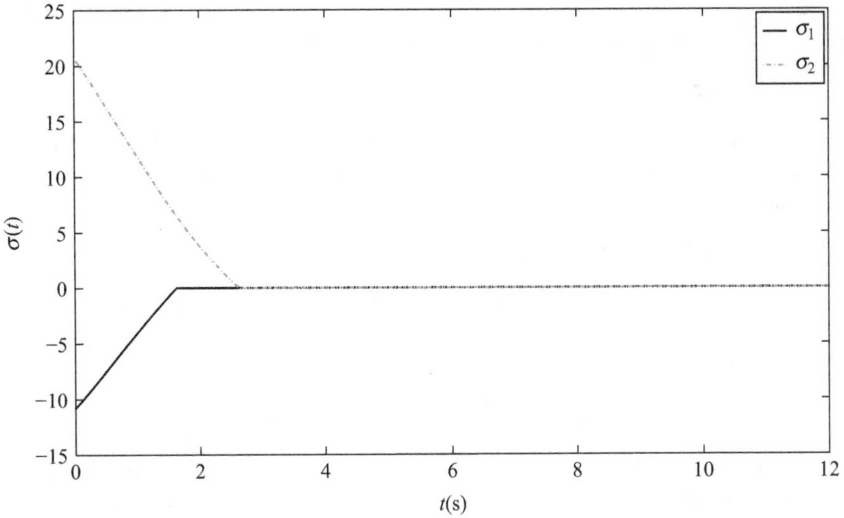

Figure 1.7 Sliding variables of the linear system driven by SMC

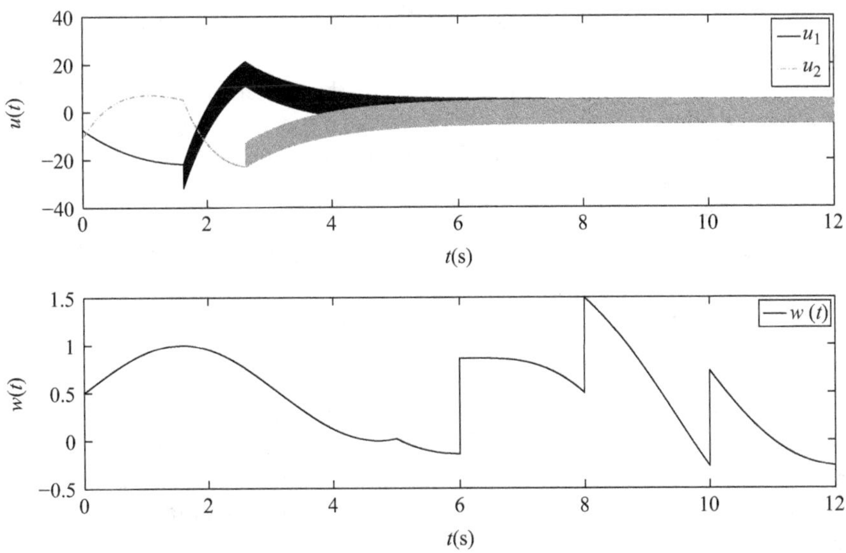

Figure 1.8 Control signal of the linear system driven by SMC (above). Disturbance $w(t)$ applied to the linear system driven by SMC (below)

robustness against matched disturbances. A sliding mode is defined *integral* if the system, while in sliding, is of the same order as the original system, see [17].

Consider again the SISO system:

$$\dot{x} = f(x,t) + B(x,t)v \tag{1.41}$$

where $x \in \mathbb{R}^n$ is the state and $u \in \mathbb{R}$ is the control variable. Then, the ISM approach requires to split the control variable $v(t)$ into two parts, i.e.,

$$v(t) = u(t) + u_{\mathrm{ISM}}(t) \tag{1.42}$$

where $u(t)$ is generated by a suitable controller designed relying on the nominal plant, and $u_{\mathrm{ISM}}(t)$ is a discontinuous control action designed to compensate the uncertainties affecting the system. The *integral sliding manifold* is instead defined as

$$\Sigma(t) = \sigma(t) - \varphi(t) = 0 \tag{1.43}$$

where Σ is an auxiliary sliding variable, σ is the actual sliding variable, and φ is a transient function given by

$$\varphi(t) = \sigma(t_0) + \int_{t_0}^{t} \frac{\partial \sigma}{\partial x}\{f(x,\zeta) + B(x,\zeta)u\}d\zeta \tag{1.44}$$

with the initial condition $\varphi(t_0) = \sigma(x(t_0))$. Since $\varphi(t)$ is chosen such that $\varphi(t_0)$ is equal to the sliding variable σ at t_0, this implies that, by applying the discontinuous control law:

$$u_{\mathrm{ISM}}(t) = -U\mathrm{sign}(\Sigma(t)), \tag{1.45}$$

the controlled system is in the sliding mode on the manifold $\Sigma(t) = 0$ since the initial time instant [2,17].

Note that, as shown in [17], in order to reduce the chattering phenomenon, which is the major drawback of traditional sliding mode, it is possible to use the equivalent value $u_{\mathrm{ISM_{eq}}}(t)$ of the discontinuous control (see [2], for a definition). Since it depends on the uncertain terms affecting the system, it can be obtained at the output of a first-order linear filter with the real discontinuous control as input. In this case, in order to ensure $\Sigma(t) = 0, \forall t \geq t_0$, the transient function φ must be redesigned as follows:

$$\varphi(t) = \sigma(t_0) + \int_{t_0}^{t} \frac{\partial \sigma}{\partial x}\{f(x,\zeta) + B(x,\zeta)(u + u_{\mathrm{ISM_{eq}}} - u_{\mathrm{ISM}})\}d\zeta, \tag{1.46}$$

with the initial condition always being equal to $\varphi(t_0) = \sigma(x(t_0))$.

This methodology has been widely used for the compensation of linear systems [18] with uncertain conditions, for example, the methodology has been applied in nonlinear systems affected by matched and unmatched uncertainties [19], in the context of robust model predictive control [20] and has been applied to different applications (see, for example, [21,22]).

1.4.3 Integral sliding modes for linear systems

Let us consider again the linear system (1.30), used in Section 1.4.1, and let us design the controller by following equations (1.31)–(1.34), the reaching phase of the sliding motion can be suppressed by the addition of an integral term in the surface. With this aim, the nonlinear term now takes the following form:

$$u_{nl} = -\alpha\mathrm{sign}(\Sigma) \tag{1.47}$$

where α is a positive constant which satisfies $\alpha \geq ||\gamma(t)||_\infty$ and

$$\Sigma = \sigma + \varphi(t) \tag{1.48}$$

is now the integral sliding variable with:

$$\sigma = \begin{bmatrix} -L & I_m \end{bmatrix} Tx \tag{1.49}$$

and $L \in \mathbb{R}^{m\times(n-m)}$ be selected such as $\bar{A}_{11} + \bar{A}_{12}L$ is a Hurwitz matrix; $\varphi(t)$ is a function of the initial conditions x_0 and the state given as follows:

$$\varphi(t) = \begin{bmatrix} -L & I_m \end{bmatrix} T \left(x_0 + \tilde{A} \int_{t_0}^{t} x(\zeta)d\zeta \right) \tag{1.50}$$

where sign(σ) represents the vector form of the signum function given in (1.38).

Notice that now the variable sigma, which was used as the switching function in Subsection 1.4.1, is now part of (1.48), however it still defines the sliding surface.

Let us illustrate the design methodology using the numerical example presented in Section 1.4.1. With this aim, the sliding variable, which now takes the form presented in (1.48) is given by:

$$\Sigma = \begin{bmatrix} 1 & 1 & 0 & 2.2 & 3.2 \\ -1 & -0.475 & 1 & -3.047 & -4.797 \end{bmatrix} \left(x(t) - x_0 \right.$$
$$\left. - \begin{bmatrix} -5.211 & -6.217 & -4.016 & -3.558 & -4.757 \\ 2 & 3 & 4 & 1 & 0 \\ 0.430 & -0.286 & -2.389 & 0.455 & 1.925 \\ 0 & 0 & 0 & 1 & 2 \\ 1 & 1 & 0 & 0 & 0 \end{bmatrix} \int_0^t x(\zeta)d\zeta \right)$$

The exponential convergence of the state vector is shown in Figure 1.9. Notice that the state trajectories are continuous, they do not exhibit switchings. The last is a consequence of the new term $\varphi(t)$, introduced in the surface to suppress the reaching phase, it means that no transition between reaching and sliding phases does exist in the motion, notice as well that any discontinuity can be appreciated in Figure 1.10. Finally, in accordance with the no existence of reaching phase, the control signals presented in the upper graph of Figure 1.11 exhibits from the first instant of time the high-frequency behavior that is characteristic of the sliding mode controllers when the surface has been reached. Moreover, this behavior allows us to guarantee the insensitivity to the matched disturbances from the first instant of time.

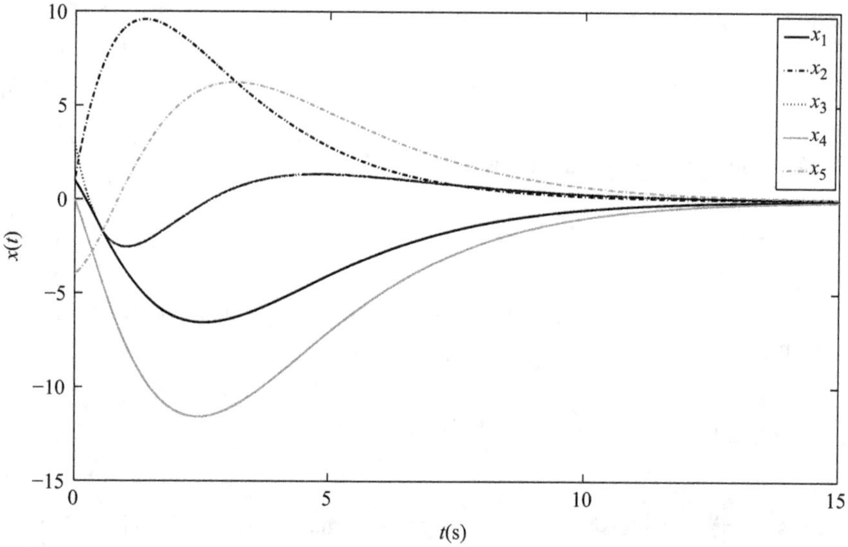

Figure 1.9 Linear system state driven by ISMC

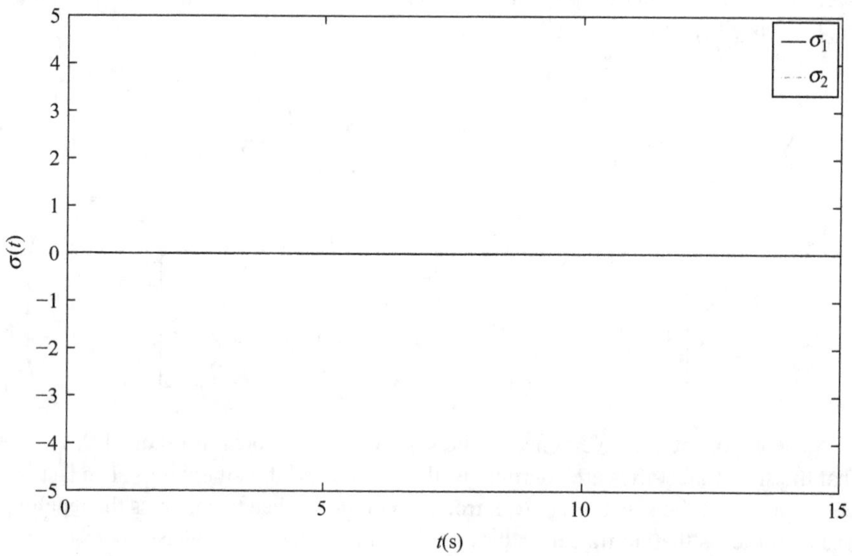

Figure 1.10 Sliding variables of the linear system driven by ISMC

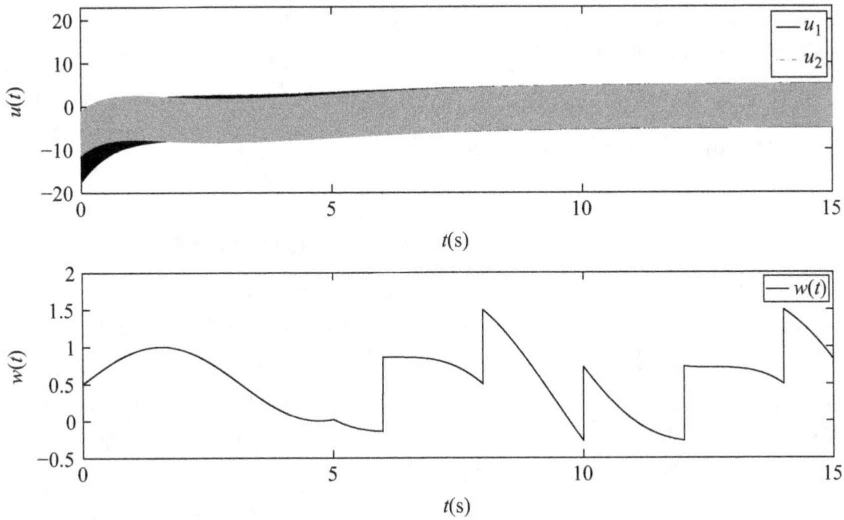

Figure 1.11 Control signal of the linear system driven by ISMC (above). Disturbance w(t) applied to the linear system driven by ISMC (below)

1.5 Second-order sliding-modes

1.5.1 The concept of high order sliding-mode

In a similar way as presented for the first-order sliding mode, and using the *Filippov's definition* [15], now it is necessary to introduce an analogous definition for the rth-order sliding mode.

Let us consider the system:

$$\dot{x} = a(t,x) + b(t,x)u \tag{1.51}$$

$$\sigma = \sigma(x,t) \tag{1.52}$$

where $x \in \mathbb{R}^n$ is the state, $u \in \mathbb{R}$ is the control input and the function $\sigma \in \mathbb{R}$ is an output to be exactly stabilized in finite time, i.e., $\sigma = 0$. Let us consider that the first r total derivatives of σ are continuous functions of the state variable, define the relative degree of σ as the smallest constant value r such that the following equalities are satisfied:

$$\frac{\partial}{\partial u}\sigma^{(j)} = 0,\ i = 1,\ldots,r-1; \quad \frac{\partial}{\partial u}\sigma^{(r)} \neq 0 \tag{1.53}$$

Then, the rth derivative of the output σ satisfies an equation of the form:

$$\sigma^{(r)} = h(x,t) + g(x,t)u \tag{1.54}$$

where the nonlinear functions $h(x,t)$ and $g(x,t)$ are bounded by known positive constants C, K_m and K_M as

$$h(t,x) \in [-C, \quad C], \quad g(t,x) \in [K_m, \quad K_M], \tag{1.55}$$

Let the rth-order sliding set $\mathscr{S}_r$ be defined as

$$\mathscr{S}_r = \{x \mid \sigma(x) = \dot{\sigma}(x) = \cdots = \sigma^{(r-1)}(x) = 0\} \tag{1.56}$$

Therefore, the following definitions, which are equivalent, can be stated.

Definition 1.2. *The set $\mathscr{S}_r$ is said to contain an rth order sliding mode if it consists of integral curves of the system in Filippov's sense and there exists a vicinity $\mathscr{N} \subseteq \mathbb{R}^n$ of $\mathscr{S}_r$ in which the shift operator is not invertible.*

Definition 1.3. *The set $\mathscr{S}_r$ is said to contain an rth order sliding mode if there exists a vicinity $\mathscr{N} \subseteq \mathbb{R}^n$ of $\mathscr{S}_r$ such that for any initial condition in $\mathscr{N}$ the trajectories of the system (in Filippov's sense) converge to $\mathscr{S}_r$ in finite-time.*

In view of the convergence of the trajectories to the sliding set $\mathscr{S}_r$, the derivative of $\sigma^{(r-1)}$ should be equal to zero to guarantee that the movement on the surface is maintained. Then, system (1.54) has to satisfy the following equality:

$$0 = h(x,t) + g(x,t)u_{eq} \tag{1.57}$$

where u_{eq} is the control signal applied to ensure that (1.57) is satisfied, and it takes the name of equivalent control. Therefore, given that the control values is known, the following equality is also satisfied:

$$u_{eq}(t) = -\frac{h(x,t)}{g(x,t)} \tag{1.58}$$

Theoretically, the equivalent output injection is the result of an infinite switching frequency of the discontinuous term u_{eq}. Nevertheless, the realization of the observer produces high (finite) switching frequency making necessary the application of a filter. To eliminate the high frequency component we will use the filter of the form:

$$\tau\dot{\bar{u}}_{eq}(t) = -\bar{u}_{eq}(t) + u_{eq}(t) \tag{1.59}$$

where $\tau \in \mathbb{R}$ and $\delta \ll \tau \ll 1$, where δ is a sampling step.

It is possible to rewrite u_{eq} as result of filtering process in the following form:

$$u_{eq}(t) = \bar{u}_{eq}(t) + \varepsilon(t) \tag{1.60}$$

where $\varepsilon(t) \in \mathbb{R}^n$ is the difference caused by the filtration and $\bar{u}_{eq}(t)$ is the filtered version of $u_{eq}(t)$.

Nevertheless, as has been shown in [2,23]

$$\lim_{\tau \to 0, \delta/\tau \to 0} \bar{u}_{eq}(\tau, \delta) = u_{eq}(t),$$

then, it is possible to assume that the equivalent output injection is equal to the output of the filter.

1.5.2 Twisting second-order sliding-modes algorithm

Let us consider that the output σ in (1.52) satisfies the equation for $r = 2$:

$$\ddot{\sigma} = h(x, t) + g(x, t)u \tag{1.61}$$

where $u, x \in \mathbb{R}$ and the nonlinear functions $h(x, t)$ and $g(x, t)$ satisfy the inequalities (1.55). For this relative degree two system, the Twisting algorithm takes the form:

$$u = -b_1 \mathrm{sign}(\sigma) - b_2 \mathrm{sign}(\dot{\sigma}) \tag{1.62}$$

where b_1 and b_2 are the parameters of the controller that should be properly chosen [5] according to

$$b_1 > b_2 > \frac{C}{K_m} \tag{1.63}$$

Under these assumptions, there exists $T > 0$ such that for all $t > T, \sigma(t) = \dot{\sigma}(t) = 0$. Thus, the Twisting algorithm is said to be a SOSM control algorithm since it provides the existence of a (stable) second order sliding mode at the origin $(\sigma, \dot{\sigma}) = (0, 0)$.

It is possible to conclude from Figure 1.12, which presents the vector field of the trivial second order system, with $h(x, t) = 0$ and $g(x, t) = 1$, driven by the twisting controller, that the solutions of system (1.61) and (1.62) cross the axis $\sigma = 0$ in a

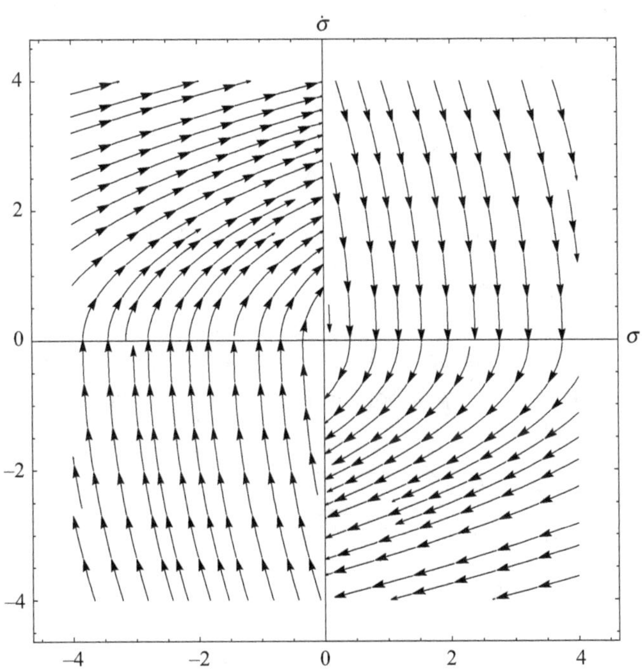

Figure 1.12 Twisting algorithm vector field

finite time. Let us now analyze the trajectories assuming, without loss of generality, that the initial values at $t = 0$ are $\sigma(0) = 0$, $\dot{\sigma}(0) = \dot{\sigma}_0 > 0$. The trajectories of (1.61) and (1.62) are given by

$$\frac{d(\dot{\sigma})}{d\sigma} = \frac{-b_1 - b_2}{\dot{\sigma}} \text{ with } \sigma > 0,\, \dot{\sigma} > 0 \tag{1.64}$$

$$\frac{d(\dot{\sigma})}{d\sigma} = \frac{-b_1 + b_2}{\dot{\sigma}} \text{ with } \sigma > 0,\, \dot{\sigma} < 0 \tag{1.65}$$

For $\sigma > 0$, the solution of (1.61) and (1.62) from the point $(\sigma_1, 0)$ takes the following forms:

$$\sigma = \sigma_1 - \frac{\dot{\sigma}^2}{2(b_1 + b_2)} \text{ with } \dot{\sigma} > 0 \tag{1.66}$$

$$\sigma = \sigma_1 - \frac{\dot{\sigma}^2}{2(b_1 - b_2)} \text{ with } \dot{\sigma} \leq 0 \tag{1.67}$$

Let $\dot{\sigma}_M$ be the second point where the trajectory cuts the axis $\dot{\sigma}$. For the point $\sigma(0) = 0$, $\dot{\sigma}(0) = \dot{\sigma}_0$ and $\sigma(0) = 0$, $\dot{\sigma}(0) = \dot{\sigma}_M$ one has $2(b_1 + b_2)\sigma_1 = \dot{\sigma}_0^2$, and $2(b_1 - b_2)\sigma_1 = \dot{\sigma}_1^2$ correspondingly. Then, consequently we obtain that

$$\frac{\dot{\sigma}_1}{\dot{\sigma}_0} = \sqrt{\frac{b_1 - b_2}{b_1 + b_2}} = q < 1 \tag{1.68}$$

Extending the trajectory into the left half-plane $\sigma < 0$, and after a similar reasoning, the successive crossings of the axis $\sigma = 0$ satisfy the inequality:

$$\frac{|\dot{\sigma}_{i+1}|}{|\dot{\sigma}_i|} = q < 1 \tag{1.69}$$

therefore, is possible to conclude that the algorithm converges to the origin.

In order to compute the time of convergence, let us analyze the arc segments for $\sigma > 0$ and $\sigma \leq 0$, that compose the real trajectory. Let us estimate the time t_1^+ in the segment $\dot{\sigma}_0 \to \sigma_1$. Taking the integral of (1.61) and (1.62) it is obtained that $\dot{\sigma}(t) = -(b_1 + b_2)t + \dot{\sigma}_0$. Then, considering $\dot{\sigma}(t_1^+) = 0 \to t_1^+ = \dot{\sigma}_0/(b_1 + b_2)$. Now, the time taken to reach from the point σ_1 the point $\dot{\sigma}_1$ is computed from $\sigma(t) = -\frac{(b_1 - b_2)t^2}{2} + \sigma_1$, thus taking $\sigma(t_1^-) = 0$ is obtained that

$$t_1^- = \sqrt{\frac{2\sigma_1}{b_1 - b_2}} = \sqrt{\frac{1}{(b_1 - b_2)(b_1 + b_2)}}\dot{\sigma}_0 \tag{1.70}$$

which means that the time interval taken for the trajectory $\dot{\sigma}_0 - \sigma_1 - \dot{\sigma}_1$ can be computed as:

$$t_1 = t_1^+ + t_1^- = \eta\dot{\sigma}_0, \text{ where } \eta = \frac{1}{b_1 + b_2} + \sqrt{\frac{1}{(b_1 - b_2)(b_1 + b_2)}} \tag{1.71}$$

Consequently, the time interval required to pass from the points $\dot{\sigma}_{i-1} - \sigma_i - \dot{\sigma}_i$ takes the following form:

$$t_i = \eta|\dot{\sigma}_{i-1}| = \eta q^{i-1}\dot{\sigma}_0 \tag{1.72}$$

Therefore, the total time of convergence is computed as:

$$T = \sum_{i=1}^{\infty} t_i = \sum_{i=1}^{\infty} \eta|\dot{\sigma}_{i_1}| = \sum_{i=1}^{\infty} \eta q^{i-1}\dot{\sigma}_0 = \frac{\eta\dot{\sigma}_0}{1-q} \tag{1.73}$$

It is important to remark that with the use of the Twisting algorithm, the sliding surface design, i.e., σ, is no longer needed for one degree of freedom mechanical systems. Moreover, the Twisting algorithm collapses the dynamics of such systems to zero in finite time.

1.5.3 Super-twisting second-order sliding-modes algorithm

Consider the first-order nonlinear system:

$$\dot{x} = f(x,t) + u \tag{1.74}$$

where $x \in \mathscr{X} \subset \mathbb{R}$ is the state, $u \in \mathscr{U} \subset \mathbb{R}$ is a scalar input to be defined and $f(x,t)$ is a C^1 function with a bounded first order derivative, i.e.,

$$||\dot{f}(t,x)||_\infty \leq \Gamma_1, \quad \Gamma_1 \in \mathbb{R}^+/0, \quad \Gamma_1 < \infty \tag{1.75}$$

The sliding function is defined as $\sigma = x$. The super-twisting controller takes the following form:

$$\begin{aligned} u &= u_1 + u_2 \\ u_1 &= -\alpha_1|\sigma|^{1/2}\text{sign}(\sigma) \\ \dot{u}_2 &= -\alpha_2\text{sign}(\sigma) \end{aligned} \tag{1.76}$$

where α_1 and α_2 are positive constants that are chosen such that:

- $\alpha_2 > \Gamma_1$,
- and α_1 is chosen such that, for positive definite matrices $P = P^T > 0$ and $Q = Q^T > 0$, the following Lyapunov equation has a solution:

$$PM + M^TP = -Q \tag{1.77}$$

where

$$M = \begin{bmatrix} -\alpha_1 & 1 \\ -2(\alpha_2 - \Gamma_1) & 0 \end{bmatrix} \tag{1.78}$$

According to [5], if the controller (1.76) is applied to the system (1.74), which satisfies the inequality (1.75), then the sliding variable σ and its first-order derivative are enforced to converge to zero after a finite time transient.

The previous sentence could be proven by writing system (1.74) in terms of the sliding function $\sigma = x$ and the controller (1.76):

$$\begin{aligned}\dot{\sigma} &= -\alpha_1|\sigma|^{1/2}\mathrm{sign}(\sigma) + v \\ \dot{v} &= \dot{f} - \alpha_2\mathrm{sign}\sigma\end{aligned} \tag{1.79}$$

Introducing the auxiliary variable ζ, which is defined as

$$\zeta = \begin{bmatrix} |\sigma|^{1/2}\mathrm{sign}(\sigma) \\ v \end{bmatrix} \tag{1.80}$$

and computing its derivative along the trajectories of (1.79), $\dot{\zeta}$ is written as

$$\dot{\zeta} = \frac{1}{2}|\sigma|^{-1/2}\begin{bmatrix} -\alpha_1 & 1 \\ -2\alpha_2 & 0 \end{bmatrix}\zeta + \begin{bmatrix} 0 \\ \dot{f}(x) \end{bmatrix} \tag{1.81}$$

Using the strong Lyapunov function [24]:

$$V = \zeta^T P \zeta \tag{1.82}$$

it is possible to prove the finite-time convergence to zero of the variable ζ and, in consequence of its dependence on σ, the convergence of σ to zero. The derivative of (1.82) along the trajectories of (1.80) takes the following form:

$$\dot{V} = 2\zeta^T P\left(\frac{1}{2}|\sigma|^{-1/2}\begin{bmatrix} -\alpha_1 & 1 \\ -2\alpha_2 & 0 \end{bmatrix}\zeta + \begin{bmatrix} 0 \\ \dot{f}(x) \end{bmatrix}\right) \tag{1.83}$$

$$\begin{aligned}\dot{V} \leq \frac{1}{2}|\sigma|^{-1/2}\zeta^T &\left(P\begin{bmatrix} -\alpha_1 & 1 \\ -2(\alpha_2 - \dot{f}(x)\mathrm{sign}\sigma) & 0 \end{bmatrix}\right. \\ &\left. + \begin{bmatrix} -\alpha_1 & 1 \\ -2(\alpha_2 - \dot{f}(x)\mathrm{sign}\sigma) & 0 \end{bmatrix}^T P\right)\zeta\end{aligned} \tag{1.84}$$

Notice that with a proper selection of parameters the next inequality is satisfied:

$$\dot{V} \leq -\frac{1}{2}|\sigma|^{-1/2}\zeta^T Q\zeta \tag{1.85}$$

Therefore, it has been proven that the variable ζ converges to zero.

The performance of the controller is illustrated by the academic system defined by

$$\dot{x} = 3\sin(x) + w(t) + u(x) \tag{1.86}$$

where x is the scalar state and $w(t)$ is an external input. For simulation purposes, let us consider that initial conditions are set as $x(0) = 0.5$ and the unknown external input is given as follows:

$$w(t) = 8 + t + \sin(4t) \tag{1.87}$$

Assuming that the system is driven by the super-twisting controller (1.76) with gains $\alpha_1 = 8.8$ and $\alpha_2 = 1.5(8)^{1/2}$, the state of the system is shown in Figure (1.13), where

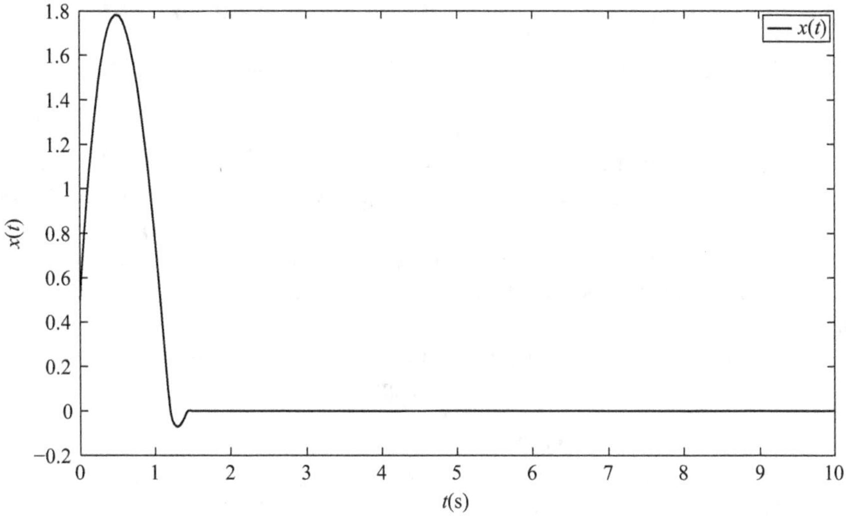

Figure 1.13 State of the nonlinear system driven by super-twisting controller

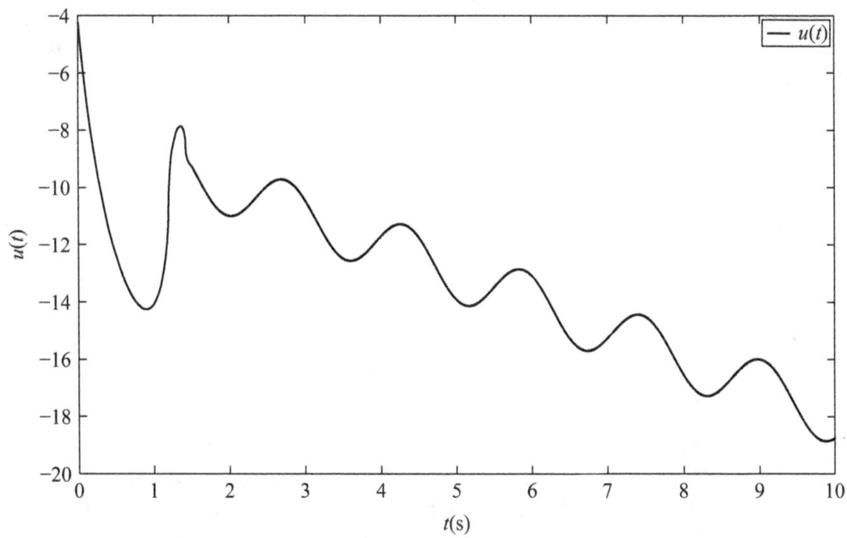

Figure 1.14 Super-twisting control signal

it is possible to see that the state reaches the origin in a finite time in spite of the growing unknown input $w(t)$ and the oscillating behavior of the system. In Figure 1.14 the continuous control signal is presented, notice that no high frequency switching is present in the control, even when the states reaches the surface $x = 0$ in time $t = 1.52$ s, that is one of the main properties of the super-twisting controller, its strong robustness provided by means of a differentiable control signal.

1.5.4 Suboptimal second-order sliding-mode algorithm

Consider the single-input system affine in the control variable:

$$\dot{x}(t) = f(x(t), t) + g(x(t), t)u(t) \tag{1.88}$$

where $x \in \mathscr{X} \subset \mathbb{R}^n$ is the state vector, the value of which at the initial time instant t_0 is $x(t_0) = x_0$, and $u \in \mathscr{U} \subset \mathbb{R}$ is a scalar input, while $f(x(t), t) : \mathscr{X} \to \mathbb{R}^n$ and $g(x(t), t) : \mathscr{X} \to \mathbb{R}^n$ are uncertain functions of class C^1. Let $\sigma(x(t), t) : \mathscr{X} \to \mathbb{R}$ be the sliding variable, which is the output function, sufficiently smooth and of class C^2. Moreover, $\sigma(x(t))$ has to be such that if $u(t)$ in (1.88) is designed so that, in a finite time t_r (ideal reaching time), $\sigma(x(t_r)) = 0 \ \forall x_0 \in \mathscr{X}$ and $\sigma(x(t)) = 0 \ \forall t > t_r$, then $\forall t \geq t_r$ the origin is an asymptotically stable equilibrium point of (1.88) constrained to $\sigma(x(t)) = 0$. Consider the input–output map:

$$\begin{cases} \dot{x}(t) &= f(x(t), t) + g(x(t), t)u(t) \\ y(t) &= \sigma(x(t)) \\ x(t_0) &= x_0. \end{cases} \tag{1.89}$$

Assume that (1.89) is complete in $\mathscr{X}$ and the relative degree of the system, i.e., the minimum order r of the time derivative $\sigma^{(r)}$ of the sliding variable in which the control u explicitly appears, is considered well defined, uniform, time invariant, and equal to 2. In the following, the dependence of σ on $x(t)$ and of all the variables on t is omitted in some cases, when it is obvious, for the sake of simplicity.

Assume that system (1.89) admits a global normal form in $\mathscr{X}$, i.e., there exists a global diffeomorphism of the form $\Phi(x) : \mathscr{X} \to \mathbb{R}^n$:

$$\Phi(x) = \begin{pmatrix} x_r \\ \sigma(x) \\ \dot{\sigma}(x) \end{pmatrix} \tag{1.90}$$

such that

$$\begin{cases} \ddot{\sigma}(t) = \phi(x_r, \sigma, \dot{\sigma}, t) + \gamma(x_r, \sigma, \dot{\sigma}, t)u(t) \\ \dot{x}_r(t) = \psi(x_r, \sigma, \dot{\sigma}, t), \end{cases} \tag{1.91}$$

where $\gamma \neq 0, \forall (x_r, \sigma(x), \dot{\sigma}(x)) \in \Phi_{\mathscr{X}}$, $\Phi_{\mathscr{X}}$ being a compact set.

In the literature, see for instance [25], making reference to previous system, the internal state $x_r \in \mathbb{R}^{n-2}$ is BIBO stable, while ϕ and γ are continuous functions such that

$$\exists F > 0 : \ |\phi(x_r, \sigma, \dot{\sigma}, t)| \leq F \ \forall (x_r, \sigma(x), \dot{\sigma}(x)) \in \Phi_{\mathscr{X}} \tag{1.92}$$

$$\exists G_{\max} > 0 : \ \gamma(x_r, \sigma, \dot{\sigma}, t) \ \leq G_{\max} \ \forall (x_r, \sigma(x), \dot{\sigma}(x)) \in \Phi_{\mathscr{X}}. \tag{1.93}$$

Moreover, assume that

$$\exists G_{\min} > 0 : \ \gamma(x_r, \sigma, \dot{\sigma}, t) \ \geq G_{\min} \ \forall (x_r, \sigma(x), \dot{\sigma}(x)) \in \Phi_{\mathscr{X}}. \tag{1.94}$$

System (1.91) is a perturbed chain of integrators built starting from the sliding variable and its time derivatives. Thus, the original control objective, attained in conventional SMC by zeroing the sliding variable in finite time, is transformed into force the system state to reach in finite time the subspace named 2-*sliding manifold* $\sigma = \dot{\sigma} = 0$ and there remains for any subsequent time instant. Note that, if the relative degree of the auxiliary system is such that $r = \hat{r} + 1$, $\hat{r}$ being the relative degree of the original system (1.88), i.e., an artificial increment of the relative degree is performed to build the auxiliary system, then an auxiliary control $w = \dot{u}$ is introduced, and the function ϕ also depends on u. According to [5], conditions (1.92)–(1.94) hold locally and specific control actions are considered to maintain the system within a boundedness region. In [26], these conditions are relaxed and it is proved that if the modulus of ϕ is upperbounded by a known function affine in $|\dot{\sigma}|$, the solution is guaranteed by a set of constant control parameters. Otherwise, if ϕ is upperbounded only by known function of σ and $\dot{\sigma}$, without assumption on $|\dot{\sigma}|$, the convergence is guaranteed by suitable adaptation of control parameters.

In the following, assume that the information about the bounds of ϕ and γ are available and holds globally, and the original dynamical system (1.91) implies the differential inclusion [27]:

$$\ddot{\sigma} \in [-F, F] + [G_{\min}, G_{\max}]\, u. \tag{1.95}$$

In the literature, it has been shown that the problem of making the 2-sliding manifold associated with (1.95) finite-time attractive can be solved by a SOSM controller of the type (see for instance [12,28–33]):

$$\begin{cases} u(t) = -\alpha(t) U \operatorname{sign}(\sigma - \beta\sigma_{\max}) \\ \alpha(t) = \begin{cases} 1 & \text{if } (\sigma - \beta\sigma_{\max})\sigma_{\max} \geq 0 \\ \alpha^* & \text{if } (\sigma - \beta\sigma_{\max})\sigma_{\max} < 0 \end{cases} \\ \beta \in [0;\ 1) \end{cases} \tag{1.96}$$

where $U > 0$ is the *control gain*, $\alpha^* > 1$ is the *modulation factor*, β is the *anticipation factor* and $\sigma_{\max}$ is the extremal value of the sliding variable σ. The latter can be found by setting at t_0 $\sigma_{\max} = \sigma(t_0)$ and then $\forall\, t > t_0$ by inspection of the past three values or two values of the sliding variable $\sigma(t)$ [12,28]. Moreover, in order to enforce the sliding mode it is necessary to tune the parameters according to the following *dominance* and *convergence* conditions, respectively:

$$\begin{cases} U > \frac{F}{G_{\min}} \\ \alpha^* \in [1;\ +\infty) \cap \left(\frac{2F + (1-\beta)G_{\max}U}{(1+\beta)G_{\min}U};\ +\infty\right). \end{cases} \tag{1.97}$$

These conditions imply twisting trajectories around the origin of the auxiliary state space. Instead, if one considers:

$$\begin{cases} U > \frac{F}{G_{\min}} \\ \alpha^* \in [1;\ +\infty) \cap \left(\frac{F + (1-\beta)G_{\max}U}{(\beta)G_{\min}U};\ +\infty\right) \end{cases} \tag{1.98}$$

a monotonic trajectory to zero can be obtained. Both the previous conditions allow to steer the sliding variable to zero in a finite time t_r such that

$$t_r \leq t_{\max_1} + \max\{t_{r_1}, t_{r_2}\} \tag{1.99}$$

with

$$\begin{aligned} t_{r_1} &= U\frac{\alpha^* G_{\min}+G_{\max}}{\alpha^* G_{\min}U-F}\sqrt{\frac{2(1-\beta)|\sigma_{\max_1}|}{G_{\max}U+F}} \times \frac{1}{1-\frac{\sqrt{|F+((1-\beta)G_{\max}-\alpha^*\beta G_{\min})U|}}{\sqrt{\alpha^* G_{\min}U-F}}} \\ t_{r_2} &= U\frac{\alpha^* G_{\min}+G_{\max}}{\alpha^* G_{\min}U+F}\sqrt{\frac{2(1-\beta)|\sigma_{\max_1}|}{G_{\max}U-F}} \times \frac{1}{1-\frac{\sqrt{|F-((1-\beta)G_{\min}-\alpha^*\beta G_{\max})U|}}{\sqrt{\alpha^* G_{\min}U+F}}} \end{aligned} \tag{1.100}$$

where $t_{\max_1}$ is the time instant at which the first extremal value of σ, i.e., $\sigma_{\max_1}$ occurs. Note that larger U, the shorter the convergence time, while if β tends to 1, the time to converge becomes infinite.

Suboptimal Second Order Sliding Mode (SSOSM) control is a particular case of SOSM control law previously described. Making reference to the generic formulation (1.96), the SSOSM control is obtained by posing $\beta = 1/2$, i.e.,:

$$\begin{cases} u(t) = -\alpha(t)U\text{sign}\left(\sigma - \frac{1}{2}\sigma_{\max}\right) \\ \alpha(t) = \begin{cases} 1 & \text{if } \left(\sigma - \frac{1}{2}\sigma_{\max}\right)\sigma_{\max} \geq 0 \\ \alpha^* & \text{if } \left(\sigma - \frac{1}{2}\sigma_{\max}\right)\sigma_{\max} < 0 \end{cases} \end{cases} \tag{1.101}$$

with conditions on the control parameters as

$$\begin{cases} U > \frac{F}{G_{\min}} \\ \alpha^* \in [1;\ +\infty) \cap \left(\frac{4F+G_{\max}U}{3G_{\min}U};\ +\infty\right) \end{cases} \tag{1.102}$$

or in the monotonic case as

$$\begin{cases} U > \frac{F}{G_{\min}} \\ \alpha^* \in [1;\ +\infty) \cap \left(\frac{2F+G_{\max}U}{G_{\min}U};\ +\infty\right). \end{cases} \tag{1.103}$$

It can be proved that, with the previous constraints (1.102) or (1.103), the control law (1.101) enforces the generation of a sequence of states with coordinates featuring a contraction property which implies the convergence in a finite time to the origin of the auxiliary state space (see [9,25] for the details of the proof).

1.6 Robust exact arbitrary order differentiator

To generate a control signal by means of the Twisting second-order sliding mode algorithm, the derivative of the sliding variable is needed. Usually, this derivative has to be obtained online by means of a numerical method, in this order we propose the robust exact differentiator [30] as a tool to obtain derivatives of arbitrary signals.

Let an input signal $f(t)$ be a function defined on $[0, \infty)$ and consisting of a bounded Lebesgue-measurable noise with unknown features and an unknown base signal $f_0(t)$ having a nth order derivative with known Lipschitz constant $L > 0$. The problem is to find, after a finite-time transient, the estimations of $f_0(t), \dot{f}_0(t), \ldots, f^{(n)}(t)$. The arbitrary-order exact robust differentiator takes the following form:

$$\begin{aligned}
\dot{z}_0 &= v_0 = z_1 - \kappa_n |z_0 - f_0(t)|^{\frac{n}{n+1}} \operatorname{sign}(z_0 - f_0(t)), \\
\dot{z}_1 &= v_1 = z_2 - \kappa_{n-1} |z_1 - v_0|^{\frac{n-1}{n}} \operatorname{sign}(z_1 - v_0), \\
&\vdots \\
\dot{z}_i &= v_i = z_i - \kappa_{n-i} |z_i - v_{i-1}|^{\frac{n-i}{n-i+1}} \operatorname{sign}(z_i - v_{i-1}), \\
&\vdots \\
\dot{z}_n &= -\kappa_1 \operatorname{sign}(z_n - v_{n-1})
\end{aligned} \tag{1.104}$$

for suitable positive constant coefficients κ_i to be chosen recursively large in the given order. In particular, a possible selection of the differentiator parameters is $\kappa_1 = 1.1L$, $\kappa_2 = 1.5L^{1/2}$, $\kappa_3 = 2L^{1/3}$, $\kappa_4 = 3L^{1/4}$, $\kappa_5 = 5L^{1/5}$, $\kappa_6 = 8L^{1/6}$ for $n \leq 6$.

Let the input noise satisfy the inequality $|f(t) - f_0(t)| \leq \varepsilon$. Then the following inequalities are established in finite time for some positive constants μ_i, ν_i depending exclusively on the parameters of the differentiator

$$\left| z_i - f_0^{(i)}(t) \right| \leq \mu_i \varepsilon^{(n-i+1)/(n+1)}, \; i = 0, \ldots, n$$

$$\left| v_i - f_0^{(i+1)}(t) \right| \leq \nu_i \varepsilon^{(n-i)/(n+1)}, \; i = 0, \ldots, n-1$$

It is important to remark that the right-hand sides of the above given inequalities are zero in the absence of noise. Thus, we can state that being the parameters of (1.104) chosen properly, the following equalities are fulfilled in the absence of noises and after a finite-time transient process:

$$z_0 = f_0(t); \quad z_i = v_{i-1} = f_0^{(i)}, \; i = 1, \ldots, n. \tag{1.105}$$

Now, the sampled case is considered, when $z_0(t_j) - f(t_j)$ is substituted for $z_0 - f(t)$ with $t_j \leq t < t_{j+1}$, $t_{j+1} - t_j = \tau > 0$.

Let $\tau > 0$ be the constant input sampling interval in the absence of noises. Then the following inequalities are established in finite time:

$$|z_i - f_0^{(i)}(t)| \leq \mu_i \tau^{n-i+1}, \quad i = 0, \ldots, n$$

$$|v_i - f_0^{(i+1)}(t)| \leq \nu_i \tau^{n-i}, \quad i = 0, \ldots, n-1$$

Let consider the noise-free function $f(t) = 2\cos(t(1 + \sin(0.1t)))$, which second order derivative has a Lipschitz constant equals $L = 29$. For this particular signal, the second-order differentiator is tuned as follows:

$$\begin{aligned}
\dot{z}_0 &= v_0 = z_1 - 2L^{1/3} \, |z_0 - f_0(t)|^{\frac{2}{3}} \operatorname{sign}(z_0 - f_0(t)), \\
\dot{z}_1 &= v_1 = z_2 - 1.5L^{1/2} \, |z_1 - v_0|^{\frac{1}{2}} \operatorname{sign}(z_1 - v_0), \\
\dot{z}_2 &= -1.1L \operatorname{sign}(z_2 - v_1)
\end{aligned} \tag{1.106}$$

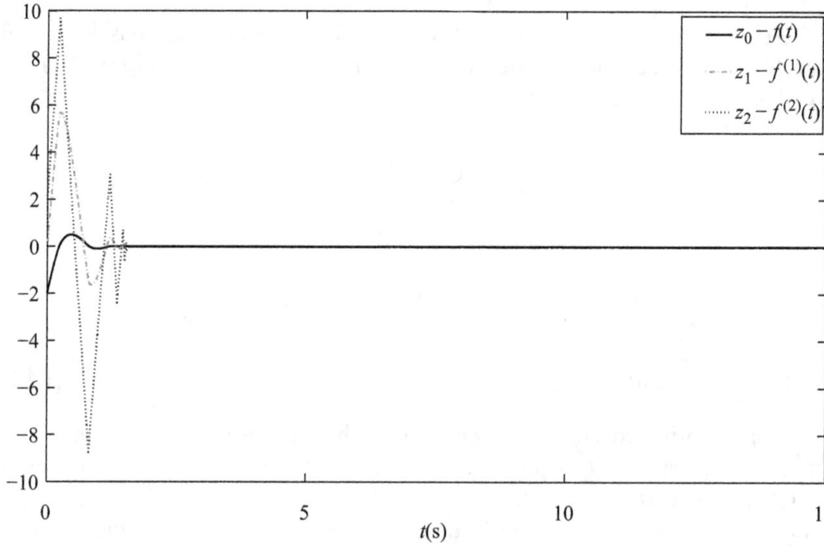

Figure 1.15 Estimation errors

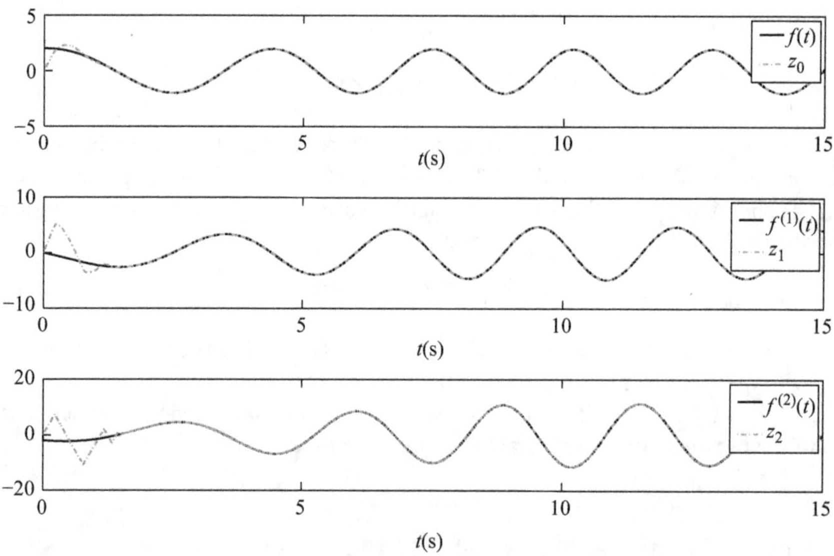

Figure 1.16 Real signals and their estimations

In Figure 1.15 the estimation errors are presented. It is clearly observable that all of them converge to zero after a finite-time transient. Therefore, the estimated signal and derivatives converge to the real values of $f(t)$ and its derivatives, as could be seen in Figure 1.16.

1.7 Conclusions

In this chapter, the main concepts used along the book have been introduced. Its content has been designed as a tutorial material that can help to understand the advanced techniques presented along the book. Even when the authors have made their best effort to produce self-contained material, the interested readers are encouraged to expand their knowledge by reading the right material, with this aim in the following lines is presented a list of recommended works.

The literature related to conventional sliding modes is extensive to name some of the books in the field let us cite [2,34,35]. In the recent book [14] the analysis for the high-order sliding-modes is also included.

Integral sliding modes have been extended to high-order sliding sets in [36]. In particular, in [37] the integral sliding modes idea has been extended to second-order sliding sets and applied for robot manipulators. A nested algorithm that guarantees the exponential tracking in the presence of matched and unmatched disturbances is presented in [38].

The application of second-order sliding mode techniques has been presented in [12]. The super-twisting second order algorithm has been applied to state observation in the articles [39,40], in which the problem of state estimation under the presence of unknown inputs is also studied.

The High-Order Sliding-Mode Differentiator structure has been used to reconstruct the states of linear systems affected by unknown inputs in [41–43], while its application to non-linear systems has been presented in [44,45].

Acknowledgments

Jorge Dávila and Leonid Fridman gratefully acknowledge the financial support by Programa de Apoyo a Proyectos de Investigacion e Innovacion Tecnologica (UNAM) 113216, DGAPA PASPA Program, SIP-IPN 20171734 and the support of CDA-IPN. Antonella Ferrara gratefully acknowledges the financial support by ITEAM Project, under the Marie Skodowska-Curie grant agreement No. 675999.

References

[1] V. Utkin, "First stage of VSS: People and events," in *Variable Structure Systems: Towards the 21st Century*, ser. Lecture Notes in Control and Information Science, X. Yu and J. Xu, Eds. London: Springer, 2002, no. 274, pp. 1–32.

[2] V. Utkin, *Sliding Modes in Control and Optimization*, ser. Communication and Control Engineering. Berlin: Springer-Verlag, 1992.

[3] S. Drakunov, D. Izosimov, A. Loukyanov, V.A. Utkin, and V.I. Utkin, "Block control principle," *Automation and Remote Control*, vol. 51, no. 5, pp. 601–609, 1990.

[4] A. Bondarev, S. Bondarev, N. Kostylyeva, and V. Utkin, "Sliding modes in systems with asymptotic state observers," *Automatica i Telemechanica (Automation and Remote Control)*, vol. 46, no. 5, pp. 679–684, 1985.

[5] A. Levant, "Sliding order and sliding accuracy in sliding mode control," *International Journal of Control*, vol. 58, no. 6, pp. 1247–1263, Dec. 1993.

[6] A. Levant, "Robust exact differentiation via sliding mode technique," *Automatica*, vol. 34, no. 3, pp. 379–384, 1998.

[7] L. Fridman and A. Levant, "Higher order sliding modes as the natural phenomena of control theory," in *Workshop Variable Structure and Lyapunov Technique*, 1994, pp. 302–309.

[8] L. Fridman and A. Levant, "Higher order sliding modes as the natural phenomena of control theory," in *Robust Control Variable Structure and Lyapunov Techniques*, ser. Lecture Notes in Control and Information Science, F. Garofalo and G. Glielmo, Eds. London: Springer, 1996, no. 217, pp. 107–133.

[9] G. Bartolini, A. Ferrara, and E. Usai, "Output tracking control of uncertain nonlinear second-order systems," *Automatica*, vol. 33, no. 12, pp. 2203–2212, Dec. 1997.

[10] W. Newman, "Robust sub-optimal control," *IEEE Transactions on Automatic Control*, vol. 35, no. 7, pp. 841–844, 1990.

[11] G. Bartolini, A. Ferrara, and E. Punta, "Multi-input second-order sliding-mode hybrid control of constrained manipulators," *Dynamics and Control*, vol. 10, no. 3, pp. 277–296, 2000.

[12] G. Bartolini, A. Pisano, E. Punta, and E. Usai, "A survey of applications of second-order sliding mode control to mechanical systems," *International Journal of Control*, vol. 76, no. 9–10, pp. 875–892, Jan. 2003.

[13] A. Levant, "Universal single-input-single-output (SISO) sliding-mode controllers with finite-time convergence," *IEEE Transactions on Automatic Control*, vol. 46, no. 9, pp. 1447–1451, 2001.

[14] Y. Shtessel, C. Edwards, L. Fridman, and A. Levant, *Sliding Mode Control and Observation*. New York, USA: Birkhauser, 2013.

[15] A. F. Filippov, *Differential Equations with Discontinuous Right-Hand Sides*, ser. Mathematics and its Applications. The Netherlands: Kluwer Academic Publishers, 1988, vol. 18.

[16] A. Polyakov and L. Fridman, "Stability notions and Lyapunov functions for sliding mode control systems," *Journal of the Franklin Institute*, vol. 351, no. 4, pp. 1831–1865, 2014.

[17] V. I. Utkin and J. Shi, "Integral sliding mode in systems operating under uncertainty conditions," in *Proceedings of the 35th IEEE Conference on Decision and Control*, vol. 4, Kobe, Japan, Dec. 1996, pp. 4591–4596.

[18] F. Castanos and L. Fridman, "Analysis and design of integral sliding manifolds for systems with unmatched perturbations," *IEEE Transactions on Automatic Control*, vol. 51, no. 5, pp. 853–858, May 2006.

[19] M. Rubagotti, A. Estrada, F. Castanos, A. Ferrara, and L. Fridman, "Integral sliding mode control for nonlinear systems with matched and unmatched perturbations," *IEEE Transactions on Automatic Control*, vol. 56, no. 11, pp. 2699–2704, Nov. 2011.

[20] M. Rubagotti, D. M. Raimondo, A. Ferrara, and L. Magni, "Robust model predictive control with integral sliding mode in continuous-time sampled-data

nonlinear systems," *IEEE Transactions on Automatic Control*, vol. 56, no. 3, pp. 556–570, Mar. 2011.

[21] L. Fridman, A. Poznyak, and F. Bejarano, Eds., *Robust Output LQ Optimal Control via Integral Sliding Modes*, ser. Systems & Control: Foundations & Applications. Berlin: Birkhäuser Basel, 2014.

[22] G. P. Incremona, A. Ferrara, and L. Magni, "Asynchronous networked MPC with ISM for uncertain nonlinear systems," *IEEE Transactions on Automatic Control*, doi: 10.1109/TAC.2017.2653760.

[23] L. Fridman, "The problem of chattering: an averaging approach," in *Variable Structure, Sliding Mode and Nonlinear Control*, ser. Lecture Notes in Control and Information Science, K. Young and U. Ozguner, Eds. London: Springer-Verlag, 1999, no. 247, pp. 363–386.

[24] J. Moreno and M. Osorio, "Strict Lyapunov functions for the super-twisting algorithm," *IEEE Transactions on Automatic Control*, vol. 57, no. 4, pp. 1035–1040, 2012.

[25] G. Bartolini, A. Ferrara, and E. Usai, "Chattering avoidance by second-order sliding mode control," *IEEE Transactions on Automatic Control*, vol. 43, no. 2, pp. 241–246, Feb. 1998.

[26] G. Bartolini, A. Ferrara, A. Pisano, and E. Usai, "On the convergence properties of a 2-sliding control algorithm for non-linear uncertain systems," *International Journal of Control*, vol. 74, no. 7, pp. 718–731, 2001.

[27] J. P. Aubin and A. Cellina, *Differential Inclusions*. Berlin: Springer-Verlag, 1984.

[28] G. Bartolini, A. Ferrara, V. I. Utkin, and T. Zolezzi, "A control vector approach to variable structure control of nonlinear systems," *International Journal of Robust and Nonlinear Control*, vol. 7, no. 4, pp. 321–335, Apr. 1997.

[29] G. Bartolini, A. Ferrara, E. Usai, and V. I. Utkin, "On multi-input chattering-free second-order sliding mode control," *IEEE Transactions on Automatic Control*, vol. 45, no. 9, pp. 1711–1717, Sep. 2000.

[30] A. Levant, "Higher-order sliding modes, differentiation and output-feedback control," *International Journal of Control*, vol. 76, no. 9–10, pp. 924–941, Jan. 2003.

[31] T. Floquet, J.-P. Barbot, and W. Perruquetti, "Higher-order sliding mode stabilization for a class of nonholonomic perturbed systems," *Automatica*, vol. 39, no. 6, pp. 1077–1083, Jun. 2003.

[32] A. Levant, "Quasi-continuous high-order sliding-mode controllers," *IEEE Transactions on Automatic Control*, vol. 50, no. 11, pp. 1812–1816, Nov. 2005.

[33] F. Dinuzzo and A. Ferrara, "Higher order sliding mode controllers with optimal reaching," *IEEE Transactions on Automatic Control*, vol. 54, no. 9, pp. 2126–2136, Sep. 2009.

[34] C. Edwards and S. Spurgeon, *Sliding Mode Control: Theory and Applications*, ser. Systems and Control. London: Taylor & Francis, 1998.

[35] W. Perruquetti and J. Barbot, *Sliding Mode Control in Engineering*. New York: Marcel Dekker, 2002.

[36] A. Levant and L. Alelishvili, "Integral high-order sliding modes," *IEEE Transactions on Automatic Control*, vol. 52, no. 7, pp. 1278–1282, Jul. 2007.

[37] A. Ferrara and G. P. Incremona, "Design of an integral suboptimal second-order sliding mode controller for the robust motion control of robot manipulators," *IEEE Transactions on Control Systems Technology*, vol. 23, no. 6, pp. 2316–2325, Nov. 2015.

[38] J. D. Sánchez-Torres, A. Navarrete-Guzman, G. Rubio-Astorga, and A. G. Loukianov, "High order integral nested sliding mode control," in *2014 13th International Workshop on Variable Structure Systems (VSS)*, June 2014, pp. 1–5.

[39] J. Davila, L. Fridman, and A. Levant, "Second-order sliding-mode observer for mechanical systems," *IEEE Transactions on Automatic Control*, vol. 50, no. 11, pp. 1785–1789, Nov. 2005.

[40] J. Davila, L. Fridman, and A. Poznyak, "Observation and identification of mechanical systems via second order sliding modes," *International Journal of Control*, vol. 79, no. 10, pp. 1251–1262, 2006.

[41] L. Fridman, A. Levant, and J. Davila, "Observation of linear systems with unknown inputs via high-order sliding-modes," *International Journal of Systems Science*, vol. 38, no. 10, pp. 773–791, 2007.

[42] F. J. Bejarano, "Partial unknown input reconstruction for linear systems," *Automatica*, vol. 47, no. 8, pp. 1751–1756, 2011.

[43] L. Fridman, J. Davila, and A. Levant, "High-order sliding-mode observation for linear systems with unknown inputs," *Nonlinear Analysis: Hybrid Systems*, vol. 5, no. 2, pp. 189–205, 2011.

[44] L. Fridman, Y. Shtessel, C. Edwards, and X. Yan, "Higher-order sliding-mode observer for state estimation and input reconstruction in nonlinear systems," *International Journal of Robust and Nonlinear Control*, vol. 18, no. 4–5, pp. 399–413, 2008.

[45] J. Davila, L. Fridman, A. Pisano, and A. Usai, "Finite-time state observation for nonlinear uncertain systems via higher order sliding modes," *International Journal of Control*, vol. 82, no. 8, pp. 1564–1574, 2009.

Chapter 2
Longitudinal vehicle dynamics control via sliding modes generation

Enrico Regolin[1], Gian Paolo Incremona[2], and Antonella Ferrara[1]

In this chapter the application of Sliding Mode Control techniques is illustrated to solve the longitudinal vehicle dynamics control problem. Sliding Mode Control is a nonlinear control method capable of offering a number of benefits, the majority of which is its robustness versus a significant class of uncertainties. It can also be profitably used to efficiently solve automotive control and observation problems, as widely testified in the literature. The aim of this chapter is to provide an overview on the longitudinal dynamics control of vehicles, in particular electric vehicles with individual motors for each wheel, focusing on recent developments based on Sliding Mode Control theory.

2.1 Introduction

The standard systems for the control of longitudinal dynamics in road vehicles are generally referred to as Antilock Braking System (ABS) and Traction Control (TC). The adoption of electronic ABS for road vehicles dates back to the early 1970s, and has ever since increased its diffusion to the point where it has become compulsory in EU regulations [1]. Dating back to the same period, TC has been used to prevent wheels skidding during acceleration or takeup, to the point where it is nowadays a typical feature of engine control units. In practice, both ABS and TC functions are implemented through wheel torque control. For this reason, in conventional vehicles, TC systems can use braking actuators as well as the driveline, so that the overall system becomes an overactuated one, with more options for the selection of the control strategy.

With the rise of electrical vehicles, and the increasingly successful variant with individually controlled electric motors (which we will refer to as E-4WD), also in the in-wheel configuration [2], the ABS and TC systems can be realized with

[1]Dipartimento di Ingegneria Industriale e dell'Informazione, University of Pavia, via Ferrata 3, 27100, Pavia, Italy

[2]Dipartimento di Elettronica, Informazione e Bioingegneria Politecnico di Milano, Piazza Leonardo da Vinci, 32, 20133, Milano, Italy

new strategies, and included in a more general framework of a generalized vehicle stability control system [3], which comprehends different aspects of vehicle dynamics: longitudinal, lateral (yaw rate control) and vertical (rollover prevention).

The problem of designing wheel-slip controllers for E-4WD vehicles is not a standalone subject, but is strictly connected with other issues, mainly:

- the design of controllers for the electric motors, with all inherent problems, including noise, vibration, and harshness (NVH) and active vibration control;
- the estimation of several unmeasurable quantities, such as vehicle velocity, forces between tire and road, road condition;
- the interaction between other active safety systems, such as Electronic Stability Control.

In this chapter, we illustrate some applications of Sliding Mode Control (SMC) to ABS devices. The considered fields of applications are split into high and low levels of control. We consider high-level control issues during the calculation of a wheel torque request, which ensures the tracking of a wheel slip reference, or the design of a model based observer for the estimation of unknown quantities, such as longitudinal wheel force. The case of low-level control case is the one where the actuators are directly controlled, so that they produce a desired torque at wheel level.

SMC is a well-known control methodology based on the use of a discontinuous control input and particularly appreciated for its robustness properties in front of a class of uncertainties affecting the systems [4–6]. SMC makes the controlled system evolve as a variable structure system and steers the system state to a pre-specified manifold, that is the so-called "sliding manifold", in a finite time so that a "sliding mode" occurs.

Yet, SMC presents a crucial drawback, the so-called chattering phenomenon, i.e., high frequency oscillations of the controlled variable which may limit its use to real actuators [7,8]. Several solutions have been studied recently in order to attenuate this phenomenon. The more effective solutions are oriented to increase the order of the sliding mode, thus producing efficient Second Order Sliding Mode (SOSM) and Higher Order Sliding Mode Control (HOSM) algorithms [9,10]. In this type of algorithms the discontinuity is imposed on the first time derivative of the effective control signal, which results in being continuous with beneficial effect for mechanical and electromechanical plants. The results of different SMC algorithms, of first and second order types and integral or nonintegral nature, evaluated on a simplified model, have been compared in [11].

The chapter is organized as follows. In Section 2.2, an overview on TC and ABS is presented, with special focus on full electric vehicles, describing some applications based on SMC techniques. In Section 2.3, the vehicle model considered for the model-based applications is introduced. In Section 2.4, we illustrate a sliding mode approach to the estimation of the longitudinal force acting between tire and road, which can be used to estimate both vehicle velocity and tire–road friction when the vehicle is in slipping conditions. Different high-level sliding mode controllers for the wheel-slip tracking are first illustrated, and then compared against a standard PID, in Section 2.5. Their performances are evaluated via simulation on IPG Carmaker SW. In Section 2.6, a First Order Sliding Mode (FOSM) controller is employed on a hydraulic ABS hardware-in-the-loop (HIL) testrig for the tracking of a target brake pressure. Finally, some conclusions are gathered in Section 2.7.

2.2 TC and ABS control systems

In the case of E-4WD vehicles, among the most common procedures for the implementation of model-based wheel-slip controllers, is the tracking of either a desired torque/force or a reference wheel slip on individual wheels. The control principles in both TC and ABS cases are similar.

In [13] the different control methods proposed in the past few years for E-4WD (and, in the ABS case, conventional) vehicles are illustrated, with special focus on those tested on HIL testrigs or prototype vehicles. The control methods are classified in different categories:

- for the TC problem, torque-based and slip-based methods;
- for ABS technology, different approaches in case of conventional ABS, electric motor control and ABS blending.

In the next few pages, we provide an overview of the current developments of TC (Section 2.2.1) in E-4WD vehicles, and ABS in both conventional and electrical configurations (Section 2.2.2). In Section 2.2.3, an example of generic control system structure is illustrated. Possible SMC technique applications are highlighted inside the general framework.

2.2.1 Traction control systems

TC systems can be separated (see [13]) in *torque-based* and *slip-based* methods. As the respective names suggest, they differentiate from one another for the reference variable adopted. In the first case the purpose of the controller is to track a certain wheel traction torque (or force), while in the latter the wheel slip-speed is the target.

In most cases, a distinct separation of the different methods in categories is not easy to accomplish. Tracking a reference torque, slip-ratio or the derivative of the tire–road friction curve $d\mu/d\lambda$ (as in [15]) often means, in practical situations, performing the same calculation from a different perspective. While in the examples provided in this chapter we focus mostly on slip-based ABS control applications, it is useful to consider the different model based approaches used for TC.

The principle behind torque-based TC controllers [12] is that the occurrence of wheel skidding is characterized by a sudden drop in the wheel estimated inertia. A feedback loop (see Figure 2.1) reduces the requested traction torque when the slip is too high: the high frequency component of the difference between actual and expected wheel velocity is used as feedback variable. The nominal vehicle model used for the determination of the expected velocity is an integrator, with proportional factor given by the overall equivalent mass of the vehicle.

A different approach to the torque-based is represented by the slip-based TC, where a reference wheel-slip is tracked via feedback control. In this configuration, a typical problem is the explicit determination of the reference slip λ^*. A way of implicitly adapting the vehicle dynamics control to the road surface, is to design a

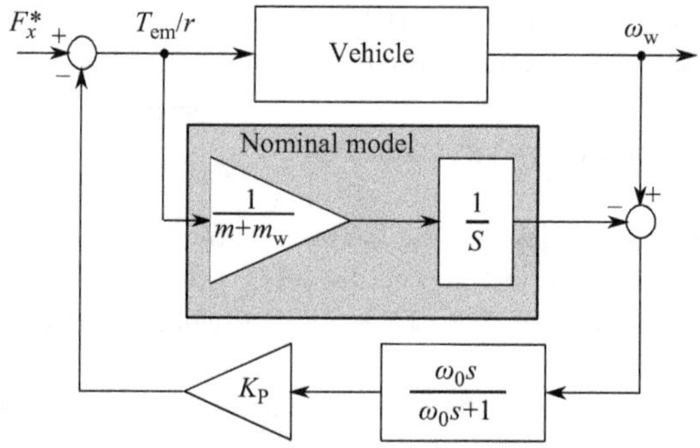

Figure 2.1 Torque-based TC concept from [12]: T_{em} is the control torque, ω_w is the rotational velocity, m_w is the mass equivalent to wheel inertia, m is the vehicle mass, and ω_0 is the cut-off frequency of the high-pass filter

feedback controller which keeps the wheel-slip in its maximum grip zone (it corresponds to the local maximum in the $\mu(\lambda)$ curve; these aspects are described in detail in Section 2.3). For this purpose, an observer for the wheel slip λ and one for the friction coefficient μ have to be designed. Based on the values $\hat{\lambda}$ and $\hat{\mu}$, the gradient $\frac{d\mu}{d\lambda}$ can be estimated, even without need for the reconstruction of the entire $\mu(\lambda)$ characteristic.

The controller in the above solution can also be implemented with SMC technique (as in [15]) for the determination of the control law for the electric motor. This configuration is similar to the one for which the different SMC controllers for ABS in Section 2.5 are designed.

2.2.2 Antilock braking systems

In general, due to the limited bandwidth of the hydraulic actuation mechanism, the ABS control algorithms for conventional vehicles are characterized by rule-based control of on–off relays with appointed thresholds. The controlled variables are typically wheel slip and vehicle acceleration: based on the control logic, a limit cycle can be generated [16], surrounding the point of peak friction.

This cyclic behavior has been investigated by Tanelli *et al.* in [14]: the ABS controller switches between the two states of the hydraulic brake, based on the deviations from constraint values of the wheel slip λ, calculated using a second-order system for the wheel-slip dynamics. The switching logic operates on the $\{\lambda,\ T_{brk}\}$ phase plane, as shown in Figure 2.2: when λ_{min}, λ_{max} are reached, the brake request switches to $T_{brk_{max}}$, $T_{brk_{min}}$ respectively, determining the limit cycle shown in the diagram. The proposed method can also estimate current road conditions in the case of additional monitoring of the cycle amplitude of the slip ratio. This approach has been investigated via simulation on a quarter-car model and showed sufficient robustness, including in realistic situations with a limited rate value of the actuator control signal.

The ABS torque blending in electrical vehicles noticeably increases the degrees of freedom of the braking system, thus allowing for optimization both in performance

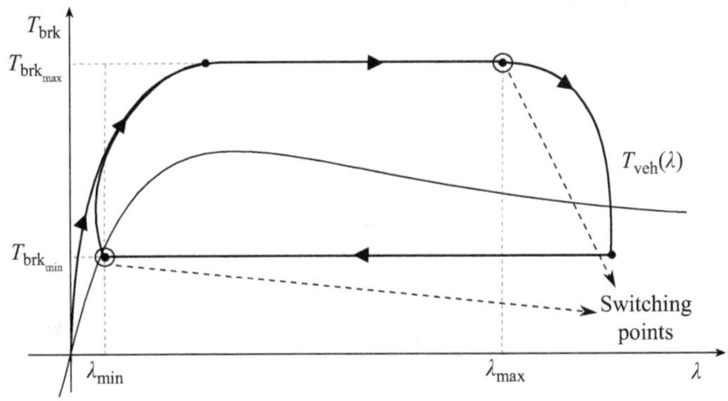

Figure 2.2 Phase plane $\{\lambda, T_{brk}\}$ of the limit cycle obtained with the strategy proposed in [14]

and efficiency of the braking phase. Several research works have shown significant benefits in terms of energy recuperation, and reduced response time, thanks to the high frequency of the electric ABS. Different strategies can be applied for the torque blending: in [17,18], the desired dynamic is obtained by regulating the electrical component via a feedback regulator, and accounting for the friction brakes dynamics via feedforward compensation. Other variants [19] assign different combinations of the braking effort depending on the surface friction: electric braking at low friction, combinations of electric and friction brakes at higher tire–road friction levels.

Pure electrical ABS systems have been investigated as well in recent years, thanks to the diffusion of brake-by-wire technology, as well as in-wheel electrical motors in E-4WD vehicles. In particular, in the case of E-4WD vehicles, this solution allows for the integration of the TC/ABS control strategies in a unique Fastest Acceleration/Deceleration Control (FADC) problem [20]. The merged TC/ABS solution has already been implemented successfully on prototype vehicles. In [21], the calculated control torque has a base component, calculated from the driver's inputs (accelerator/brake pedal), and a reactive component, which allows the tracking of a reference slip during the controlled phases.

2.2.3 Overview of sliding mode applications in wheel-slip control

In the practical examples illustrated in the next sections, we will consider closed loop control schemes for the slip control of individual wheels, such as the one schematized in Figure 2.3. We assume the Active Safety devices add their contribution to the calculation of the individual wheel torque demands, which come from the generic Electronic Control Unit (ECU), and therefore from the driver.

The Active Safety block is assumed to be a centralized high-level controller, which provides reference values for the wheel slip-ratio of each wheel. These requests can come from considerations on actual slip occurring for the implementation of the TC or ABS functions, or for other purposes, such as the generation of a certain yaw moment for the stability control. Since an electrical vehicle with individual actuators

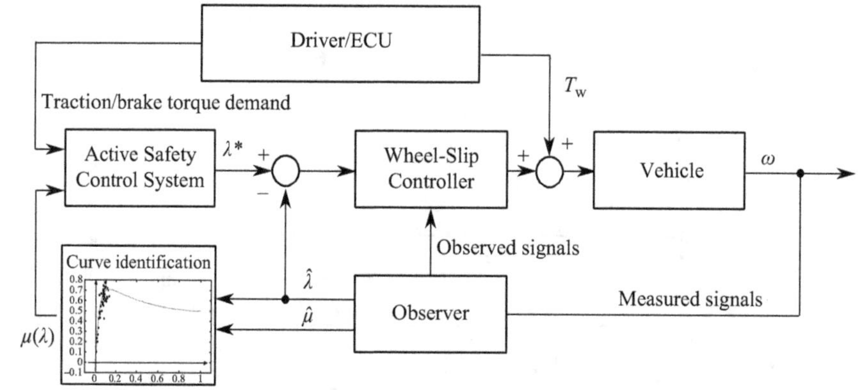

Figure 2.3 Example of control system for the control of vehicle longitudinal dynamics

for each wheel is an over-actuated system, in most cases the generation of the wheel slip references is the solution of an optimization problem, which can be solved either online, offline or with hybrid techniques. Once the reference wheel slip is known, a closed loop control scheme can be implemented, as the ones that will be illustrated in Section 2.5. Also an observer for the wheel slip λ and possibly other necessary unmeasurable quantities can be found, as the one which will be presented in Section 2.4. The observer is necessary, since while the vehicle is skidding, it is not possible to derive the vehicle speed directly from the wheel speed. One can see then, how the slip is always the result of a calculation of the instantaneous vehicle velocity v, which can be obtained from accelerometers on the vehicle and/or model based estimation. Finally, a low-level actuation control needs to be implemented, in order to ensure the proper torque $T^*_{\mathrm{brk/drv}}$, calculated by the feedback controller, is actually applied to the wheel. In Section 2.6 a demonstrative example of low-level actuation is presented, concerning the tracking of a brake pressure reference in a conventional ABS setup.

The different examples of applications proposed, which can be inserted within the frame of the scheme in Figure 2.3, are all implemented via application of SMC techniques. Since the approach to observation and control is model based, the commonly used model for the wheel dynamics is illustrated in Section 2.3, starting from the description of the tire–road interaction.

2.3 Modeling of vehicle longitudinal dynamics

In order to design a model-based control system for the wheel-slip, it is essential to define a mathematical model for the longitudinal dynamics of the single wheel. In this section, after some initial analytical considerations on the static tire–road friction models (Section 2.3.1), we illustrate a basic second-order representation of the wheel dynamics (Section 2.3.2).

2.3.1 Longitudinal tire model

The forces, which determine the vehicle dynamics, depend on the interaction between road and wheel. Pneumatic tires have the function of supporting the vehicle and controlling its trajectory. The modeling of such interaction is essential for the design of a model-based control of the longitudinal dynamics. Note that, although in this chapter we only focus on the characterization of the longitudinal forces acting at the contact point between road and tire, in practice the lateral and vertical dynamics also affect the longitudinal forces.

During the rolling of the pneumatic tire on the road, there is a deformation occurring on both surfaces, as none of them is an ideal rigid body. Such deformation, for what concerns the pneumatic, depends on inflation pressure, rigidity, wear, temperature and other factors. The deformation of the tire and of the road produces an energy loss which is known as rolling resistance. This deformation varies depending on the external momentum applied to the wheel (free rolling, acceleration or braking). In the case of acceleration torque present on a wheel, the part preceding the contact zone is compressed, therefore the peripheral velocity v of the tread band in the leading zone of the contact is lower than the product ωr, r being the wheel radius of the undeformed wheel. This phenomenon increases with the acceleration torque, leading to an increment in the gap between v and ωr, while its counterpart happens when a braking torque is applied.

Having defined the effective rolling radius r_e as the one satisfying the following equation:

$$v = \omega r_e \tag{2.1}$$

and the deviation δ of the wheel radius, determined by the acceleration ($T_{W_{drv}}$) or braking ($T_{W_{brk}}$) torque, we can distinguish the following situations:

- $r_e = r - \delta(T_{W_{drv}})$ in the case of external acceleration moment;
- $r_e = r + \delta(T_{W_{brk}})$ in the case of external braking moment.

Following these considerations, it is now possible to define a "longitudinal slip ratio" λ, as a measure depending on the variation of r_e, as follows:

$$\lambda = \frac{\omega r - v}{\max(\omega r, v)}. \tag{2.2}$$

The slip ratio defined by (2.2) is positive for driving conditions and negative for braking, and it is limited to the $[-1, 1]$ interval. Note that, even if $|\lambda| > 0$, it does not necessarily mean that the entire tire–road contact zone is slipping. Typically, for low slip values only a small portion at the tail end of the contact zone is actually slipping. The portion of tire actually slipping increases with λ, until the value λ_p is reached, which corresponds to pure sliding.

The longitudinal force F_x, which the wheel exchanges with the road, is a function of λ. Ideally, when no forces are acting on the wheel (free rolling conditions), there

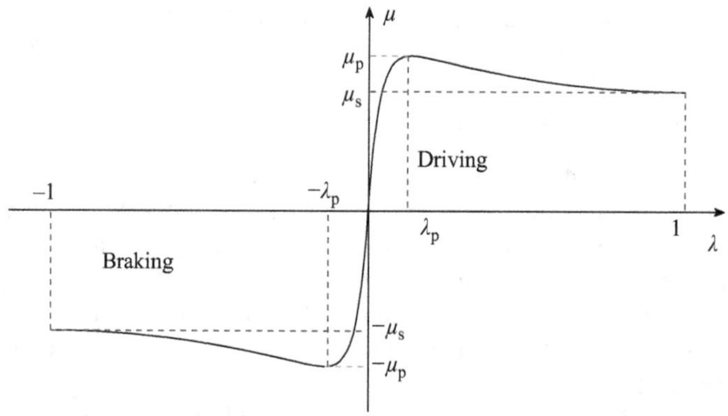

Figure 2.4 Qualitative representation of the λ(μ) curve, in the braking and driving case

is no slip (that is $\lambda = 0$), although in reality the rolling resistance causes a very small negative slip. The force F_x changes linearly with the slip until a peak value $\lambda = \lambda_p$ is reached ($\lambda = -\lambda_p$ for negative slips). Outside this range, the force F_x decreases in absolute value until the extremes $\lambda = \pm 1$ are reached. Given a constant slip λ, F_x can be considered roughly proportional to the load F_z, therefore the longitudinal friction coefficient μ is defined as

$$\mu = \frac{F_x}{F_z}. \tag{2.3}$$

The qualitative behavior of $\mu(\lambda)$ is reported in Figure 2.4. The values μ_p and μ_s identified are respectively the peak traction/braking coefficient, and the sliding traction/braking coefficient. Since the curves usually shows symmetric behavior, μ_p and μ_s are considered equal in absolute value in the traction and in the braking case. The area included between the two peaks is generally referred to as the "linear zone": outside this interval, the wheel is in an unstable condition. This can be intuitively understood, in the breaking case, by considering the wheel equation of motion during a braking session. Keeping in mind that the wheel speed ω cannot have negative values, and having assigned positive sign to the acceleration forces/torques acting on the wheel, and negative to the braking ones, we have that

$$J\dot{\omega} = |F_{x\text{brk}}|r - |T_{W\text{brk}}| = |\mu(\lambda)|r - |T_{W\text{brk}}| \tag{2.4}$$

where J is the inertia. For the sake of simplicity, we can assume the vehicle speed v to be constant, as its response to the braking action is slower than that of the wheel in case of slip. During braking, the decrease in wheel speed causes the absolute value of the slip $|\lambda|$ to increase. If, at the same time, the absolute value of the friction coefficient $|\mu|$ decreases, a sudden reduction in the braking torque $T_{W\text{brk}}$ is then also needed, to compensate the decrease in $|F_{x\text{brk}}|$, in order to avoid locking of the wheel.

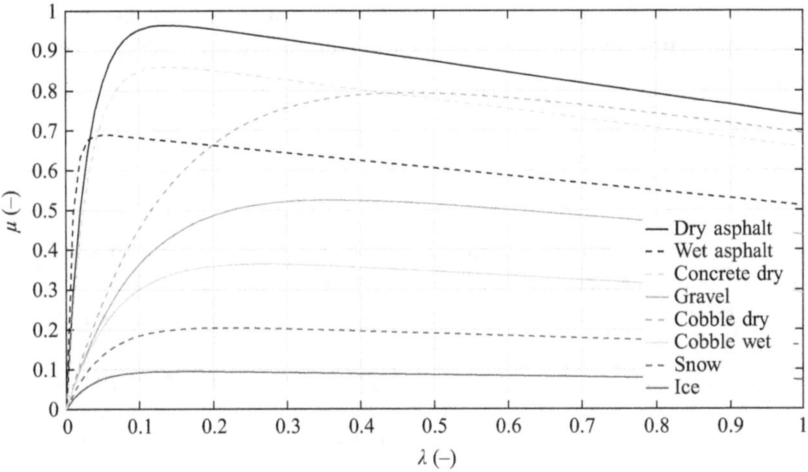

Figure 2.5 Tire–road friction curves $\mu(\lambda)$ for eight different surfaces, obtained from values in Table 2.1

This action cannot be performed sufficiently fast by the driver, who must be aided by *active safety* devices. Traditional ABS absolves this function by decreasing the hydraulic brake pressure when excessive slip is detected, and increasing it again once the wheel-slip is again in the linear zone.

The shape of the curve $\mu(\lambda)$ changes considerably, depending on various parameters, such as type of tire, road conditions, speed, and lateral tire force F_y. Different examples of $\mu(\lambda)$ curves obtained in different conditions are reported in Figure 2.5.

2.3.1.1 Tire model equation

For the mathematical representation of the $\mu(\lambda)$ curve in Figures 2.4 and 2.5, several static and dynamic models have been investigated in the past (see [22]), including those which combine longitudinal and lateral dynamics. Among the static models, Pacejka's "Magic Formula", Burckhardt's and Dugoff's models are generally adopted in slip-control applications. Since the control devices should be able to perform efficiently in different road conditions, the choice of the model should also consider its properties in terms of parameter identification capabilities. In this subsection, an analytic approach to the selection of the curve $\mu(\lambda)$ is illustrated, as proposed in [23] within the scope of parameter identification.

In general, models have several parameters which can introduce redundancy when identification is performed, therefore, it is preferable to keep the model dimension low (if possible 3 parameters or less). Since we are mostly interested in modeling the relation between λ and μ in the stable area $\lambda \leq \lambda_p$, it is sufficient to use a reduced version of the Burckhardt's model, already adopted in other works [24,25]:

$$\mu_{\text{exp}}(\theta, \lambda) = \text{sign}(\lambda) \cdot (\theta_1(1 - e^{-\theta_2|\lambda|}) - \theta_3|\lambda|) \tag{2.5}$$

with $\theta = (\theta_1, \theta_2, \theta_3)^T$ being the vector of model parameters. Additionally, we consider the rational function representation similar to the one proposed in [26], and evaluated in [25,27] that is

$$\mu_{\text{ratfun}}(p, \lambda) = \frac{p_1 \lambda}{1 + p_2|\lambda| + p_3 \lambda^2} \tag{2.6}$$

with $p = (p_1, p_2, p_3)^T$ being the vector of model parameters. Since all possible models for $\mu(\lambda)$ are odd functions, in order to simplify the notation we will focus only on the positive λ case. As a consequence, the sign$(\cdot)$ and $|\cdot|$ functions can be removed in (2.5) and (2.6).

In most applications, the main interest lays in the identification of the maximum available friction coefficient μ_{p}. Therefore, it is especially important, when estimating the entire curve, to detect the $\lambda = \lambda_{\text{p}}$ at which the peak occurs. From (2.5) and (2.6), the following analytical forms of the slip corresponding to maximum grip are found:

$$\lambda_{\text{exp,p}} = -\frac{1}{\theta_2} \log\left(\frac{\theta_3}{\theta_1 \theta_2}\right) \approx -\frac{1}{\theta_2} \log\left(\frac{\text{const}}{\theta_2}\right) \tag{2.7}$$

and

$$\lambda_{\text{ratfun,p}} = \frac{1}{\sqrt{p_3}}. \tag{2.8}$$

The approximation $\frac{\theta_3}{\theta_1} \approx \text{const}$ was made, based on the empirical values from Table 2.1. In both models the value λ_{p} appears to be influenced by a single parameter. Due to the shape of functions (2.7) and (2.8), one can expect, during parameter identification, more accurate convergence for low parameter values. For θ_2 this means good accuracy for $\theta_2 < 30$ and worse performances for $\theta_2 > 40$.

Table 2.1 Nominal θ values in model (2.5) for different surfaces

Road type	$\boldsymbol{\theta_1}$	$\boldsymbol{\theta_2}$	$\boldsymbol{\theta_3}$
Dry asphalt	1.01	45	0.3
Wet asphalt	0.7	100	0.2
Dry concrete	0.9	35	0.25
Gravel	0.6	10	0.17
Dry cobblestone	0.95	7	0.28
Wet cobblestone	0.4	15	0.1
Snow	0.22	20	0.05
Ice	0.1	30	0.01

The candidate model has to be characterized by a high sensitivity to the variation of parameters θ_2, p_3, in the slip region where the samples (λ_i, μ_i), i being the ith sample, are most likely to be found. The parameter sensitivity functions are defined as follows:

$$\left|\frac{\partial}{\partial p_i}\mu(p,\lambda)\right|, \quad i = 1, \ldots, m \tag{2.9}$$

which yields

$$\left|\frac{\partial}{\partial \theta_2}\mu_{\text{exp}}(\theta,\lambda)\right| = \theta_1 \lambda e^{-\theta_2 \lambda} \tag{2.10}$$

and

$$\left|\frac{\partial}{\partial p_3}\mu_{\text{ratfun}}(p,\lambda)\right| = \frac{p_1 \lambda^3}{(1 + p_2|\lambda| + p_3\lambda^2)^2}. \tag{2.11}$$

As apparent in Figure 2.6, the normalized sensitivity of the parameter θ_2 in model (2.5) is preferable to the one of parameter p_3 in (2.6) for $\lambda \le \lambda_p$, which makes the selection of model (2.5) the obvious choice. Different $\mu(\lambda)$ curves for various surfaces, obtained through (2.5), are shown in Figure 2.5. It is evident how, compared to the dry asphalt condition, the nonlinear parameter θ_2 affects the shape of the curves in case of cobblestone and gravel, whereas, in case of wet asphalt, one only sees a reduction

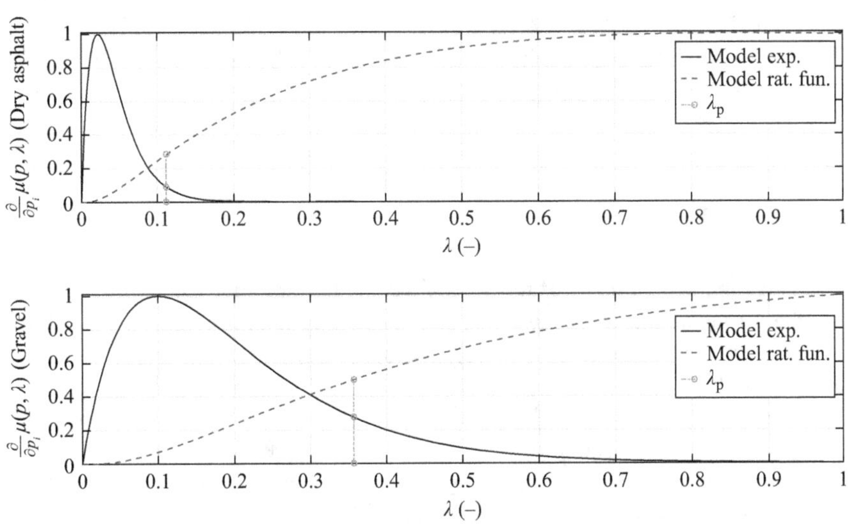

Figure 2.6 Normalized sensitivity of the parameter influencing the friction coefficient peak for models (2.5) and (2.6)

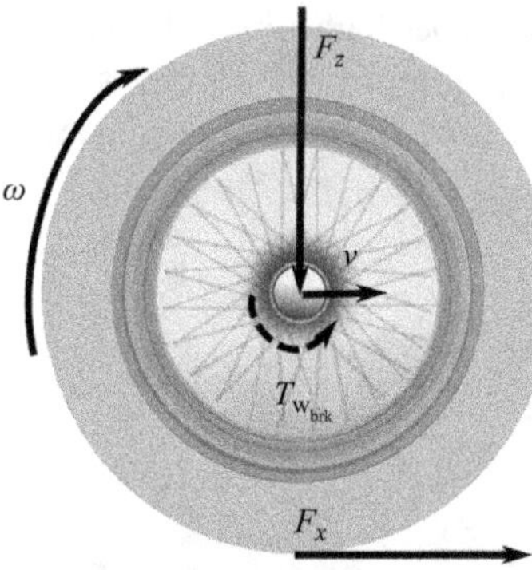

Figure 2.7 Graphical representation of the single-corner model (2.12)

of maximum grip, but no shifting of the maximum grip slip-ratio. In case of extreme low-grip surfaces such as ice and snow, the slope in the micro-slip region is lower, mainly due to the drop in θ_1.

2.3.2 Vehicle model

Considering from the $\mu(\lambda)$ curve derived in the previous section (equation (2.5)), one needs to select a sufficiently accurate vehicle model to be used for the development of proper observers and controllers. The usual choice is the single-corner model, which is often employed for traction and braking control system design [28]. The single-corner model, which describes the dynamics of a single wheel, represented in Figure 2.7, is as follows:

$$\begin{cases} J\dot{\omega} = -rF_x + T_{\mathrm{w}} \\ m_{\mathrm{W_{eq}}}\dot{v} = F_x \end{cases} \tag{2.12}$$

where ω is the angular speed of the wheel, v is the longitudinal speed of the vehicle, T_{w} is the torque acting on the wheel, F_x is the longitudinal tire–road contact force, J is the wheel moment of inertia, $m_{\mathrm{W_{eq}}}$ the quarter-car mass (or wheel equivalent mass), and r the wheel radius.

Model (2.12) describes the longitudinal dynamics, under the assumption that the wheel side-slip angle is equal to zero, and no lateral forces are acting on the wheel, i.e., the driver is performing a pure longitudinal maneuver. Under these conditions, we can use (2.3) to approximate the longitudinal force F_x as

$$F_x = F_z\mu(\lambda) \tag{2.13}$$

where $\mu(\lambda)$ is the tire–road friction curve (2.5). Model (2.12) can be extended to the full vehicle, including aerodynamic e friction losses, as follows [29]:

$$\begin{cases} m_{\text{veh}}\dot{v} = \sum_{i,j} F_{x_{i,j}} - F_{\text{loss}_i}(v) & \text{(2.14a)} \\ J_{i,j}\dot{\omega}_{i,j} = T_{\text{w}_{i,j}} - r_{i,j}F_{x_{i,j}} & \text{(2.14b)} \\ F_{\text{loss}}(v) = F_{\text{air}}(v) + F_{\text{roll}} = c_x v^2 \operatorname{sign}(v) + f_{\text{roll}} m_{\text{veh}} g & \text{(2.14c)} \\ F_{x_{i,j}} = \mu_{i,j}(\lambda_{i,j})F_{z_{i,j}} & \text{(2.14d)} \\ F_{z_{\text{f},j}} = \dfrac{l_{\text{r}} m_{\text{veh}} g - l_{\text{h}} m_{\text{veh}} \dot{v}}{2(l_{\text{f}} + l_{\text{r}})} & \text{(2.14e)} \\ F_{z_{\text{r},j}} = \dfrac{l_{\text{f}} m_{\text{veh}} g + l_{\text{h}} m_{\text{veh}} \dot{v}}{2(l_{\text{f}} + l_{\text{r}})} & \text{(2.14f)} \end{cases}$$

where the subscripts $i = \text{f, r}$ and $j = \text{l, r}$ distinguish, respectively, front/rear and left/right wheels. F_{air} is the air drag, and F_{roll} is the rolling resistance. Moreover, m_{veh} is the vehicle mass, c_x is the longitudinal wind drag coefficient, f_{roll} is the rolling resistance coefficient, l_{f} is the distance from the front axle to the center of gravity (CoG), l_{r} is the distance from the CoG to the rear axle, and l_{h} is the vertical distance to the CoG. Note that (2.14e) and (2.14f) can be adopted for the estimation of $m_{\text{w}_{\text{eq}}}$ for each wheel in the single-corner model.

By using a more accurate description of the forces acting on the vehicle/wheel system, which includes air and rolling resistance losses, we can improve the quality of the observation/control devices. It should be noted, though, that other external influences are neglected, such as roll and yaw moments, lateral and vertical motions, brake, throttle, steering actuators and manifold dynamics. The effects of these approximations are reflected in loss of quality of the estimation, when the algorithms are run in the loop with a full vehicle simulator. As for the c_x coefficient, in order to verify the effectiveness of the proposed control schemes in the presence of uncertainties, this can vary over time. Also the mass of the vehicle and the tire–road friction coefficient are time-varying and represent the unmatched uncertainties affecting the system.

By properly substituting (2.12) in (2.2), and assuming the vehicle velocity to be constant, the wheel slip dynamics can be reduced to a first order differential equation, which changes in case of driving/braking:

$$\begin{aligned} \dot{\lambda} &= -\frac{r}{Jv}\left[\psi_{\text{brk}}(\lambda) + T_{\text{w}}\right] \\ &= f_{\text{brk}} + b_{\text{brk}} T_{\text{w}} \end{aligned} \tag{2.15}$$

$$\psi_{\text{brk}}(\lambda) = \left(-\frac{J(1-\lambda)}{m_{\text{w}_{\text{eq}}} r} + r\right) F_z \mu(\lambda) \tag{2.16}$$

$$\dot{\lambda} = \frac{(1-\lambda)^2}{Jv}[\psi_{\text{drv}}(\lambda) + T_{\text{w}}] \\ = f_{\text{drv}} + b_{\text{drv}}T_{\text{w}} \tag{2.17}$$

$$\psi_{\text{drv}}(\lambda) = -\left(\frac{J}{m_{\text{w}_{\text{eq}}}r(1-\lambda)} + r\right)F_z\mu(\lambda) \tag{2.18}$$

where f_{brk}, b_{brk}, f_{drv} and b_{drv} are easily obtainable functions, which will be indicated with f and b in Section 2.5.

It can be proven, by substituting typical values for J, r, $m_{\text{w}_{\text{eq}}}$, that $\psi_{\text{brk}}(\lambda) > 0$, for $\lambda < 0$ and $\psi_{\text{drv}}(\lambda) < 0$, for $\lambda > 0$. This implies that, with no external influence ($T_{\text{w}} = 0$), the systems are naturally stable, and the slip vanishes asymptotically. By braking/accelerating ($T_{\text{w}} \neq 0$), we change the dynamics of λ in the direction of reducing the natural damping effect, and increasing the absolute value of λ.

2.4 Sliding mode observer for tire–road friction and velocity

In this section, we illustrate a SMC-based observer for the tire–road friction coefficient μ and the vehicle velocity v.

The observer, based on the algorithm proposed in [24], estimates the longitudinal acceleration of each individual wheel, provided that values of vehicle mass and inertia are known, and accurate measurements of wheel speed and driving/braking torque are available. Based on these estimates, a logic is used to reconstruct actual vehicle velocity during slipping, as well as the friction coefficient μ (as defined in (2.3)) of each individual wheel.

The algorithm, as presented in this section, is not in general suited to be applied to E-4WD vehicles, but only to two-wheel drive vehicles, either electrical or conventional. In order to adapt this observer to the E-4WD vehicle, an improvement of the logic illustrated in the next section would be necessary.

2.4.1 Sliding mode tire–road friction observer

The observation law, which is used for the estimation of the traction force F_x acting on the single wheel, is based on the dynamic model (2.12) illustrated in Section 2.3.2.

Based on the first equation in (2.12), a first-order sliding mode observer is designed to estimate the rotational speed ω of the wheel, which is actually a measurable quantity. Define the sliding variable as:

$$s_{\omega} = \omega - \hat{\omega}. \tag{2.19}$$

From this, the dynamics of the wheel speed estimate $\hat{\omega}$ are given by

$$\dot{\hat{\omega}} = \frac{1}{J}(\Omega + T_{\text{w}}) \tag{2.20}$$

where the control signal of the sliding mode observer is

$$\Omega = \Omega_{\max}\,\text{sign}\,(s_{\omega}) \tag{2.21}$$

with $\Omega_{\max}$ being a positive constant. By differentiating (2.19), and considering the single-corner model (2.12), one obtains:

$$\dot{s}_\omega = \dot{\omega} - \dot{\hat{\omega}} = \frac{1}{J}(rF_x + T_w - \Omega - T_w) \tag{2.22}$$

$$= \frac{1}{J}(rF_x - \Omega_{\max}\operatorname{sign}(s_\omega)). \tag{2.23}$$

The condition to ensure the convergence to zero of the sliding variable (2.19) can be found by adopting the Lyapunov function:

$$V = \frac{1}{2}s_\omega^2 \tag{2.24}$$

and design the control such that its time derivative is negative definite:

$$\dot{V} = s_\omega \dot{s}_\omega < 0. \tag{2.25}$$

Note that, given the initial value of the sliding variable $|s_\omega(0)|$, one has that

$$t_r \leq \frac{|s_\omega(0)|}{\eta}, \tag{2.26}$$

with $\eta \in \mathbb{R}^+$, if the following reaching condition is satisfied:

$$\dot{V} = s_\omega \dot{s}_\omega \leq -\eta |s_\omega| \tag{2.27}$$

which is more restrictive than (2.25) and implies it. Proof of this can be obtained by integrating both terms of (2.27), both in case of positive and negative values of s_ω (see [30]).

Thus, if the gain $\Omega_{\max}$ is chosen such that (2.27) is satisfied, the controlled system state evolution will be confined to the sliding manifold for $t \geq t_r$. Developing (2.27), by substituting (2.23) into it, we obtain:

$$s_\omega \dot{s}_\omega \leq \frac{1}{J}(rF_x - \Omega_{\max}\operatorname{sign}(s_\omega))s_\omega \leq -\eta|s_\omega| \tag{2.28}$$

which means that $\Omega_{\max}$ should satisfy:

$$\Omega_{\max} > r\max_\lambda(F_x) = r\max_\lambda(\mu(\lambda)\cdot F_z) \tag{2.29}$$

Then, choosing $\Omega_{\max}$ as in (2.29), the reaching condition (2.27) is satisfied and the motion reaches the sliding surface s_ω in finite time, which can be upper-bounded as in (2.26).

The control signal Ω in (2.21) switches at a high, theoretically infinite, frequency. However, in reality, the impossibility of having such ideal control signal, makes the state oscillate at finite frequencies, with both high and slow components [5].

It has been proven [4] that, on the sliding manifold, the output of a proper low-pass filter with the discontinuous control signal (2.21) as input, converges asymptotically to the so-called equivalent control, which represents the continuous control signal that would maintain the system trajectory onto the sliding surface $s_\omega = 0$, and that is equivalent from the point of view of its effects to the discontinuous control law actually applied [31]. It should be noted that the equivalent control is only defined on the sliding manifold: by filtering the control signal during the reaching phase (which is in theory a continuous control action), a distorted version of the control signal is obtained.

In our case, the equivalent control, denoted as Ω_{eq}, can be computed from (2.19) and (2.20) by letting:

$$\dot{s}_{\omega} = \frac{1}{J}(rF_x - \Omega_{\mathrm{eq}}) = 0 \tag{2.30}$$

which yields immediately:

$$\Omega_{\mathrm{eq}} = rF_x, \tag{2.31}$$

so that the equivalent control gives indication of the current value of the longitudinal force exerted by the tire. Therefore, by considering the chassis dynamics given in the second equation of (2.12) and assuming that the wheel radius r is known, i.e., equal to its static value, we get that an estimate of the vehicle deceleration $\hat{\dot{v}}$ can be computed as

$$\hat{\dot{v}} = \hat{a} = -\frac{1}{rm_{\mathrm{w}}}\Omega_{\mathrm{eq}}. \tag{2.32}$$

Finally, the approximation of the equivalent control Ω_{eq} obtained by filtering out the high frequency component of the actual control signal Ω is

$$\tau_{\mathrm{obsv}}\dot{\hat{\Omega}} + \hat{\Omega} = \Omega \tag{2.33}$$

$$\Omega_{\mathrm{eq}} \approx \hat{\Omega} \tag{2.34}$$

where τ_{obsv} is the filter time constant. The filter time constant should be chosen sufficiently small to preserve the informative components of the control Ω undistorted, but large enough to eliminate the high frequency noise.

2.4.1.1 Vehicle velocity and tire–road friction estimation

The vehicle acceleration estimation $\hat{a}$ can be used to compute the velocity $\hat{v}$ when wheel slip occurs during braking maneuvers. The method proposed in [32] considers a vehicle configuration with two driving wheels, so that the average peripheral speed $\overline{v}_{\mathrm{ND}}$ of the non-driving ones can be effectively used for vehicle speed estimation in acceleration conditions. The average peripheral speed $\overline{v}$ of all wheels is used to determine the vehicle velocity at low speed and when external torques acting on wheels are small. Finally, the estimated vehicle acceleration $\hat{a}$ is integrated when a significant braking action is detected. The average speed values previously defined are

$$\overline{v}(t) = \frac{1}{4}\sum_{i=1}^{4} v_i(t) \tag{2.35}$$

$$\overline{v}_{\mathrm{ND}}(t) = \frac{1}{2}\sum_{i=1}^{2} v_{\mathrm{ND}_i}(t). \tag{2.36}$$

Having defined a minimum velocity threshold value $v_{\min}$, and proper acceleration thresholds β, δ, the strategy for the calculation of the velocity estimate $\hat{v}$ is summarized in Table 2.2, where t_0 is the time instant defining the start of the braking phase.

Table 2.2 Vehicle velocity estimation in two wheel drive vehicles

Driving condition	**Detection**	**v calculation**
Low speed	$\overline{v}(t) < v_{\min}$	$\hat{v}(t) = \overline{v}(t)$
Acceleration	$\overline{v}(t) > v_{\min} \wedge \hat{a}(t) > \delta$	$\hat{v}(t) = \overline{v}_{\mathrm{ND}}(t)$
Slow variations	$-\beta \leq \hat{a}(t) < \delta \wedge \overline{v}(t) > v_{\min}$	$\hat{v}(t) = \overline{v}(t)$
Braking	$\hat{a}(t) < -\beta \wedge \overline{v}(t) > v_{\min}$	$\hat{v}(t) = \hat{v}(t_0) + \int_{t_0}^{t} \hat{a}(t)dt$

Note that, when trying to extend this method to E-4WD vehicles, there should be a replacement for the strategy adopted in case of low vehicle speed and acceleration, since it is possible that, during acceleration, skidding occurs for all wheels.

The individual wheel-slip is derived from the estimated vehicle velocity and the measured wheel slip, by using (2.2), in order to obtain slip estimates samples $\hat{\lambda}_{i,j}$ for each wheel.

With the longitudinal force estimation $\hat{F}_x$ derived from (2.31), the tire–road friction coefficient can also be computed via (2.3). During pure longitudinal maneuvers, the normal force F_z can be assumed to be identical for wheels on the same axles (front or rear), and depends on the geometrical configuration of the vehicle body, as well as on its instantaneous acceleration. We can assume that the normal forces for the front and rear axles F_{z_i}, $i = \mathrm{f}$, r, are the ones calculated in (2.14e) and (2.14f).

Substituting $\hat{F}_x$, $\hat{F}_z$ in (2.3), we obtain the estimated value for the instantaneous friction coefficient:

$$\hat{\mu} = \frac{\hat{F}_x}{\hat{F}_z} = \frac{2l_{\mathrm{tot}}}{m_{\mathrm{veh}}} \left[\frac{rg(l_{\mathrm{tot}} - l_i)}{\Omega_{\mathrm{eq}_{i,j}}} - \frac{l_{\mathrm{h}} s(i)}{m_{\mathrm{W_{eq}}}} \right]^{-1} \tag{2.37}$$

with the distance between the two axles $l_{\mathrm{tot}} = l_{\mathrm{f}} + l_{\mathrm{r}}$, and the sign of the second addend defined by

$$s(i) = \begin{cases} s(i) = -1 & \text{for} \quad i = \mathrm{f} \\ s(i) = 1 & \text{for} \quad i = \mathrm{r}. \end{cases} \tag{2.38}$$

Note that the friction coefficient estimation depends entirely on the equivalent control $\Omega_{\mathrm{eq}_{i,j}}$ and constant values defining the vehicle mechanical properties. This implies that a correct representation of the vehicles geometry and weight is crucial for an accurate estimation of the road surface properties.

2.4.2 Observer evaluation

The observer has been validated on IPG Carmaker SW. The vehicle adopted is an Audi R8, with four independent electric motors (one for each wheel), of which only the front ones are used for traction, in order to run the velocity estimation algorithm (see Table 2.2) while simulating a front-wheel drive. The purpose of the test is to evaluate the algorithm capability of estimating the wheel slip-ratio and friction coefficient for each individual wheel. For this reason a simulation test has been set up,

Figure 2.8 Snapshot, from the IPG movie feature, of the driving sequence on the μ-split surface

which includes an acceleration phase followed by a braking phase on a μ-split surface (as it can be inferred from the different colors of the asphalt in Figure 2.8). The goal is to identify the friction coefficients of the two different surfaces, which have a maximum friction coefficient value μ_p equal to 0.8 and 0.4 for the left and right sides, respectively.

In order for the vehicle to avoid losing lateral stability, an Integral Sliding Mode (ISM) based (see Section 2.5.1) ABS system is implemented with an adaptive slip reference as the one mentioned in Section 2.2.1. The slip λ_p corresponding to peak adherence is reached implicitly by correcting the constant reference λ^* by using the estimate of the derivative $d\mu/d\lambda$. A small additional compensation is required by the driver, in order to offset the difference in longitudinal force applied on the left and right side, by acting on the steering wheel.

The result of the estimation of the vehicle velocity is reported in Figure 2.9. As it can be seen in the first graph, the overall vehicle acceleration, calculated as the average of the single wheel acceleration estimates obtained from (2.32), is tracked accurately (the estimate only shows a delay smaller than $0.05s$ during the transients). In the second graph, the steps of the algorithm explained in Table 2.2 are shown, which guarantee the accurate vehicle velocity estimation reported in the third graph. Note that, in order to run the algorithm successfully, proper hysteresis have to be introduced for the transition between different steps in the vehicle velocity estimation algorithm. Additionally, the wheel radius r has to be accurately estimated (for this purpose, when working with a real vehicle, the tire pressure should also be considered), in order to avoid offsets in the vehicle speed calculation, via (2.36), outside the braking phase.

Based on the vehicle velocity estimate during slipping, the slip ratio calculations for the front wheels are reported in Figure 2.10(a), plotted against the actual values available from the simulator. This comparison provides a significant measure of the

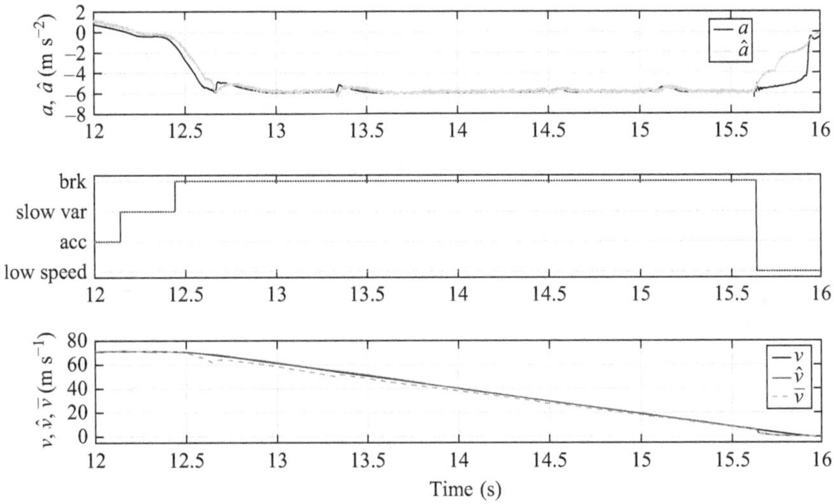

Figure 2.9 From the top: vehicle acceleration tracking, vehicle velocity estimation algorithm sequence, and vehicle velocity estimation result

quality of the estimation, since small deviations in the vehicle velocity might result in a significant error in the slip-ratio. The maximum estimation error of the slip-ratio is less than 0.02 during the braking sequence, the error showing a small but steady increase, due to the integral nature of the calculation of $\hat{v}$ in Table 2.2.

For the calculation of the friction coefficient μ, good estimations of the longitudinal and normal forces acting on each wheel are necessary. An alternative is to use directly the explicit formulation from (2.37). The results of the estimation is shown in Figure 2.10(b), when compared to the value obtained by computing the ratio in (2.3) with the actual force values F_x, F_z provided by the simulator. After a transition phase during the interval 12.5–12.8 s, the estimation is accurate for both high grip (front left wheel) and low grip (front right wheel) surfaces.

Such accuracy is reflected on the scatter plot of the samples $\{\hat{\lambda}, \hat{\mu}\}$, compared to the actual ones provided by the simulator $\{\lambda, \mu\}$ (Figure 2.11). Once the braking phase starts, the ISM controller ensures the tracking of the target slip reference $\lambda^*_{i,j}$ for each wheel. During the transition phase, for slip-ratio values in the 0.02–0.05 range, the inaccuracy of the estimation of the friction coefficients reflects in an underestimation of that portion of the $\mu(\lambda)$ curve. This distortion is less evident for slip values above 0.05. This is an expected result, considering that the base target slip is $\lambda^* = 0.1$ for all wheels, regardless of the specific $\mu(\lambda)$ shape. The majority of the samples in this region represent correctly the tire–road friction curve, while a few of them underestimate it, again due to the transient phase. From Figure 2.11 one can also notice, how the small slip-ratio offset has limited impact on the $\{\lambda, \mu\}$ phase plane representation. Note that, as an input to the observer, the actual brake

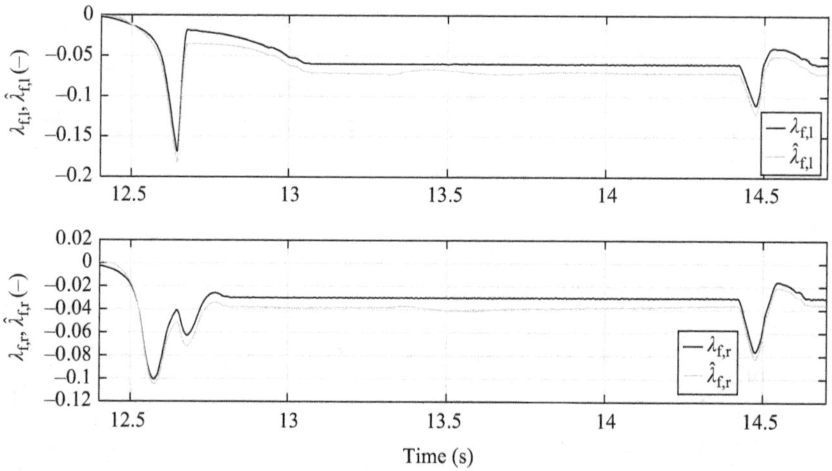

(a) Wheel slip estimation result

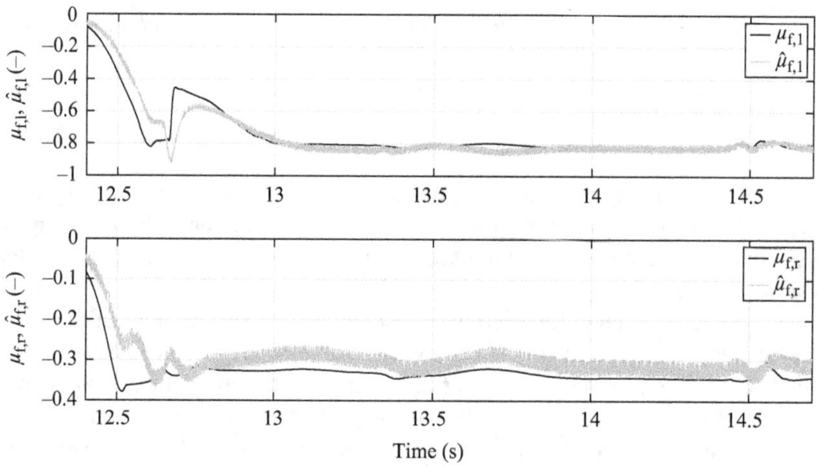

(b) Friction coefficient estimation result

Figure 2.10 Evaluation of the observer performance for front wheels: (a) wheel slip-ratio and (b) friction coefficient

torque $T_{w_{brk}}$ obtained from the simulation platform has been used. This is a major drawback of the illustrated technique, especially in case of conventional hydraulic braking system, due to the difficulties in the estimation of the actual braking torque. One possible solution, could be to identify offline a dynamic model of the braking actuation system, for example by using the sliding mode control component from an ISM controller, by adopting the technique which will be proposed in Section 2.5.6.

The results illustrated in this section appear promising for the estimation of wheel-slip and friction coefficient via sliding modes generation. Nevertheless,

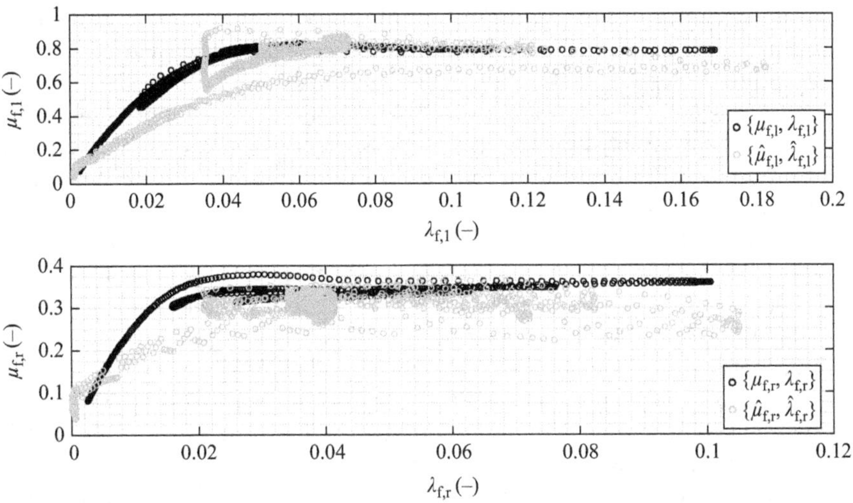

Figure 2.11 *Phase plane representation of the wheel slip and friction coefficient from Figures 2.10(a) and 2.10(b). Comparison between estimates* $\{\hat{\lambda}, \hat{\mu}\}$ *and actual values* $\{\lambda, \mu\}$ *from simulator*

the implementation of algorithms for the online identification of the parameters of the tire–road friction model, such as the one proposed in [24], would be negatively affected by the bad approximation of the slip during the transient. In order to obtain a more reliable parametric reconstruction of the $\mu(\lambda)$ curve, a time-varying reference slip signal would be needed, such as the one which will be used for the simulation in Section 2.5.6. By making the slip vary over a greater interval, including the low slip range $[0, \lambda_p]$, the observer would be estimating the longitudinal traction force F_x at low slips, while being on the sliding manifold, thus allowing us to acquire an undistorted estimation of μ.

2.5 Sliding mode control algorithms for wheel-slip control

SMC has been successfully applied to solve automotive control and observation problems, as testified by several recent publications and research projects. In particular several work have been done about traction/braking control systems [11,33]. The traction force produced by a vehicle is strongly influenced by road conditions. This is why it is necessary to design a robust traction force controller taking into account the time-varying tire/road interaction. Moreover, the designed control law has to prevent the generation of vibrations in order to avoid discomfort. Finally, the major design requirement is to make the wheel slip ratio follow a desired value, while guaranteeing that the control is stabilizing.

Typically, automotive subsystems are characterized by nonlinearities, modeling uncertainties and disturbances [29,34]. Then, SMC seems a very adequate and effective methodology to face with this kind of systems [4–6]. In the literature there are several proposals in which SMC solutions have been introduced and successfully tested on prototypes [33,35–40]. Since the implementation of a feedback control requires several quantities to be available to the controller at any sampling time instant and some of them are not always available, also SMC-based observers have been proposed [20,32,41–47]. Conventional control schemes are typically of closed loop type and mainly based on PID controllers but also robust controllers relying on sliding mode generation have been introduced [48–52].

In this chapter, in particular, different sliding mode control strategies are discussed: an ISM control [53], a Suboptimal Second Order Sliding Mode (SSOSM) control [54], a Super Twisting Sliding Mode (STSM) control [10] and the recently published Integral SSOSM (ISSOSM) control algorithm [55].

For the sake of comparison, a PID control algorithm and a FOSM control approach will be considered. In fact, since the sliding variable is set as the slip error, the relative degree is equal to 1 and a FOSM naturally applies. The basics of a classical FOSM control are given in [4]. Note that, the main difficulty with applying of this approach to solve the slip control problem is the discontinuity of the control variable which can cause the chattering phenomenon [8] which is hardly acceptable in practical cases. For the sake of simplicity, in the following subsections, the subscripts i, j are omitted, when obvious.

2.5.1 Integral sliding mode control

ISM method enables to generate an ideal sliding mode of the controlled system starting from the initial time instant t_0 [53]. The main feature of ISM control is that the control variable is split into two parts, that is

$$T_{\mathrm{w}}(t) = T_{\mathrm{w}_0}(t) + T_{\mathrm{w}_1}(t) \tag{2.39}$$

where T_{w_0} is generated by any suitably designed high level controller, for instance a PID controller as in this chapter, and T_{w_1} is a discontinuous control action designed to compensate the uncertainties affecting the system. The so-called integral sliding manifold is defined as the combination of a standard sliding variable, $\sigma = e_\lambda = \lambda^* - \lambda$, and an integral term, φ, as follows:

$$\Sigma(t) = \sigma(t) + \varphi(t). \tag{2.40}$$

More specifically, the integral term φ, for instance in the acceleration case, is

$$\varphi(t) = \sigma(t_0) + \int_{t_0}^{t} \frac{\partial \sigma}{\partial e_\lambda} \left(\dot{\lambda}^* - f(\zeta) - b(\zeta) T_{\mathrm{w}_0}(\zeta)\right) d\zeta \tag{2.41}$$

with the initial condition $\varphi(t_0) = \sigma(t_0)$. Then, the ISM control law is defined as

$$T_{\mathrm{w}_1}(t) = -U_{\max} \operatorname{sign}(\Sigma(t)) \tag{2.42}$$

with $U_{\max} > 0$ to enforce the sliding mode. By virtue of the choice of $\varphi(t)$ and $\varphi(t_0)$, it is apparent that the controlled system is in sliding mode on the manifold $\Sigma(t) = 0$ since the initial time instant. Moreover, under suitable assumptions on the auxiliary sliding variable, it is possible to show that the unmatched uncertainties are not amplified [53].

Note that, signal (2.42) is discontinuous so that a chattering phenomenon can occur. An alternative way to (2.42) in order to alleviate chattering has been presented in [53]. It is based on the so-called "equivalent control" concept (see [4] for a definition). Since the equivalent control, by definition, depends on the uncertain terms affecting the system it cannot be directly computed. However, it has been proved that it can be obtained as the output of a first-order linear filter as follows:

$$\tau_{\text{filt}} \dot{T}_{\text{w1,eq}} + T_{\text{w1,eq}} = T_{\text{w1}} \tag{2.43}$$

where τ_{filt} is the time constant of the filter suitably chosen in order to not distort the slow component of the control action. Since the equivalent control computed in such a way is an approximation, the rejection of the disturbance terms is guaranteed by suitably selecting the integral term of the sliding manifold as follows:

$$\varphi(t) = \sigma(t_0) + \int_{t_0}^{t} \frac{\partial \sigma}{\partial e_\lambda} \left(\dot{\lambda}^* - f(\zeta) - b(\zeta)(T_{\text{w}}(\zeta) - T_{\text{w}_1}(\zeta)) \right) d\zeta \tag{2.44}$$

where $T_{\text{w}}(t) = T_{\text{w}_0}(t) + T_{\text{w}_{1,\text{eq}}}(t)$. This implies that the mismatch between the actual equivalent control and its average value is intrinsically compensated in the sliding manifold.

2.5.2 *Higher order sliding mode control*

Another effective solution proposed in the literature to alleviate the chattering phenomenon is to solve an Higher Order Sliding Mode (HOSM) control problem. This is based on the definition of an "auxiliary system" associated with the original system, that is a perturbed chain of integrators built starting from the sliding variable and its time derivatives. The original control objective of steering to zero the sliding variable in a finite time, is transformed into the aim of finite time regulating the auxiliary system. This means, for any rth order sliding mode control (r being the relative degree of the auxiliary system), to force the system state to reach in finite time the subspace named *r-sliding manifold* $\sigma = \dot{\sigma} = \cdots = \sigma^{(r-1)} = 0$ and there remain for any subsequent time instant. The time derivative $\sigma^{(r)}$ is the bounded function which can be expressed as

$$\sigma^{(r)}(t) = \psi(t) + b(t)w(t) \tag{2.45}$$

where $\psi = \sigma^{(r)}|_{w=0}$ and $b = (\partial \sigma^{(r)} / \partial w) \neq 0$ are unknown functions and $w = \dot{u}$ is the so-called auxiliary control variable. Moreover, one has that there exist positive constants Ψ, $b_{\min}$, $b_{\max}$, such that

$$|f(x,t)| \leq \Psi \tag{2.46}$$

$$0 < b_{\min} \leq b(x,t) \leq b_{\max}. \tag{2.47}$$

Finally, in the literature, it has been shown that the HOSM control problem can be solved by any r-sliding mode controller of the type:

$$w(t) = U_{\max} \Xi \left(\sigma, \dot{\sigma}, \ldots, \sigma^{(r-1)} \right) \tag{2.48}$$

(see for instance [10,56–60]), where Ξ is a discontinuous function, and $U_{\max} > 0$ in order to ensure the finite time convergence of the state trajectories.

2.5.3 Suboptimal second order sliding mode control

The considered automotive system, with the selected sliding variable as the slip rate error with respect to a reference value, has a relative degree equal to 1. Since the use of a discontinuous control law is not appropriate for an electromechanical application, because of the chattering effect, it can be convenient to artificially increase the relative degree of the auxiliary system as suggested in [61]. To this end, by defining $\xi_1(t) = \sigma(t)$ and $\xi_2(t) = \dot{\sigma}(t)$, it yields:

$$\begin{cases} \dot{\xi}_1(t) = \xi_2(t) \\ \dot{\xi}_2(t) = \psi(t) + b(t)w(t) \end{cases} \tag{2.49}$$

where $\xi_2(t)$ is assumed to be unmeasurable, functions ψ and b have the bounds indicated in (2.46) and (2.47), $w(t)$ is the auxiliary control law which has to be designed to regulate $\xi_1(t)$ and $\xi_2(t)$ thus enforcing a second order sliding mode.

Then, by virtue of the artificial increment of the relative degree, the control actually fed into the plant is continuous since it is the output of an integrator having in input $w(t)$. So, given the auxiliary system (2.49), the control law can be expressed as

$$T_w(t) = -\int_{t_0}^{t} \alpha W_{\max} \operatorname{sign}\left(\xi_1(\zeta) - \tfrac{1}{2}\xi_{\max}\right) d\zeta \tag{2.50}$$

where $\xi_{\max}$ is the local minimum or maximum of the sliding variable, while the control parameters $\alpha = \alpha^*$ and $W_{\max}$ are chosen such that

$$W_{\max} > \max\left(\frac{\Psi}{\alpha^* b_{\min}}; \frac{4F}{3b_{\min} - \alpha^* b_{\max}}\right) \tag{2.51}$$

$$\alpha^* \in (0, 1] \cap \left(0, \frac{3b_{\min}}{b_{\max}}\right). \tag{2.52}$$

It can be proved (see [61]) that, with the constraints (2.51) and (2.52), the control law (2.50) enforces the generation of a sequence of states with coordinates featuring the contraction property expressed by $|\xi_{\max_{i+1}}| < |\xi_{\max_i}|$, $i \in \mathbb{N}^+$, which occurs when the switching line:

$$\xi_1 = -\frac{\xi_2|\xi_2|}{2W_{\max}} \tag{2.53}$$

is crossed.

Note that the SSOSM algorithm requires the control actually fed into the plant is continuous, which is highly appreciable in case of mechanical plants, and the knowledge of the first time derivative of the sliding variable is not necessary, since the computation of its extremal values, through the methods described in [54], is sufficient.

2.5.4 Integral suboptimal second order sliding mode control

In this subsection, the recently introduced ISSOSM control methodology is recalled [55]. This represents an extension of the SSOSM control algorithm with improved robustness properties against the uncertainties affecting the system. The idea is to reduce to a minimum the so-called reaching phase [4], during which the controlled system is not insensitive to the disturbances. Consider the auxiliary system (2.49) and define a *transient function* as

$$\begin{cases} \varphi(t) = (t - t_r)^2(c_0 + c_1(t - t_0)), & \forall t,\ t_0 \le t \le t_r \\ \varphi(t) = 0, \quad \forall t > t_r \end{cases} \tag{2.54}$$

where c_0 and c_1 are

$$c_0 = \sigma(t_0)T_p^{-2} \tag{2.55}$$

$$c_1 = \dot{\sigma}(t_0)T_p^{-2} + 2\sigma(t_0)T_p^{-3} \tag{2.56}$$

while $T_p = t_r - t_0$ is the so-called “prescribed time”, which allows one to steer the sliding variable σ to zero at the time t_r. Note that, from (2.55) and (2.56), the transient function is realized such that the initial conditions are

$$\sigma(t_0) = \varphi(t_0) \tag{2.57}$$

$$\dot{\sigma}(t_0) = \dot{\varphi}(t_0). \tag{2.58}$$

Then, the auxiliary sliding manifold is defined as in (2.40), where Σ is an auxiliary sliding variable such that $\xi_1 = \Sigma$ and $\xi_2 = \dot{\Sigma}$, while the control law is of the same form of (2.50) with constraints as in (2.51) and (2.52). The finite time convergence of the sliding variable in front of matched uncertainties can be proved relying on the results presented in [55].

2.5.5 Super twisting sliding mode control

STSM control is another particular case of second order sliding mode control in which, similarly to the SSOSM algorithm, the knowledge of the first time derivative of the sliding variable is not required [10]. The STSM control law can be expressed as follows:

$$\begin{aligned} T_w(t) &= v_{STSM}(t) - W_{max}\sqrt{|\xi_1(t)|}\ \text{sign}\,(\xi_1(t)) \\ \dot{v}_{STSM}(t) &= -V_{max}\ \text{sign}\,(\xi_1(t)) \end{aligned} \tag{2.59}$$

where W_{max} and V_{max} are suitably chosen in order to ensure the sliding mode generation [10]. Note that, STSM generates the continuous control function that steer the sliding variable and its derivative to zero in a finite time in the presence of matched disturbances, the bounds of which are assumed to be known. The main drawback of STSM control is related to the need to know the boundaries of the disturbance gradient. The overestimation of these values can imply to have larger than necessary control gains.

2.5.6 Simulations

The sliding mode controllers previously discussed are now evaluated in a simulation test, where the wheel-slip is regulated to a desired time varying value. Only results relevant to the front-left wheel are illustrated, since the performances obtained for the other wheels are comparable. The tests have been performed by using IPG Carmaker SW, as anticipated in Section 2.2.1, thus consolidating the results presented in [11], which were only evaluated on a simplified model. The simulation consists of a braking session, from an initial speed of $80\,\mathrm{km\,h^{-1}}$, which features a friction coefficient variation from $\mu_{\max_1} = 0.4$ to $\mu_{\max_2} = 0.8$, which, for the front wheels, occurs at approximately time instant 16.6 s. The purpose of the test is not to optimize the braking distance, but rather to display its capability to track a time varying wheel-slip reference λ^*, which has a sinusoid like oscillation between the extreme values -0.4 and 0.

Note that in the ISM design a nominal controller is required to be included, whereas it is not necessary in the other SMC algorithms. The nominal control is the same PID algorithm used in the standalone experiment. All controller parameters have been tuned in order to obtain an adequate response for each algorithm.

More specifically, in order to assess the performances of the SMCs, the control results have been compared with those of a PID, the gains of which are varied according to a Gain-Scheduling technique, and are tuned based on the Ziegler–Nichols method [62]. To tune the PID, the response of a feedback model, based on (2.15), to a step signal of amplitude 0.2 has been considered. According to the standard Ziegler–Nichols procedure, the feedback gain $\bar{K}_{\mathrm{P}}$ and the oscillation period T_{P}, corresponding to the stability limit, are calculated at different working points, determined by different values of velocity v and friction coefficient μ. Based on the calculated values of $\bar{K}_{\mathrm{P}}$, T_{P}, adaptive gain matrices are derived as follows:

$$K_{\mathrm{P}} = 0.6\bar{K}_{\mathrm{P}} \qquad T_{\mathrm{I}} = 0.5T_{\mathrm{P}} \qquad T_{\mathrm{D}} = 0.125T_{\mathrm{P}} \tag{2.60}$$

for the PID expressed in the form

$$u(t) = K_{\mathrm{P}}\left(e(t) + \frac{1}{T_{\mathrm{I}}}\int_0^t e(\tau)d\tau + T_{\mathrm{D}}\frac{de(t)}{dt}\right) \tag{2.61}$$

As far as the FOSM controller is concerned, to avoid feeding the discontinuous signal directly to the actuators, a first-order linear filter is used, having as input the discontinuous signal with amplitude 2,500 Nm, and time constant equal to $\tau_{\mathrm{FOSM}} = 0.05$ s.

In Figure 2.12 the response of FOSM, PID (top), SSOSM, ISSOSM and STSM (bottom) controllers is shown. It can be observed that FOSM gives rise to a controlled slip-ratio with oscillations. As a matter of fact, the presence of the filter, which is intended to alleviate the chattering effect, prevents the attainment of an ideal sliding mode. The sliding variable evolves in a close vicinity of the sliding manifold. This implies that the controlled system is not robust in front of possible matched uncertainties. The corresponding braking torque presents an oscillatory behavior which could make the adoption of this control algorithm critical in real cases.

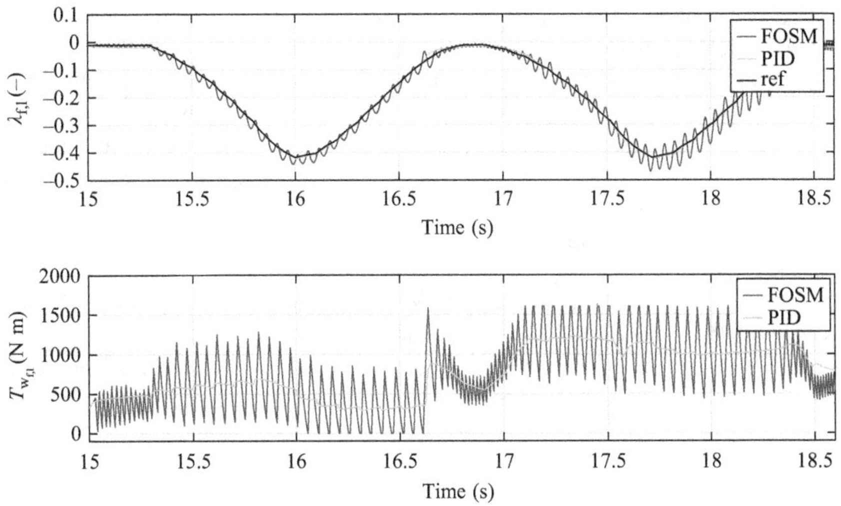

(a) FOSM, PID

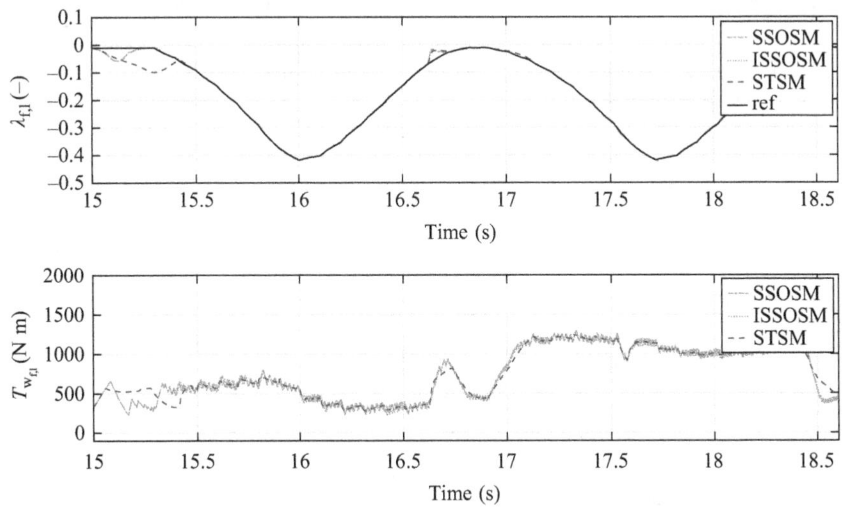

(b) SSOSM, ISSOSM and STSM

Figure 2.12 Time evolution of the front-left wheel-slip and of the braking torque when the considered controllers are applied to the vehicle: (a) FOSM, PID controllers and (b) SSOSM, ISSOSM and STSM controllers

As for the other controllers, during the first transient interval, SSOSM and ISSOSM algorithms guarantee better behaviors. The main difference between the two second-order algorithms, is due to the fact that the ISSOSM control law provides improved robustness properties of the controlled system, as mentioned in Section 2.5.1, and proved in [55]. In fact, the SSOSM controller, as any SMC not of integral type, makes the plant unsensitive to the matched uncertainties only after the so-called reaching phase.

The same considerations about robustness hold for the STSM controller. Note that the control torques of all second-order SMCs are continuous without the need for the low-pass filter, so that they result in being valid solutions for field implementation. The root mean square (RMS) tracking error σ_{RMS} and the RMS value of the control signal T_{RMS} for each controller are reported in Table 2.3.

Figure 2.13 shows the results in case of ISM controller, compared to the PID one. As discussed in Section 2.5.1, one of the main features of the ISM control algorithm is that the control law is split into two different parts: the nominal controller, which in this case is a PID, and the sliding mode component. In fact, performance and control signal are similar to that of the PID standalone (as well as to those of second-order SMCs, see Table 2.3), as the filtered control signal (2.43), designed according to

Table 2.3 Performance indices

RMS	PID	FOSM	SSOSM	ISSOSM	STSM	ISM
σ_{RMS}	0.0052	0.0260	0.0080	0.0081	0.0183	0.0130
T_{RMS}	817.918	854.119	804.571	804.356	804.796	816.958

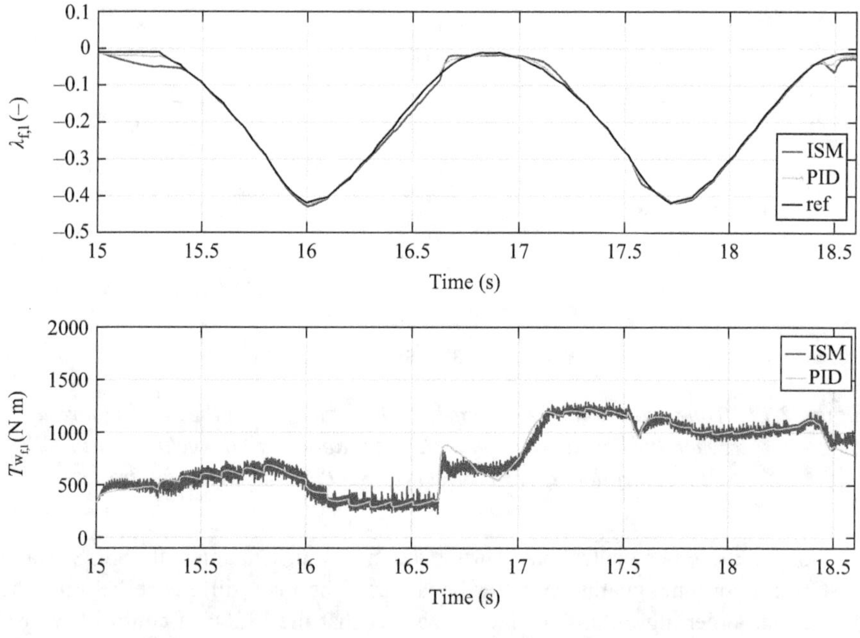

Figure 2.13 Time evolution of the front-left wheel-slip and of the braking torque when PID and ISM controllers are applied to the vehicle

transient function (2.44), is applied. This implies that the original discontinuous part of the controller becomes continuous, with consequential chattering alleviation. The so-called "equivalent control" is exploited with the beneficial effect that the sliding mode component plays the role of disturbance estimator, as it corresponds exactly to the matched uncertainties affecting the system.

In the applications illustrated in this chapter, the ABS is applied to electric motors with fast response, which can be approximated as first-order low-pass filters, with small time constant $\tau_{\mathrm{em}} \approx 0.002$ s. In case a hydraulic braking system is adopted, the transfer function from brake torque request to actual brake torque is less straightforward. Investigations have shown that such relation can be approximated as second-order linear dynamics [63]. The distortion introduced by the braking dynamics, in addition to other external influences not considered in model (2.12), makes the correct estimation of the braking torque affecting the wheel a challenging task. One possible way of estimating the actual braking torque T_{brk} on the wheel is to use the aforementioned disturbance estimation capability of the ISM technique.

Let us assume that the unknown component of the braking dynamics $T_{\mathrm{brk}_{\mathrm{dyn}}}$ corresponds to the matched disturbance $T_{\mathrm{w_d}}$, i.e.,

$$T_{\mathrm{brk}} = T^*_{\mathrm{brk}} + T_{\mathrm{brk}_{\mathrm{dyn}}} = T^*_{\mathrm{brk}} + T_{\mathrm{w_d}}. \tag{2.62}$$

We can get a demonstration of the validity of such an assumption, by running two separate simulations of a braking sequence: one equal to those reported in Figures 2.12 and 2.13, and a second one which contains an additional sinusoidal signal added as a matched disturbance to the brake torque request T^*_{brk}. The results of the simulations are reported in Figure 2.14, which shows how accurately the sliding mode component

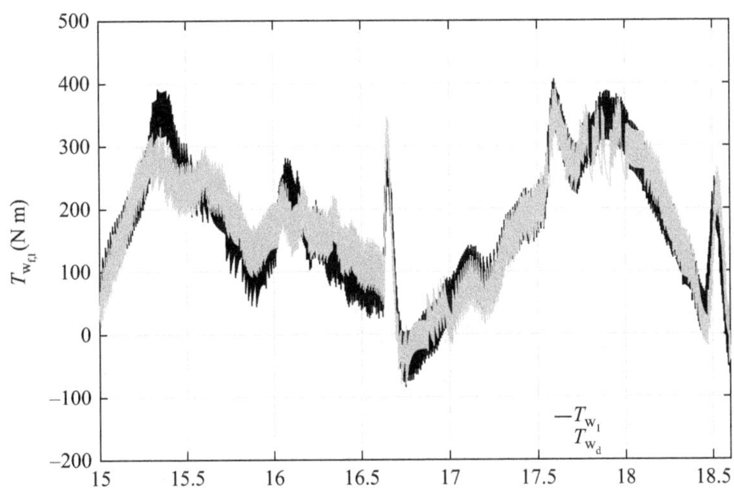

Figure 2.14 Time evolution of the ISM control signal when a sinusoidal matched disturbance affects the system

T_{w1} from the braking sequence matches signal T_{w_d}, the latter being the sum of the sliding mode component from test and the known additional sinusoidal disturbance.

Note that ISM control is a model-based algorithm. In a practical implementation, it may happen that a reliable estimation of F_x is not available. In this case the model is less accurate, and, consequently, the estimation of T_{brk} deteriorates as well.

2.6 Sliding mode low level actuation in conventional ABS

The ABS controllers illustrated in the previous sections are designed to track a desired wheel-slip ratio, by requesting a braking torque from the electrical actuator. In principle, these controllers work thanks to the quickness of response and accuracy of the electric actuator in a E-4WD. The ABS technology employed in conventional vehicles is not as naturally suited for such a task. The typical configuration of ABS/ESP systems includes a set of valves and pumps that can either reduce the pressure on disks generated by the brake pedal action, or artificially increase the pressure on selected wheels. Typically, a rule-based ABS algorithm [64] realizes pressure modulation, by going through three phases: *increase* (or apply), *decrease* (or dump) and *hold*.

The working principle, in the case of a hydraulic braking system, can be explained as follows. The ABS is activated as soon as the wheel slip reaches a certain threshold. In order to reduce the wheel slip, by producing an acceleration of the wheel, the pressure in the corresponding brake caliper is released until a certain system state is reached, corresponding to the reaching of either a wheel slip or acceleration threshold. After the new state is reached, pressure can be increased again. The frequency of these control loops is about 1–3 Hz [64], as long as significant deviations between the actual speed and the reference wheel speed occur.

This type of control approach is widely used in industry owing to its high robustness and relatively low demands to the control units. Due to the on–off nature of the control strategy, to the deviations in the caliper pressure estimation, to the delays in the actuation chain, and to several other factors, an operation of the ABS outside the linear area of the $\{\lambda, \mu\}$ diagram (Figure 2.4) is inevitable.

In order to practically apply this strategy, the specific hydraulic configuration of the ABS/ESP system has to be considered. As a result of decades of development of ABS technology, several configurations have been designed, which have the common trait of controlling selected wheels between two states, high and low pressure. Due to this "binary" characteristic, the generation of a continuous torque/pressure on the disk calipers is not an easy task, with its degree of difficulty depending on the configuration of the hydraulic circuit. Decoupled branches for each brake allow for an easier tracking of the different pressures independently, while coupled configurations introduce strong disturbances, as the pressures in the different chambers tend to even out during a dynamic transient in which the brake pressure targets are different, due to noncontrollable one-way valves and safety fluid outlets. A simplified representation of the coupled hydraulic circuit, as the one in Figure 2.15, can be used for the description of the system which was used in the demonstrative example in this section.

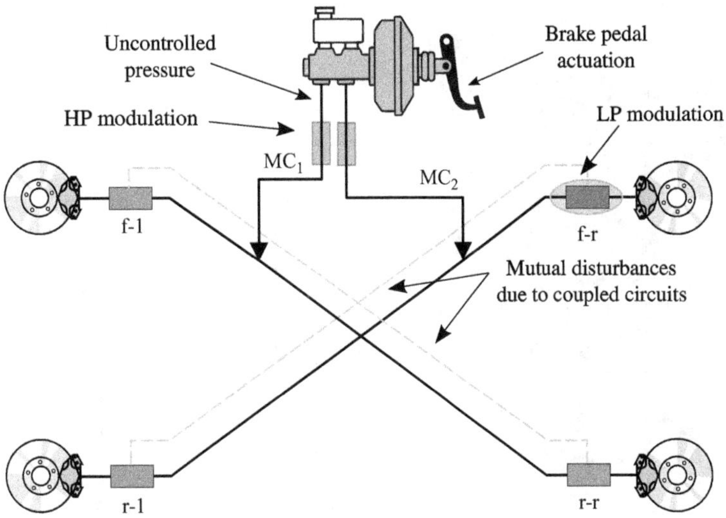

Figure 2.15 Schematic representation of the hydraulic model in "X" configuration, used for the design of the control system

Due to the considerations expressed above, the approach of continuously controlling the pressure on the brake discs has not found wide application in the industry yet [64]. Nevertheless, the high actuation frequency of the hydraulic valves could make it a natural field of application for a traditional FOSM controller, in a similar way to how a relay system is controlled in switching mode. In fact, some of the most common application field for SMC are electrical and electromechanical systems (see for instance [5,65] and the references therein), where the conversion of electrical energy is performed by power electronics. Typically, power electronic devices work in switching mode: semiconductor elements are switched on and off, as does the control voltage, resulting in smoother variations in the load current.

A detailed explanation of the functioning principle of ABS/ESP systems in their different configurations can be found in [66]. In the next subsection we describe briefly how the on/off principle we just mentioned practically works in the so-called "X configuration", and how the system can be steered in order to replicate a brake-by-wire system. In Section 2.6.2, we present an SMC-based approach for the control of such system, and, finally, in Section 2.6.3 we illustrate the results of such control system, evaluated on a HIL testrig.

2.6.1 HIL braking system configuration

A HIL setup (Figure 2.16(a)) was used to test the FOSM control law for the low level actuator. It consists of an ABS/ESP brake system with X layout: this layout involves a diagonal split configuration where each circuit operates one front brake and its

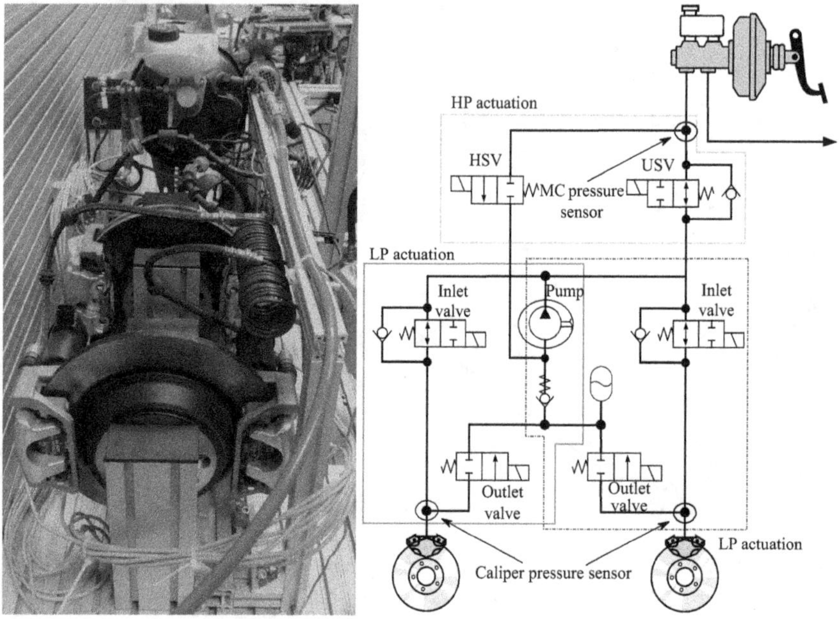

Figure 2.16 HIL testrig and hydraulic schematic of one branch in the X configuration: (a) HIL testrig: brake discs and calipers and (b) hydraulic modulator scheme: single branch

diagonally opposed rear brake. The X configuration is widely used in commercial and light duty vehicles because it involves the minimum amount of brake piping, hoses, hydraulic joints and static or dynamic seals. Moreover the X configuration is preferable on vehicles with higher forward weight distribution (generally front wheel drive vehicles) in order to meet the safety requirements in case of failure of one branch of the circuit [66].

The most important component of an ABS/ESP system is the hydraulic modulator. It incorporates 12 solenoid valves that can open or close the hydraulic circuits connecting the master cylinder (MC) and the brakes (Fig. 2.16(b)). For the sake of simplicity, we divide the hydraulic circuit of the modulator into a High Pressure (HP) side, which is used to control the MC pressure, and a Low Pressure (LP) side, whose actuators regulate the brake calipers pressure. The ABS system is designed to be driven by the LP actuators alone. During an EPS intervention, in order to increase the MC pressure without the driver's direct intervention, the HP actuators are also operated.

In the LP side, the inlet valves, between the master cylinder and the brakes, control the pressure application. The outlet valves, between the brakes and the return pump, control the pressure release. Under normal conditions, the solenoid valves in

Table 2.4 LP hydraulic configurations

LP mode	Inlet valve	Outlet valve
Increase	Open	Closed
Hold	Closed	Closed
Decrease	Closed	Open

Table 2.5 HP hydraulic configurations

HP mode	USV	HSV	Pump
Increase	Closed	Open	On
Hold	Closed	Closed	Off
Decrease	Open	Closed	Off

the hydraulic modulator are set in apply mode (*increase*), which means the inlet valve is open. The hydraulic modulator then forms a straight-through connection between the master cylinder and the brakes. Consequently, the brake pressure generated in the master cylinder when the brakes are applied is transmitted directly to the brakes at each wheel.

As the degree of brake slip increases due to braking on a slippery surface or heavy braking, the risk of the wheels locking up also increases: in these conditions the ABS is triggered. The solenoid valves are switched to the pressure to maintain (*hold*) setting: the connection between the master cylinder and the brakes is shut off (inlet valve is closed) so that any increase of pressure in the master cylinder does not lead to a pressure increase at the brakes. If the degree of slip of any of the wheels increases further despite this action, the pressure in the considered brake(s) must be reduced. To achieve this, the solenoid valves are switched to the pressure release setting (*decrease*). The inlet valve is still closed, and in addition, the outlet valve opens to allow the return pump integrated in the hydraulic modulator to draw brake fluid from the brake(s) in a controlled manner. The pressure in the relevant brake(s) is thus reduced so that wheel lock-up does not occur. The combinations of inlet/outlet valves corresponding to the desired modes are summarized in Table 2.4.

The ESP intervention may require to increase the brake pressure specified by the driver in order to stabilize the vehicle. To do this, the USV (Under-pressure Switching Valve) is closed and the HSV (High-pressure Switching Valve) is opened. Similarly to the ABS case, a DC motor driven plunger pump is used, with the difference that in the ESP case the pump can generate pressure without requiring the primary pressure at the MC produced by the driver pressing the brake pedal. The combinations of the HP actuators states and the associated MC pressure modes are reported in Table 2.5.

Six pressure sensors are installed in order to perform measurements of brake pressure in each circuits as well as in each brake caliper. For the hydraulic circuit

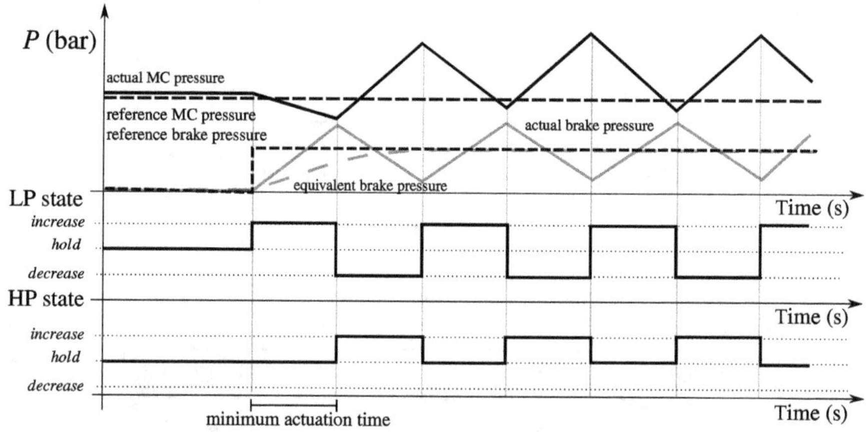

Figure 2.17 Theoretical time domain behavior of the pressure signals in the case of continuous reference tracking via on–off switching

branches $k = 1, 2$, and the brake positions i,j defined in the previous sections, the following quantities are defined:

- Brake Caliper (BC) reference and actual pressure: $P^*_{\mathrm{brk}_{i,j}}(t)$, $P_{\mathrm{brk}_{i,j}}(t)$;
- Master Cylinder (MC) reference and actual pressure: $P^*_{\mathrm{MC}_k}(t)$, $P_{\mathrm{MC}_k}(t)$;
- Low Pressure (LP) and High Pressure (HP) actuators mode request: $u_{\mathrm{brk}_{i,j}}(t)$, $u_{\mathrm{MC}_k}(t)$.

By properly controlling the pump and the USV, HSV valves, the ESP system can be operated so that it behaves like a brake-by-wire braking system: the difference with the standard ABS setting is that the MC pressure is not generated by the driver pressing the brake pedal, but by a control logic operating the HP hydraulic actuators. Due to the finite actuation frequency of the valves, it is a difficult task to reach a precise pressure level, and then keep it by isolating the brake caliper from the MC. The behavior which is expected by switching the actuators at their peak sustainable frequency is shown in Figure 2.17. We assume the MC pressure $P_{\mathrm{MC}_k}(t)$ is within a certain tolerated distance from its reference value $P^*_{\mathrm{MC}_k}(t)$, which is greater than the target value for the brake pressure $P^*_{\mathrm{brk}_{i,j}}(t)$, and generally is equal to the maximum of the reference values for the two calipers connected to the MC. Assuming the brake caliper pressure initial condition is $P^*_{\mathrm{brk}_{i,j}}(0) = 0$, by switching the LP modulation valves into *increase* mode for the minimum time allowed by the actuation physical limits, the pressures of the MC and the one at the brake caliper tend to even out. Therefore, in the next interval, in order to increase the MC pressure above its reference value and decrease the brake pressure towards its own, the valve states are switched. This sequence brings the system into a state similar to the starting one, so that the sequence repeats itself for as long as the target brake pressure $P^*_{\mathrm{brk}_{i,j}}$ is at an intermediate level between the current MC pressure $P_{\mathrm{MC}_k}(t)$ and 0 bar.

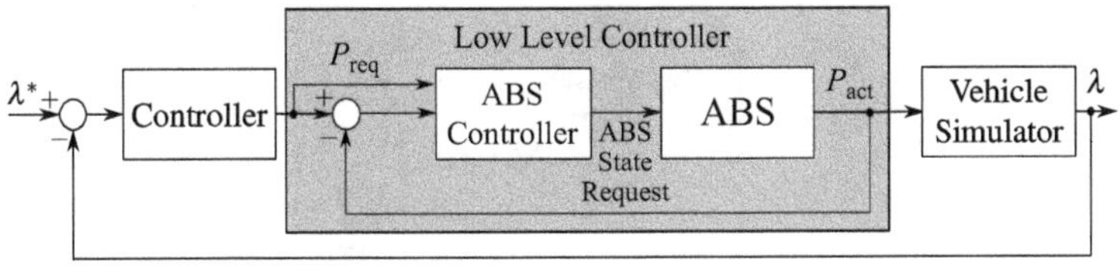

Figure 2.18 Schematic representation of the low level controller in the context of ABS control system

2.6.2 Low level control algorithm

The rule-based pressure control illustrated in the previous section has the purpose of tracking a continuous target brake pressure by using a discontinuous control action. The finite actuation states correspond to specific configurations of the valves and pumps (see Tables 2.4 and 2.5). Ideally, if the control states could switch at infinite frequency, the control algorithm illustrated above would be a conventional FOSM control.

Unfortunately, the valves have finite commutation time, and the pump should not be activated above a threshold frequency. Additionally, in case of the LP actuation mechanism, the switching between different operation modes (*increase* to *decrease* and vice versa) requires the transition through the intermediate state *hold*, which causes the actuation time to double up. Finally, the fast response of the pressure inside the calipers, when the valves are switched, does not allow for the natural dumping in the pressure response, which would be necessary for the attenuation of the pressure oscillations. Albeit all these limitations, we can consider that, for our purpose of wheel-slip tracking, the braking effect on the brake disc is equivalent to the continuous component of the actual brake pressure in Figure 2.17. Note that the pressure sensor is located before the caliper, therefore it is reasonable to assume that the high frequency components of the measured signal are attenuated in some measure by the mechanical action of the calipers on the disk.

The low level control principle that we illustrate in this section can be framed in the bigger picture of wheel-slip control as it is shown in Figure 2.18. We assume for simplicity that the set of torque demands on the brake disks are generated with one of the control laws presented in Section 2.5, and that the transfer function between brake torque and brake disks pressure is a constant $\bar{G}$, so that the following approximations hold:

$$T_{\mathrm{brk}_{i,j}}(t) \approx \bar{G} \cdot P_{\mathrm{brk}_{i,j}}(t) \tag{2.63}$$

The pressure demands have to be translated into a set of requests for the HP and LP actuators. As was explained in Section 2.6.1, during the brake-by-wire actuation:

- the LP side is controlled between the states *increase* and *decrease*;
- the HP side is controlled between the states *increase* and *hold*, as the decrease in pressure inside the hydraulic circuit is a consequence of the increase in the final caliper pressure. Since the *decrease* phase is only entered when a global release of

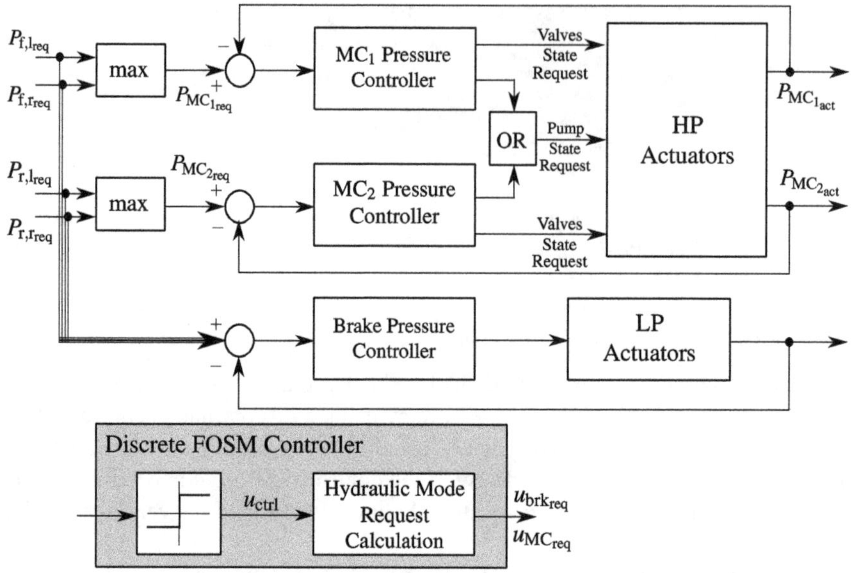

Figure 2.19 Low level control scheme for the HP and LP actuators

the calipers is requested, this case is not covered in the SMC approach illustrated here.

The FOSM control scheme for the HP and LP controlled variables, which reflects the considerations above, is the one represented in Figure 2.19. The sliding variables, for HP and LP actuators, are the pressure tracking errors, respectively:

$$\sigma_{HP} = P^*_{MC_k} - P_{MC_k} \tag{2.64}$$

$$\sigma_{LP} = P^*_{brk_{i,j}} - P_{brk_{i,j}}. \tag{2.65}$$

For $\sigma = \sigma_{HP}, \sigma_{LP}$, the FOSM control law is then implemented, following the "Boundary Layer Solution" described in [5], as

$$u_{ctrl} = \text{sat}\left(\frac{\sigma}{\sigma_{tol}}\right) \tag{2.66}$$

where σ_{tol} is the tolerated pressure error for each specific actuation, which can also be set differently for *increase* and *decrease* sequences. The hydraulic circuit modes are derived from u_{ctrl}, for the HP and LP actuators, as follows:

$$u_{brk} = \begin{cases} \textit{increase} & \text{if } u_{ctrl} = 1 \\ \textit{hold} & \text{if } |u_{ctrl}| < 1 \\ \textit{decrease} & \text{if } u_{ctrl} = -1 \end{cases} \tag{2.67}$$

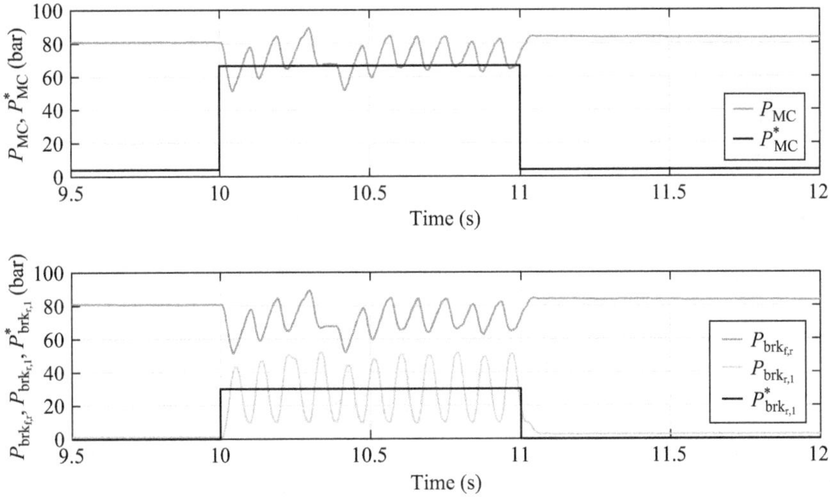

Figure 2.20 Single brake (r-l) step response, while the coupled brake (f-r) is in its default mode increase

and

$$u_{MC} = \begin{cases} \textit{increase} & \text{if } u_{ctrl} = 1 \\ \textit{hold} & \text{if } u_{ctrl} < 1. \end{cases} \tag{2.68}$$

The computation of u_{brk} and u_{MC} has to consider the minimum commutation time between the *increase*, *hold* and *decrease* modes.

Note that, in case of the MC pressure control, the purpose of the closed loop control is not an accurate tracking of the reference signal, but rather the maintenance of a pressure level that ensures the tracking of the reference at LP level. For this reason, the references $P^*_{MC,k}$ can also be corrected via feedback laws based on the low frequency components of the LP tracking errors.

2.6.3 Low level control results

In the reported experiments, valves and pump are actuated at a maximum frequency of 20 Hz, a conservative value selected to avoid possible damages to the mechanical components. The typical benchmarking test which is performed for the evaluation of a pressure tracking device in an ABS/ESP system is the response to a step input. The results of this test, performed on the HIL testrig, are shown in Figures 2.20 and 2.21, for the second branch, corresponding to front-right (f-r) and rear-left (r-l) brakes.

In Figure 2.20, the step response of the r-l brake caliper pressure is displayed, when the coupled front-right brake is kept in its default *increase* mode. This is the base control test to be performed, as it introduces the least amount of interaction between the two coupled hydraulic circuits. The f-r pressure is already at its maximum

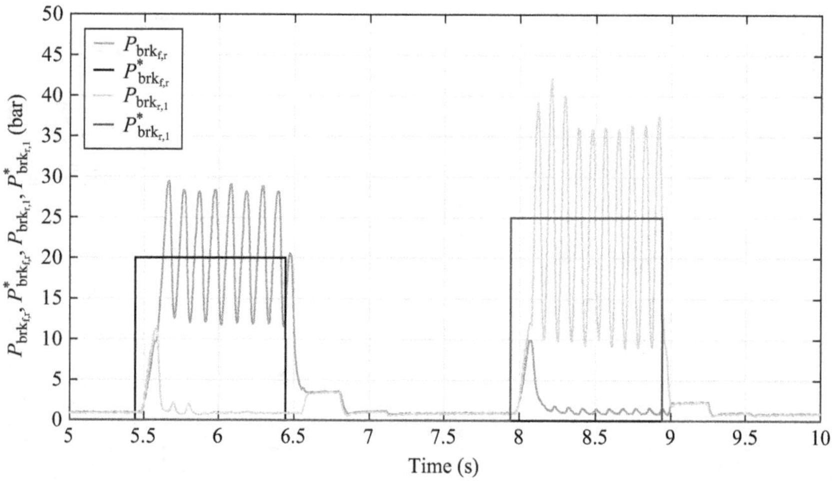

Figure 2.21 Step response of both brakes, while the noncontrolled one is in decrease mode

value $P_{brk_{fr}} = P_{MC_2}$, and the effect of the control on it is to generate an oscillating behavior which, however, does not deviate too much from the prescribed value. The r-l brake caliper pressure, on the other hand, reacts instantaneously to the step input, its continuous component reaching the reference value in approximately 0.05 s both to the rising and to the falling step.

A second step response test is displayed in Figure 2.21: this time the uncontrolled brake caliper is set to *release* state. The tracking of the target pressure is accomplished also in this case, for both brakes in the branch, by the average values of the controlled variables. Nevertheless, the coupling of the hydraulic circuits has some negative effects on the control performance. When considering the step response of the first brake actuated (f-r), two undesired behaviors are noticed:

- the pressure increase in r-l after the step command: when the state request for the HP circuit switches to *increase*, both LP valves sets are still in their default mode (*increase*), which causes the overshoot of $P_{brk_{rl}}$, since some time is required to make them switch to the *release* mode. Around the time instant 5.55 s, after the commutation of the valves is completed, the r-l pressure decreases again;
- both f-r and r-l pressure stabilize at approximately 4 bar after the pressure request $P^*_{brk_{fr}}$ drops to 0 bar. In this case the undesired effect is due to the r-l outlet valve, which, when opening, allows the pressure in the two coupled calipers to equalize. The resulting pressure of approximately 4 bar in the lower part of the circuit is equal to the remaining pressure P_{MC_2} in the master cylinder, which makes it necessary for the HP circuit to enter *release* state, as the only way to bring $P_{brk_{fr}}$, $P_{brk_{rl}}$ to zero.

Same considerations can be made for step response of r-l (approximately 8–9 s).

2.7 Conclusions

In this chapter, an overview on the longitudinal dynamics control of vehicles, focusing on the application of Sliding Mode Control theory, has been provided. The basics of traction control are recalled and a suitable vehicle model for the longitudinal dynamics has been derived making reference to the more used identification approaches reported in the literature. The second part of the chapter has been devoted to observation problems. More specifically, Sliding Mode-based observers are presented to estimate the friction coefficient, forces and velocities needed to find an accurate estimate of the effective wheel slip values, which represent the controlled variables in the considered traction control problem. Then, in the third part of the chapter, a discussion on the main Sliding Mode Control approaches, suitable to solve the traction control problem, has been reported. They have been applied to the considered vehicle dynamics in order to assess their beneficial properties in front of unavoidable modeling uncertainties and external disturbances. All the controllers have been evaluated in a very realistic simulation scenario by using IPG Carmaker SW, with satisfactory results. Finally, the last part of the chapter includes the description of a first-order low level sliding mode controller, implemented for an ABS HIL Testrig at the Technology University of Ilmenau. The technological aspects related to the application of Sliding Mode Control have been emphasized, showing the beneficial effects of this control method, very adequate to be used even in field implementation.

Acknowledgments

The project leading to this application has received funding from European Union's Horizon 2020 research and innovation programme under the Marie Skodowska-Curie grant agreement No. 675999. Authors would like to express thanks to Prof. Valentin Ivanov, Dr. Dzmitry Savitzky and M.Sc. Vincenzo Ricciardi (Technology University of Ilmenau) for their support with the setup and analysis of the tests on the ABS HIL Testrig.

References

[1] "ECE/TRANS/WP.29/2007/106, Proposal for the 11 series of amendments to Regulation no. 13 (Braking)," 2007.

[2] M. Kamachi, K. Walters, and H. Yoshida, "Improvement of vehicle dynamic performance by means of in-wheel electric motors," *Mitsubishi Motors Technical Review*, no. 18, pp. 106–112, 2006.

[3] D. A. Crolla and D. Cao, "The impact of hybrid and electric powertrains on vehicle dynamics, control systems and energy regeneration," *Vehicle System Dynamics. International Journal of Vehicle Mechanics Mobility*, vol. 50, no. 1, pp. 95–109, 2012.

[4] V. Utkin, *Sliding Modes in Control and Optimization*. Communication and Control Engineering, Heidelberg: Springer-Verlag Berlin, 1992.

[5] V. I. Utkin, J. Guldner, and J. Shi, *Sliding Model Control in Electromechanical Systems*. Boca Raton: Taylor & Francis Group, 1999.

[6] C. Edwards and S. Spurgeon, *Sliding Mode Control: Theory and Applications*. Systems and Control, London: Taylor & Francis Group, 1998.

[7] I. Boiko, L. Fridman, A. Pisano, and E. Usai, "Analysis of chattering in systems with second-order sliding modes," *IEEE Transactions on Automatic Control*, vol. 52, pp. 2085–2102, Nov. 2007.

[8] A. Levant, "Chattering analysis," *IEEE Transactions on Automatic Control*, vol. 55, pp. 1380–1389, Jun. 2010.

[9] G. Bartolini, A. Ferrara, A. Levant, and E. Usai, "On second order sliding mode controllers," in *Variable Structure Systems, Sliding Mode and Nonlinear Control* (K. D. Young and Ü. Özgüner, eds.), Lecture Notes in Control and Information, pp. 329–350, London, UK: Springer-Verlag, 1999.

[10] A. Levant, "Higher-order sliding modes, differentiation and output-feedback control," *International Journal of Control*, vol. 76, pp. 924–941, Jan. 2003.

[11] G. P. Incremona, E. Regolin, A. Mosca, and A. Ferrara, "Sliding mode control algorithms for wheel slip control of road vehicles," in *Proceedings of the American Control Conference*, vol. – (Seattle, WA, USA), pp. –, May. 2017.

[12] T. Kawabe, "Slip suppression of electric vehicles using model predictive PID controller," *International Journal of World Academy Science, Engineering and Technology*, vol. 67, pp. 524–529, Jul. 2012.

[13] V. Ivanov, D. Savitski, and B. Shyrokau, "A survey of traction control and antilock braking systems of full electric vehicles with individually controlled electric motors," *IEEE Transactions on Vehicular Technology*, vol. 64, pp. 3878–3896, Sep. 2015.

[14] M. Tanelli, G. Osorio, M. di Bernardo, S. M. Savaresi, and A. Astolfi, "Limit cycles analysis in hybrid anti-lock braking systems," in *Proceedings of the IEEE 46th Conference on Decision and Control* (New Orleans, LA, USA), pp. 3865–3870, Dec. 2007.

[15] V. D. Colli, G. Tomassi, and M. Scarano, "Single wheel longitudinal traction control for electric vehicles," *IEEE Transactions on Power Electronics*, vol. 21, pp. 799–808, May 2006.

[16] W. Pasillas-Lpine, "Hybrid modeling and limit cycle analysis for a class of five-phase anti-lock brake algorithms," *Vehicle System Dynamics*, vol. 44, pp. 173–188, Feb. 2006.

[17] S.-I. Sakai and Y. Hori, "Advantage of electric motor for anti skid control of electric vehicle," *European Power Electronics Drives Journal*, vol. 11, pp. 26–32, Sep. 2001.

[18] J. L. Zhang, C. L. Yin, and J. W. Zhang, "Improvement of drivability and fuel economy with a hybrid antiskid braking system in hybrid electric vehicles," *International Journal of Automotive Technology*, vol. 11, pp. 205–213, Apr. 2010.

[19] C. Song, J. Wang, and L. Jin, "Study on the composite ABS control of vehicles with four electric wheels," *Journal of Computers*, vol. 6, pp. 618–626, Mar. 2011.

[20] M. Amodeo, A. Ferrara, R. Terzaghi, and C. Vecchio, "Wheel slip control via second-order sliding-mode generation," *IEEE Transactions on Intelligent Transportation Systems*, vol. 11, pp. 122–131, Mar. 2010.
[21] V. Ivanov, B. Shyrokauand, D. Savitski, and J. Orus, "Design and testing of ABS for electric vehicles with individually controlled on-board motor drives," *SAE International Journal of Passenger Cars – Mechanical Systems*, vol. 7, pp. 902–913, Aug. 2014.
[22] J. Svendenius, *Tire Modeling and Friction Estimation*. Ph.D. Thesis, Lund University, Department of Automatic Control, Lund University, Box 118, SE-221 00 Lund, Sweden, Apr. 2007.
[23] E. Regolin and A. Ferrara, "SVM classification and Kalman filter based estimation of the tire–road friction curve," in *Proceedings of the 20th World Congress of the International Federation of Automatic Control (IFAC)*, vol. – (Toulouse, France), pp. –, Jul. 2017.
[24] M. Tanelli, L. Piroddi, and S. M. Savaresi, "Real-time identification of tire–road friction conditions," *IET Control Theory and Applications*, vol. 3, pp. 891–906, Jul. 2009.
[25] R. de Castro, R. E. Araujo, J. S. Cardoso, and D. Freitas, "A new linear parametrization for peak friction coefficient estimation in real time," in *Proceedings of the IEEE Vehicle Power and Propulsion Conference* (Lille, France), Sep. 2010.
[26] U. Kiencke, "Real time estimation of adhesion characteristic between tyres and road," in *Proceedings of the IFAC 12th Triennial World Congress*, 1993.
[27] S. Müller, M. Uchanski, and K. Hedrick, "Estimation of the maximum tire–road friction coefficient," *Journal of Dynamic Systems, Measurement and Control*, vol. 125, pp. 607–617, Dec. 2003.
[28] S. Savaresi and M. Tanelli, *Active Braking Control Systems Design for Vehicles*. London: Springer, 2010.
[29] R. Rajamani, *Vehicle Dynamics and Control*. New York: Springer US, 2012.
[30] J. J. E. Slotine and W. Li, *Applied Nonlinear Control*. Englewood Cliffs, NJ: Prentice-Hall, 1991.
[31] V. I. Utkin, *Sliding Modes in Optimization and Control Problems*. New York: Springer Verlag, 1992.
[32] M. Tanelli, A. Ferrara, and P. Giani, "Combined vehicle velocity and tire–road friction estimation via sliding mode observers," in *Proceedings of the IEEE International Conference on Control Applications* (Dubrovnik, Croatia), pp. 130–135, Oct. 2012.
[33] S. Drakunov, Ü. Özgüner, P. Dix, and B. Ashrafi, "ABS control using optimum search via sliding modes," *IEEE Transactions on Control Systems Technology*, vol. 3, pp. 79–85, Mar. 1995.
[34] G. Genta, *Motor Vehicle Dynamics – Modeling and Simulation*. Singapore: World Scientific, 1993.
[35] L. Fu, Ü. Özgüner, and İ. Haskara, "Automotive applications of sliding mode control," *IFAC Proceedings Volumes*, vol. 44, no. 1, pp. 1898–1903, 2011.

[36] İ. Haskara, H. C., and Ü. Özgüner, *Sliding Mode Compensation, Estimation and Optimization Methods in Automotive Control*, pp. 155–174. Lecture Notes in Control and Information Sciences, Berlin: Springer, 2002.
[37] Y.-W. Kim, G. Rizzoni, and V. I. Utkin, "Developing a fault tolerant power-train control system by integrating design of control and diagnostics," *International Journal of Robust and Nonlinear Control*, vol. 11, pp. 1095–1114, Sep. 2001.
[38] S. B. Choi, Y. T. Choi, and D. W. Park, "A sliding mode control of a full-car electrorheological suspension system via hardware in-the-loop simulation," *Journal of Dynamic Systems, Measurement and Control*, vol. 122, pp. 114–121, Apr. 1998.
[39] J. Ackermann, J. Guldner, W. Sienel, R. Steinhauser, and V. I. Utkin, "Linear and nonlinear controller design for robust automatic steering," *IEEE Transactions on Control Systems Technology*, vol. 3, pp. 132–143, Mar. 1995.
[40] P. Hingwe and M. Tomizuka, "Experimental evaluation of a chatter free sliding mode control for lateral control in AHS," in *Proceedings of the American Control Conference*, vol. 5 (Albuquerque, NM), pp. 3365–3369, Jun. 1997.
[41] C. Edwards, R. G. Hebden, and S. K. Spurgeon, "Sliding mode observers for vehicle mode detection," *Vehicle System Dynamics*, vol. 43, pp. 823–843, Feb. 2005.
[42] N. K. M'sirdi, A. Rabhi, L. Fridman, J. Davila, and Y. Delanne, "Second order sliding mode observer for estimation of velocities, wheel sleep, radius and stiffness," in *Proceedings of the American Control Conference* (Minneapolis, MN), pp. 3316–3321, Jun. 2006.
[43] N. Patel, C. Edwards, and S. K. Spurgeon, "Optimal braking and estimation of tyre friction in automotive vehicles using sliding modes," *International Journal of Systems Science*, vol. 38, pp. 901–912, Oct. 2007.
[44] H. Imine, L. Fridman, H. Shraim, and M. Djemai, *Sliding Mode Based Analysis and Identification of Vehicle Dynamics, Lecture Notes in Control and Information Sciences*, vol. 414. Berlin: Springer-Verlag , 2011.
[45] R. Tafner, M. Reichhartinger, and M. Horn, "Robust online roll dynamics identification of a vehicle using sliding mode concepts," *Control Engineering Practice*, vol. 29, pp. 235–246, Apr. 2014.
[46] R. Tafner, M. Horn, and A. Ferrara, "Experimental evaluation of non-linear unknown input observers applied to an EPS system," in *Proceedings of the American Control Conference* (Boston, MA), pp. 2409–2414, Jul. 2016.
[47] M. Reichhartinger, S. K. Spurgeon, and M. Weyrer, "Design of an unknown input observer to enhance driver experience of electric power steering systems," in *Proceedings of the European Control Conference* (Aalborg, Denmark), Jul. 2016.
[48] T. Acarman, Ü. Özgüner, C. Hatipoglu, and A.-M. Igusky, "Pneumatic brake system modeling for systems analysis," *Transactions Journal of Commercial Vehicles – V109-2*, pp. 1–9, Dec. 2000.

[49] J. D. Sànchez-Torres, A. G. Loukianov, M. I. Galicia, and R. Dominguez, "Robust nested sliding mode integral control for anti-lock brake system," *International Journal of Vehicle Design*, vol. 62, pp. 188–205, Jan. 2013.

[50] İ. Haskara, V. V. Kokotovic, and L. A. Mianzo, "Control of an electromechanical valve actuator for a camless engine," *International Journal of Robust and Nonlinear Control*, vol. 14, pp. 561–579, Feb. 2004.

[51] M. Reichhartinger and M. Horn, "Application of higher order sliding-mode concepts to a throttle actuator for gasoline engines," *IEEE Transactions on Industrial Electronics*, vol. 56, pp. 3322–3329, Sep. 2009.

[52] M. Yokoyama, J. K. Hedrick, and S. Toyama, "A model following sliding mode controller for semi-active suspension systems with MR dampers," in *Proceedings of the American Control Conference*, vol. 4 (Arlington, VA), pp. 2652–2657, Jul. 2001.

[53] V. I. Utkin and J. Shi, "Integral sliding mode in systems operating under uncertainty conditions," in *Proceedings of the 35th IEEE Conference on Decision and Control*, vol. 4 (Kobe, Japan), pp. 4591–4596, Dec. 1996.

[54] G. Bartolini, A. Ferrara, and E. Usai, "Chattering avoidance by second-order sliding mode control," *IEEE Transactions on Automatic Control*, vol. 43, pp. 241–246, Feb. 1998.

[55] A. Ferrara and G. P. Incremona, "Design of an integral suboptimal second order sliding mode controller for the robust motion control of robot manipulators," *IEEE Transactions on Control Systems Technology*, vol. 23, pp. 2316–2325, May 2015.

[56] G. Bartolini, A. Ferrara, V. I. Utkin, and T. Zolezzi, "A control vector approach to variable structure control of nonlinear systems," *International Journal of Robust and Nonlinear Control*, vol. 7, pp. 321–335, Apr. 1997.

[57] G. Bartolini, A. Ferrara, E. Usai, and V. I. Utkin, "On multi-input chattering-free second-order sliding mode control," *IEEE Transactions on Automatic Control*, vol. 45, pp. 1711–1717, Sep. 2000.

[58] T. Floquet, J.-P. Barbot, and W. Perruquetti, "Higher-order sliding mode stabilization for a class of nonholonomic perturbed systems," *Automatica*, vol. 39, pp. 1077–1083, Jun. 2003.

[59] A. Levant, "Quasi-continuous high-order sliding-mode controllers," *IEEE Transactions on Automatic Control*, vol. 50, pp. 1812–1816, Nov. 2005.

[60] F. Dinuzzo and A. Ferrara, "Higher order sliding mode controllers with optimal reaching," *IEEE Transactions on Automatic Control*, vol. 54, pp. 2126–2136, Sep. 2009.

[61] G. Bartolini, A. Ferrara, and E. Usai, "Output tracking control of uncertain nonlinear second-order systems," *Automatica*, vol. 33, pp. 2203–2212, Dec. 1997.

[62] J. G. Ziegler and N. B. Nichols, "Transactions of ASME," *ASME Journal of Dynamic Systems*, vol. 64, no. 759, pp. 759–768, 1942.

[63] C. F. Lee, D. Savitski, C. Manzie, and V. Ivanov, "Active brake judder compensation using an electro-hydraulic brake system," *SAE International Journal of Commercial Vehicles*, vol. 8, pp. 20–26, May 2015.

[64] D. Savitski, V. Ivanov, K. Augsburg, *et al.*, "The new paradigm of an anti-lock braking system for a full electric vehicle: experimental investigation and benchmarking," *Journal of Automobile Engineering*, vol. 230, no. 10, pp. 1364–1377, Oct. 2015.

[65] M. Cucuzzella, G. P. Incremona, and A. Ferrara, "Design of robust higher order sliding mode control for microgrids," *IEEE Journal on Emerging and Selected Topics in Circuits and Systems*, vol. 5, pp. 393–401, Sep. 2015.

[66] K. Reif, *Brakes, Brake Control and Driver Assistance Systems*. Berlin: Springer Vieweg, 2014.

Chapter 3
Sliding mode control of traction and braking in two-wheeled vehicles

Mara Tanelli[1], *Matteo Corno*[1], *and Antonella Ferrara*[2]

Four-wheeled vehicles are being equipped with many different active control systems which enhance drivers and passengers comfort and safety. In the field of two-wheeled vehicles, instead, the development of electronic control systems is following with some delay. However, the importance of active control for traction and braking has been recently recognized also for motorcycles. The motivation for this is twofold: on the one hand, in the racing context, these systems are designed to enhance vehicle performance; on the other hand, in the production context, the same control systems are intended to enhance the safety of non-professional bikers. Within such an evolving context, this chapter addresses the control problems related to the longitudinal dynamics of two-wheeled vehicles, that is traction and braking. To do this, some preliminary material introducing the dynamics of such vehicles is provided. Furthermore, the technological issues arising from the actuation dynamics and the measured signals available on board are discussed, to better highlight the challenges of this specific control problem. Then, the traction and braking problems are formulated within a sliding-mode framework, and the performance of second-order sliding mode controllers are analyzed.

3.1 Introduction

Automotive control is one of the fields where automatic control theory has the greatest public visibility. Vehicle Dynamics Control (VDC) systems are a selling point for many manufacturers. In the automotive industry, automatic control is a little less a "hidden" technology than in other fields. The industry got to this point through a long process that started in the 1971 with the anti-skid Sure-Brake system proposed by Chrysler and Bendix and from there, meeting alternate fate, got to the most recent advances in vehicle dynamics control.

[1]Dipartimento di Elettronica, Informazione e Bioingegneria, Politecnico di Milano, Piazza Leonardo da Vinci, 32, 20133, Milano, Italy
[2]Dipartimento di Ingegneria Industriale e dell'Informazione, University of Pavia, Via Ferrata 5, 27100 Pavia, Italy

The development of vehicle dynamics control systems for motorcycles started with some delay, but had a faster growth. The first automatic control system for motorcycles was, as for cars, the anti-locking braking system (ABS) introduced in 1983 by BMW. Now, the European Commission is mandating all motorcycles (over 125 cc) sold in Europe from 2016 to be equipped with ABS. The history of traction control is even more compelling. In 2008, only Ducati and BMW provided traction control-quipped motorcycles; by the end of 2012 the list of brands sporting their own version of traction control included: BMW, Ducati, Aprilia, MV-Agusta, Kawasaki, Honda, Yamaha and basically all the other big players in the field [1].

The sudden success of motorcycle dynamics control is due to two related factors: on the one hand, electro-mechanical actuators (electronic throttle-bodies, semi-active suspensions, actuated brakes) have become more cost-effective, reliable, lighter and smaller; on the other hand, the success of advanced control techniques on the racing track has promoted the image of automatic control as a performance-enhancing technology, rather than a safety-oriented one. High-end motorcycles are recreational vehicles for the thrill-seeking. Performance is a stronger selling point than safety. Once the initial investment had been faced by high-end motorcycles, the technology started to trickle down to more cost-effective vehicles where safety plays an important role as they are often used as commuter vehicles.

This chapter addresses the analysis and design of a safety-oriented rear wheel slip control system for ride-by-wire sport motorcycles, and of an active braking controller, relying on the so-called Second Order Sliding Mode (SOSM) approach. The sliding mode control methodology is interesting in the two-wheeled vehicle context because of its robustness properties, which make it particularly suitable to deal with the parameter uncertainties and the wide range of operating conditions typical of the automotive context [2–4]. Furthermore, SOSM controllers feature higher accuracy with respect to first order sliding mode controllers and generate continuous control actions, since the discontinuity, necessary to enforce sliding modes, is confined to the derivative of the control signal, thereby reducing the actuator stress and wear [3–5]. Apart from robustness against possible disturbances and parameter variations affecting the vehicle model, the sliding mode control methodology has also the advantage of producing low complexity control laws compared to other robust control approaches, which appear particularly suitable to be implemented in the Electronic Control Unit (ECU) of a controlled vehicle [3–6].

3.2 State of the art

The traction and braking dynamics are mainly determined by the so-called wheel-slip. Wheel-slip is defined as the normalized difference between the longitudinal velocity of the vehicle and the peripheral velocity of a tire:

$$\lambda = \frac{\omega r - v}{\max(v, \omega r)} \tag{3.1}$$

where r is the tire rolling radius, ω is the wheel angular velocity and v is the longitudinal velocity of the center of gravity of the vehicle. The longitudinal tire force is a

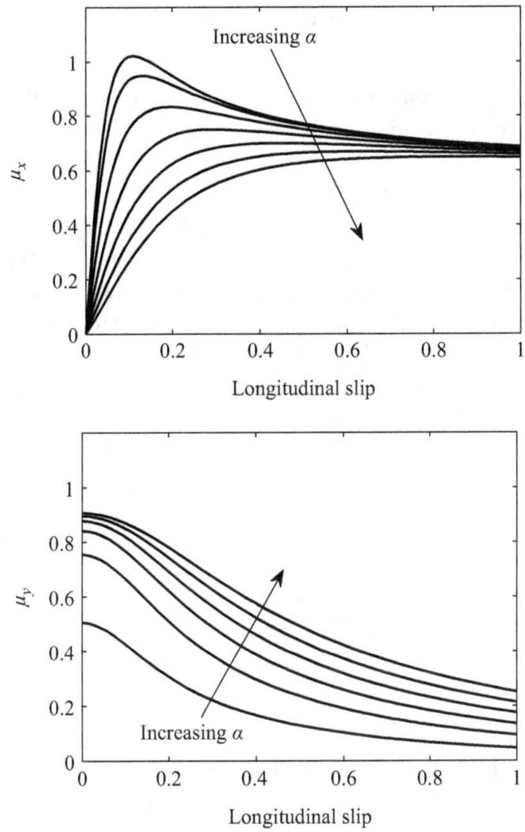

Figure 3.1 Plot of longitudinal and lateral friction coefficient as functions of wheel-slip and side slip angle

non-linear function of wheel-slip, wheel side-slip and vertical load: $F_x = \mu_x(\lambda, \alpha)F_z$ where F_z is the vertical load and μ_x is the longitudinal friction coefficient [7]. The side-slip angle, α, is the angle between a rolling wheel's actual direction of travel and the direction towards which it is pointing [7].

Figure 3.1 shows a typical plot of the longitudinal and lateral friction coefficient characteristics. All characteristics have an ascending part, a peak (at λ^*) and a descending part. From figure the importance of controlling wheel-slip is clear; an excessive slip determines a non-maximal longitudinal force. Furthermore, for values of slip above the peak, the lateral force drops, and the dynamics become open-loop unstable, [8]. As soon as the peak value is reached, if the wheel torque is not reduced wheel-slip will rapidly diverge to 1. The drop of lateral force is particularly critical in single-track vehicles because the lateral force not only allows the vehicle to steer, but it keeps it upright and stable.

The problem of traction and braking control is a well-known problem in four-wheeled vehicle dynamics control; however, research shows that the methods devised for four-wheeled vehicles are not directly applicable. The literature identifies mainly two reasons:

- motorcycles have a higher center of gravity with respect to the wheel-base than cars. This makes the load-transfer phenomena more prevalent. The analysis in [9,10] presents two approaches to the study of the effect of load-transfer on braking performance. The former uses optimization techniques to design the suspension characteristics to minimize braking distance; whereas the latter proves that, in sport-motorcycles, a controller based only on wheel-slip does not provide the optimal performance in terms of braking distance. The conclusion is that load-transfer phenomena have to be explicitly accounted for in the design of the controller. The work initiated by these two papers is further developed in [11,12].
- The tire lateral force plays a critical role in stabilizing the capsize mode of the motorcycle; the control performance requirements are therefore more stringent. An overshoot in wheel-slip that could only momentarily affect the trajectory of a car could cause the motorcycle to fall over. Furthermore, in order to provide yaw stability the ABS system for cars are designed to lock the front tires before the rear tires. If this were to happen on a motorcycle, the motorcycle would fall [13].

Traction and braking dynamics are very similar; despite this fact, research has shown that they do require different methodologies. These differences are mainly due to differences in actuator dynamics. Several options are available to control the wheel torque. For traction torque the choices are:

- Electronic throttle bodies (ETB). The air in-take valve influences engine torque [14]. ETB's can be operated quite freely without concerns of exceeding the engine capabilities (overheating, partial combustion, emission regulation); on the other hand, the dynamics from valve position and engine torque depends on the engine regime.
- Engine spark advance. By changing the spark timing in spark ignited engines, the engine torque can be directly modulated [15]. The dynamics from spark advance to engine torque is more direct than the air path dynamics. Moreover, this approach does not require additional electromechanical hardware. On the other hand, the spark advance dynamics are nonlinear and more importantly have considerable limitation in terms of engine capability. The spark advance cannot be kept far from its nominal condition for too long without affecting the engine temperature and combustion dynamics. This could eventually damage the engine.
- Cylinder cut-off. By simply not injecting fuel and not igniting a cylinder, the engine torque can be reduced [16]. This technique is a very crude method that shares all the disadvantages of engine spark advance control along with the impossibility of precisely modulating the engine torque.
- Electronic clutch. An electronically actuated clutch can be used to limit the engine torque transferred to the rear wheel [17]. Electronic clutches enable a very precise torque control, but with some limitations: they can only be employed during launch and for a limited amount of time. Clutches are wearable elements and their continuous actuation considerably limits their life.

Braking torque is more easily controlled by brakes; in particular hydraulically actuated brakes and electro-hydraulic brakes are the most common choices. In [18,19] the design of an electro-hydraulic brake is discussed showing that a pressure control bandwidth of 20 Hz is easily achieved. Braking torque control is therefore more easily achieved and the actuator dynamics are also more accurately modeled.

The different role played by the actuators in the overall dynamics calls for different methods and approaches to the development of the control system.

3.2.1 Braking control

Upon reviewing the braking control literature, one can find two main approaches: wheel-deceleration and wheel-slip based control.

Wheel-deceleration control is very common in automotive ABS braking by virtue of its robustness and the need for relatively inaccurate (and thus inexpensive) actuation. Wheel-deceleration-based (derived from classical ABS control) systems are based on letting the wheel-slip oscillate around the optimal value. The methodological difficulties associated to proving its safety and stability in the more complex motorcycle context and its intrinsic lower level of achievable performance are among the causes of the preference of the scientific community toward wheel-slip control. Nevertheless, most commercially available motorcycle ABS systems are developed by the same automotive suppliers and are probably based on adapted methodologies. No control system-oriented analysis of commercial ABS is available; however, some works assess and compare braking performance. In [20], Donovan compares the stopping distance of four different motorcycle makers. The results cover the comparison between ABS and no ABS on both wet and dry surfaces for a series of different load conditions. On dry surface, the average ABS stopping distance is 5%–7% (depending on the load) shorter than the best manual braking performance. The performance further improves when wet surface is considered yielding a reduction of the stopping distance between 5% and 15%.

The cited analyses are very useful for policy makers but do not provide information on how the current ABS system can be improved. These information are provided by the detailed analysis of the technical literature, which, very precisely points toward wheel-slip control. Matter-of-factly, there are no scientific publications that address the design of a wheel deceleration-based control for PTW. Conversely, the literature devoted to wheel-slip control is rich and diverse. The design of a wheel-slip controller can be done either based on models or completely model free. The model-based approach is applicable as the braking actuator dynamics is easily modeled and the torque precisely controlled.

Among the model-based approaches, we will consider performance-oriented systems [10,11] and safety-oriented systems [21]. The objective of a performance-oriented system is optimizing the braking distance, usually in racing settings, and thus the robustness issue is somewhat neglected. Neglecting the robustness issues has two advantages: on the one hand, the controller can achieve better performance; on the other hand, there is no need to develop a control-oriented model but the controller can be tuned directly on the complete multi-body dynamics. In [10], Bikesim (a multi-body motorcycle simulator) is used to tune an optimal braking controller. The resulting controller uses a combined wheel-slip control and load control on the front wheel whereas a slip control suffices for the rear wheel. The proposed approach is capable of detecting and keeping the motorcycle braking on the maximum deceleration limit without pitching over; furthermore, the paper quantifies the contribution of the rear wheel braking in sport motorcycles. On high friction surfaces, the rear wheel brake contributes only minimally to the overall braking deceleration (if only the in-plane dynamics are considered). The results of [10] are picked up by Sharp in [11] where the tuning of the controller is discussed in more details and a feed-forward control action (termed pre-planned control) is added. The tuning of the closed-loop controller is however carried out by trial-and-error in simulation.

The performance-oriented methods described above are useful to quantify the roles of different motorcycle parameters but are not robust enough to provide consistent performance when used on roads. In order to guarantee robustness, a control-oriented model is required that accounts for the varying parameters and guide the design of robust closed-loop controllers. In the context of this review this approach is called safety-oriented. The analysis in [21] shows that the classical single corner model does not capture the load transfer dynamics accurately enough. The authors thus propose a Linear Parameter Varying (LPV) model derived from the Jacobian linearization of the complete vehicle dynamics. This step considerably simplifies the model-reduction task and provides a system that is amenable to advanced control system design tools. The LPV framework has the capacity of describing nonlinear systems while maintaining some of the linear systems properties. LPV models, roughly speaking, can be defined as linear systems where, either the matrices of the state equations or the coefficients of the input–output relation, depend on one or more time-varying parameters. In the case at hand, the chosen time-varying parameters are velocity and wheel-slip itself, resulting in a two-dimensional family of seventh-order systems accounting for the wheel-slip, wheel-hop, bounce and pitch dynamics. Only the fully scheduled controller (accounting for both velocity and wheel-slip) achieves the required level of performance and robustness. The method discussed in [21], thanks to the availability of the LPV model, can be shown to be robust in face of unknown tire characteristics. Figure 3.2 plots a braking maneuver performed on three different surfaces.

The model-based approaches yield good control performance, but need an accurate model, which can be expensive to get. This drawback can be edged by recurring to model-free approaches (see [22–25]). Model-free approaches are based on the idea of designing the controller without any explicit model of the system dynamics. The systems and controls community has developed a plethora of methods for model-free control. Among the available choices, sliding mode control, fuzzy control and direct control synthesis approaches have been successfully applied. Lu's contribution in [22] describes a complete ABS system for a light motorcycle. The system is composed of (1) an electric motor driven pressure actuator, (2) a pressure control and (3) an ABS controller. The braking pressure actuator architecture guarantees a smooth braking pressure modulation. The ABS controller is based on a sliding mode control approach. The sliding surface is defined according to a desired wheel-slip. Simulation

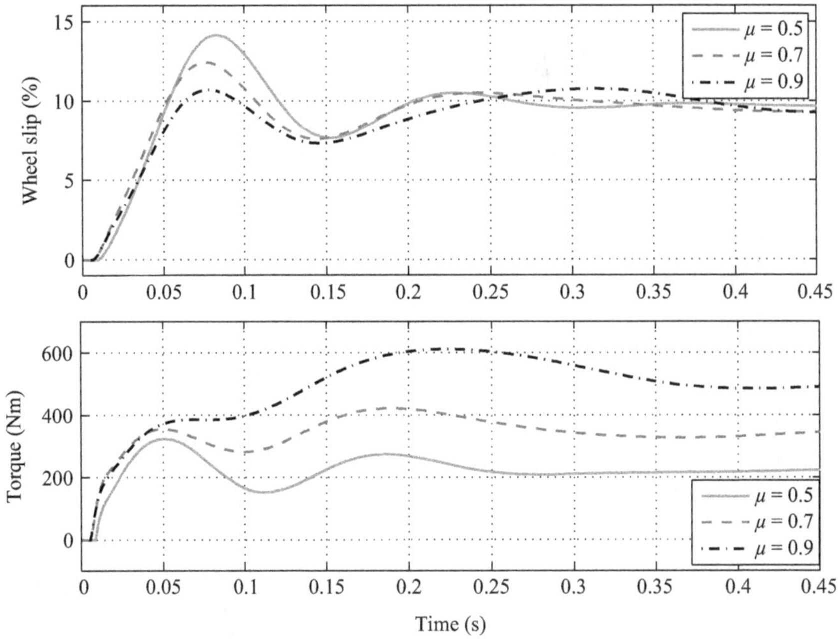

Figure 3.2 Robustness validation of the LPV-based wheel-slip control method. Adapted from [21]

and experimental results compare the proposed ABS system against an automotive derived ABS controller (based on the wheel deceleration); the wheel-slip based control achieves slightly shorter braking distances: the stopping distance from 65 km/h for the wheel-slip based control on dry slippery surface was measured at 11.63 m whereas the wheel-deceleration approach yielded 12.60 m. The paper reports braking distances, but lacks a thorough discussion of the tuning process. The authors explain that the sliding mode control approach enhances the robustness of ABS braking system, but do not provide any supporting evidence. This line of reasoning is also argued for by Tanelli and Ferrara [24]. The paper formally proves the robustness of the controller. Sliding mode controllers achieve their high level of robustness by virtue of a persistent switching of the control around an ideal sliding surface. The resulting oscillations in the control variable and wheel-slip, may affect the drivability.

This issue is addressed either by fuzzyfication or direct control design. Huang in [26] uses the same hardware as in [22] to implement a fuzzy controller. The fuzzy controller generates the reference pressure for the brakes based on a heuristic. A methodological analysis of the controller is not given, but an extensive experimental validation is discussed. Of particular interests are the tests performed on a three-phase pavement road: the light motorcycle starts braking on a dry road, transitions onto a wet road and finally returns to the dry road. Experiments show that the control system

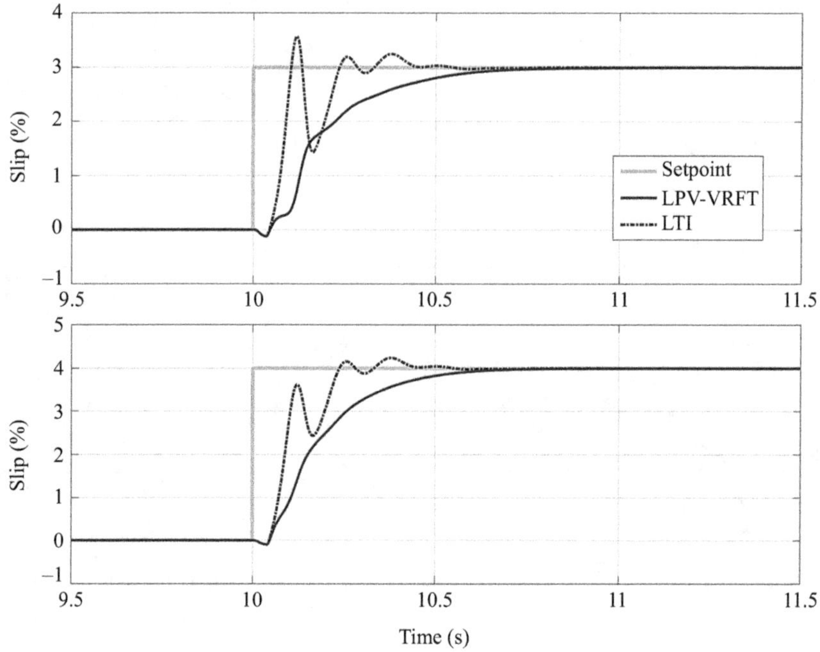

Figure 3.3 LPV and LTI slip controller during panic brake maneuvers with two different set point values

is stable and the desired braking performance is achieved without high frequency excitation of the braking torque.

Another possibility is direct control design. In [27,28] a direct LPV controller is designed. The method is a non-iterative direct data-driven technique, i.e., a gain-scheduled fixed order controller is derived from a finite number of experiments, without need of explicitly identifying the plant dynamics. This yields an easy and efficient design procedure. The method employs instrumental variables and optimal data pre-filtering to deal with measurement noise and under-parameterization of the controller. The method results in a gain-scheduled controller with the vertical load as the scheduling parameter. Figure 3.3 shows the performance of the LPV controller, compared to that of an LTI controller. Clearly the LPV controller yields a better response, and offers more robustness to the variation of the braking conditions. In Figure 3.3, two different set-point values are tested, and the LPV controller appears to offer a substantially identical behavior, while this is not true for the LTI one.

In general, model-based methods have two advantages over model-free ones. On the one hand, model-based methods can better exploit the accurate knowledge of the dynamics and provide better performance. On the other hand, model-based methods enable the formal proof of performance and robustness and are thus better suited for safety critical systems. For some of the above methods, the analysis could be carried out also for model-free methods, but a model would still be required and this would cancel out part of the advantages of adopting a model-free design.

3.2.2 Traction control

Traction Control (TC), despite having similar dynamic features as braking control, is more challenging. Accurate first-principle models of the engine torque generation are obtained with much difficult. The engine characteristics are accurately modeled in steady state with statics maps, but the dynamic behavior is more difficult. Historically, the difficulties brought by the time-varying engine dynamics were addressed with heuristics, that yield rather crude results. Subsequently, the introduction of rigorous system identification techniques provided the required model to perform model-based design.

The work by Cardinale in [29] is, to the best of the authors' knowledge, the first publicly available contribution on the subject. Commercial TC systems were available before the publication of that paper, but company policies have limited the availability of information. Cardinale and collaborators propose a spark-based control; the spark advance is controlled based on the difference between the front and rear wheel velocities. Although not explicitly stated, the control algorithm is a second-order sliding mode controller where the sliding surface is represented by null rear wheel-slip. The controller activates when the difference between the front and rear wheel velocity is above a threshold. Defining $e(k) = \omega_f(k) - \omega_r(k)$, if $e(k) > \varepsilon$ then the spark advance is determined by

$$c(k) = c(k-1) + \text{sign}(e(k-1) - e(k))\delta. \tag{3.2}$$

The two control parameters, the activation threshold (ε and the cut-off advance increment/decrement step δ) are empirically tuned. Experimental results show that the rear wheel-slip is limited during sudden accelerations, but a thorough quantitative analysis is lacking.

The lack of a dynamics model prevents any methodological analysis or tuning of the controller. The authors' work in [30] firstly addressed this issue. The authors propose an identification protocol that enables a quick and accurate modeling of the engine-to-slip dynamics from experimental data. The method is based on the open loop excitation of the control variable (in the original paper sine-sweep and step inputs were employed) and a frequency-based system identification. The method yields control-oriented models that, although not useful to understand the role of each mechanical component, are useful for control system design. The method can also be employed to compare different motorcycles and different actuation solutions. The analysis leads to the following conclusions:

- Both throttle-to-slip and spark-to-slip dynamics exhibit a resonance at 8 Hz.
- Spark advance is "faster" than throttle action. At 10 Hz, there is a 60° difference in phase: half of this loss is due to the servo-loop. This observation proves that slip control through throttle control is achievable.
- Although spark advance provides a slightly faster actuation, the response of the system is less linear, and therefore more difficult to model and control.

The method is further discussed in [1] comparing different motorcycle models. Formentin provides a further contribution in [31] addressing the issue of optimizing the identification experiment, through design of experiments.

Massaro and collaborators take on the experimental identification method in [32] extending it with a detailed discussion of the dynamics. Combining the experimental data with their multi-body simulator, they are able to provide a physical interpretation of the dynamics. Among other comments, they argue that a stiffer sprocket absorber (and in general a stiffer transmission) would make the design of the control system easier.

The availability of an accurate model opens the possibility of more advanced control systems. Chapter 8 in [1] gives a complete overview of the design process of a traction control system for a motorcycle. Several aspects are considered: reference generation, activation strategy and controller tuning among the most important ones. The TC system is wheel-slip based and implements a novel approach for the reference generation. This sets the proposed method aside from the classical threshold based TC systems. Fixed wheel-slip thresholds systems deprive the driver of any control when the TC is active. In the authors' approach, the activation threshold is fixed, but once it is crossed, the driver can modulate the wheel-slip reference through the throttle grip. The rider controls the reference slip by opening or closing the throttle; when the throttle is fully open the rider is requesting the maximum allowed slip. This mechanism guarantees better safety, robustness and controllability than the single-threshold logic. The second important element is the controller itself. The controller is tuned via classical control theory on the identified model. The resulting controller is a gain scheduled controller. The scheduling is based on the online identification of the friction surface. The design is supported by an extensive experimental validation; the proposed method is compared against a commercial system from several stand-points: acceleration time, wheel-slip and lateral stability.

As for the ABS problem, also model-free approaches have been investigated. For example, in [33] a second-order sliding model control is proposed, analyzed and validated in simulation.

The literature draws a clear picture; from the performance point of view wheel-slip based traction and braking control are superior to wheel deceleration.

In order to be effectively employed, wheel-slip methods require the knowledge of the vehicle velocity which cannot be directly measured. Vehicle velocity estimation is a critical problem on car; on motorcycle it is made even more difficult by having fewer wheels to rely upon and by the possibility of wheelies and stoppies, see [34] for details on these issues.

To put TC into a more general perspective, consider Figure 3.4, which shows a schematic view of the overall TC problem for motorbikes. As can be seen, the control problem in its most general view is comprised of three different control sub-problems. The first one (see the dashed oval box with label ① in Figure 3.4) concerns the servo-control of the Electronic Throttle Body (ETB), which is the considered actuator. Note that electronic control of throttle in motorbikes has been only recently introduced [35,36]; before the introduction of such a technology, engine torque was (and still is) mainly modulated via spark advance control. By anticipating or delaying

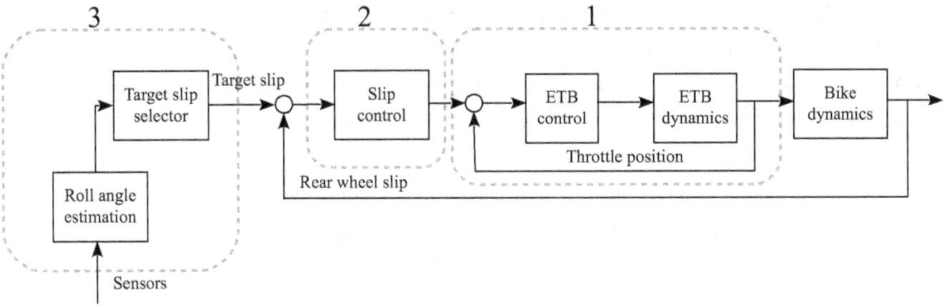

Figure 3.4 A schematic view of the overall TC problem

the spark in the cylinders, it is possible to control the generated torque. As such, the control sub-problem ① in Figure 3.4 could in principle need to address spark advance control, or, in a more sophisticated system, a combination of spark advance and ETB control, as it is commonly done in F1 cars. Once the actuator controller has been designed, an outer loop for controlling the rear wheel slip needs to be designed (dashed oval box with label ② in Figure 3.4). This step, which has been addressed for four-wheeled vehicles (see e.g., [37]) has not yet been treated in the literature for motorbikes and constitutes the main topic of this work. However, even though it constitutes the most important task to be solved for TC control, rear wheel slip control is not the last design step. In fact, as shown in Figure 3.4, the slip controller needs a target wheel slip to track. In two-wheeled vehicles, when moving in a curve, there is a trade-off between longitudinal and lateral forces (see e.g., [38]). When on a curve, in fact, a TC system should provide the largest possible amount of longitudinal force for transferring the traction torque to the ground, while guaranteeing sufficient lateral force for negotiating the curve. As such, the last building block of a TC system (dashed oval box with label ③ in Figure 3.4) is a supervisory unit which, based on a measure or an estimation of the roll angle, selects the optimal target slip. The roll angle estimation with low-cost sensors is still an open problem (first results on this topic can be found in [39]), so that the supervisory block ③ for TC systems has not yet been addressed in the literature for motorbikes and is topic of ongoing research.

It is worth noting that, as far as control systems design is concerned, dealing with motorcycle dynamics is far more subtle than dealing with for four-wheeled vehicles. In fact, it is common practice to design most active control systems for cars based on simplified dynamical models (e.g., the quarter-car model and the half-car model for braking control systems and the single-track model for active stability control, [40]), while complete vehicle models are employed mostly for testing and validation phases. In two-wheeled vehicles, instead, the presence of a single axle, together with the peculiar suspensions, steer and fork geometry, makes it difficult to devise appropriate simplified models.

As such, the effort of analyzing well-defined driving conditions seems to be the key for a comprehensive understanding of motorcycles dynamics. Such an approach is well confirmed in the scientific literature of this field (see [41–46]). Note that, although this approach is well suited for dynamic analysis, it may not be easily used for control systems design, as its results are strongly dependent on many (uncertain) system parameters, and this makes it difficult to validate them on a real vehicle.

3.3 Control-oriented modeling of the longitudinal dynamics of two-wheeled vehicles

For the preliminary design of wheel-slip control algorithms in two-wheeled vehicles, the following dynamical model can be employed, see e.g., [47–49]:

$$J_r\dot{\omega}_r = -r_rF_{x_r} + T_r \tag{3.3}$$

$$J_f\dot{\omega}_f = -r_fF_{x_f} + T_f \tag{3.4}$$

$$m\dot{v} = F_{x_r} + F_{x_f}, \tag{3.5}$$

where ω_f [rad/s] and ω_r [rad/s] are the angular speeds of the front and rear wheel, respectively, v [m/s] is the longitudinal speed of the vehicle body, T_r and T_f [Nm] are the rear and front torques, respectively, F_{x_f} and F_{x_r} [N] are the front and rear longitudinal tire–road contact forces (assumed to be positive during traction and negative during braking), J_f and J_r [kg m^2], m [kg] and r_f and $r = r$ [m] are the wheel inertias, the vehicle mass, and the wheel radii, respectively. Note that, for simplicity, the front and rear wheel inertias and the wheel radii are assumed to be equal and indicated with J and r, respectively. Further, for traction control purposes one has the driving torque $T_r > 0$ and $T_f = 0$ while, for braking control $T_i \leq 0$, $i \in \{f, r\}$.

The system is nonlinear due to the dependence of F_{x_i}, $i \in \{f, r\}$, on the state variables v and ω_i, $i \in \{f, r\}$. The expression of F_{x_i} as a function of these variables is involved and influenced by a large number of features of the road, tire and suspension; however, it can be approximated as follows (see [40]):

$$F_{x_i} = F_{z_i}\,\mu(\lambda_i, a_{i_t}; \vartheta), \quad i \in \{f, r\}, \tag{3.6}$$

where F_{z_i} is the vertical force at the tire–road contact point and $\mu(\cdot, \cdot; \vartheta)$ is a function of

- the longitudinal slip $\lambda_i \in [-1, 1]$, which is defined as

$$\lambda_i = \frac{\omega_i r - v}{\max\{\omega_i r, v\}}. \tag{3.7}$$

  Note that this definition implies that $\lambda \in [0, 1]$ in traction and $\lambda \in [-1, 0]$ during braking, where $\lambda = 0$ models a constant-speed motion, while $\lambda = \pm 1$ represents the case of locked wheels;
- the wheel side-slip angle a_{i_t}.

Further, during braking one needs to model the load transfer phenomena between the front and rear wheels. To do this, we define the vertical force as

$$
\begin{aligned}
F_{z_f} &= \frac{mgl_r}{l} - \frac{mh}{l}\dot{v} = W_f - \Delta_{F_z}\dot{v} \\
F_{z_r} &= \frac{mgl_f}{l} + \frac{mh}{l}\dot{v} = W_r + \Delta_{F_z}\dot{v},
\end{aligned}
\tag{3.8}
$$

where l is the wheelbase, l_f and l_r are the distances between the projection of the center of mass on the road and the front and rear wheel contact points, respectively, h is the height of the center of mass and g is the gravitational acceleration. Note that $\dot{v}$ is the vehicle *acceleration*, hence it is negative during braking.

Note, in passing, that from (3.6) one has that the longitudinal force produced by a wheel is bounded, i.e.:

$$|F_{xi}| \leq \Psi, \quad i \in \{f, r\}. \tag{3.9}$$

The tire model (3.6) is a steady-state model of the interaction between the tire and the road. The transient tire behavior, due to tire relaxation dynamics, yields traction forces F_{xi} with bounded first time derivative, i.e.:

$$|\dot{F}_{xi}| \leq \Gamma, \quad i \in \{f, r\}. \tag{3.10}$$

Vector ϑ in $\mu(\cdot, \cdot; \vartheta)$ (see again (3.6)) represents the set of parameters that identify the tire–road friction condition. Since for maneuvers performed along a straight line one can set the wheel side-slip angle equal to zero ($a_{i_t} = 0$), we shall omit the dependence of F_{x_i} on a_{i_t} and denote the μ function as $\mu(\cdot; \vartheta)$.

Many empirical analytical expressions for function $\mu(\cdot; \vartheta)$ have been proposed in the literature. A widely used expression (see e.g., [47]) is

$$\mu(\lambda; \vartheta) = \vartheta_1(1 - e^{-\lambda\vartheta_2}) - \lambda\vartheta_3, \tag{3.11}$$

where ϑ_i, $i = 1, 2, 3$, are the three components of vector ϑ. By changing the values of these three parameters, many different tire–road friction conditions can be modeled. In Figure 3.5(a) the shape of $\mu(\lambda; \vartheta)$ is displayed for different road conditions and for positive longitudinal forces (in the case of negative forces one has the symmetrical condition with respect to the origin). From now on, for ease of notation, the dependency of μ on ϑ will also be omitted, and the function in (3.11) will be referred to as $\mu(\lambda)$.

3.3.1 The complete motorcycle traction dynamics

By employing system (3.3), the rear wheel slip dynamics can be highlighted. To this aim, in order to use the wheel slip definition in (3.7), a measure or a reliable estimate of the vehicle speed is needed. As is discussed in [50] for the case of braking control, vehicle speed estimation for two-wheeled vehicle is an open problem.

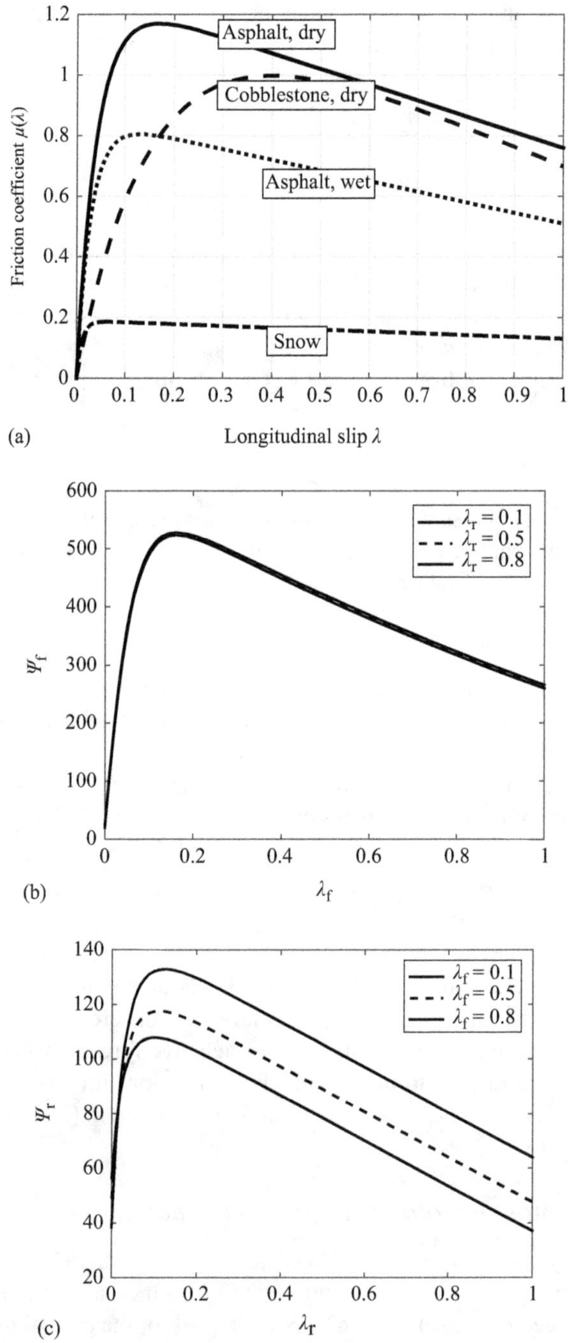

Figure 3.5 Plot of: (a) the function $\mu(\lambda; \vartheta)$ in different road conditions; (b) $\Psi_f(\cdot, \lambda_r)$ for different values of λ_r: $\lambda_r = 0.1$ (solid line), $\lambda_r = 0.5$ (dashed line) and $\lambda_r = 0.8$ (dotted line); (c) $\Psi_r(\lambda_f, \cdot)$ for different values of λ_f: $\lambda_f = 0.1$ (solid line), $\lambda_f = 0.5$ (dashed line) and $\lambda_f = 0.8$ (dotted line)

For traction control purposes, however, the problem of vehicle speed estimation is eased by the fact that the only driven wheel is the rear one, so that, in principle, the front wheel linear speed should provide a reasonable estimate of the vehicle speed. Again, the suspension and pitch dynamics, which in two-wheeled vehicles are much more coupled with longitudinal dynamics than they are in cars, should warn that the use of the front wheel speed might provide non-precise speed estimates in very strong acceleration phases. However, as in practice no alternative (or more accurate) vehicle speed estimate has been made available yet, the definition of the *relative* rear wheel slip is introduced, namely:

$$\lambda_{r,r} = \frac{\omega_r r - \omega_f r}{\omega_r r}, \tag{3.12}$$

which is nothing but (3.7) with $\omega_f r$ replacing v. This quantity is what can be actually measured on commercial motorbikes. Along this line, the quantity computed as in (3.7) using the true vehicle speed v is called *absolute* rear wheel slip $\lambda_{r,a}$.

In what follows it is assumed that the longitudinal dynamics of the vehicle (expressed by the state variable v) are significantly slower than the rotational dynamics of the wheels (expressed by the state variables λ_i or ω_i) due to the differences in inertia. Henceforth, v is considered as a slowly time-varying parameter when analyzing the evolution in time of λ_i (see e.g., [48,51]). Under this assumption, (3.5) (center of mass dynamics) is neglected, and the model reduces to that of the wheels dynamics only. Further, in system (3.3)–(3.5) the state variables are v and ω_i. As λ_i, v and ω_i are linked by the algebraic equation (3.7), it is possible to replace ω_i with λ_i as state variable. Specifically, let us analyze the *absolute* rear wheel slip $\lambda_{r,a}$. Considering (3.3) and (3.5) (the front wheel dynamics (3.4) only affect the vehicle speed equation in the driving torque to rear slip dynamic relation), and considering the absolute slip definition in (3.7) together with the longitudinal force description in (3.6), the absolute rear wheel slip dynamics can be written as

$$\begin{aligned}\dot{\lambda}_{r,a} &= \frac{v}{\omega_r^2 r}\dot{\omega}_r - \frac{1}{\omega_r r}\dot{v} = -\frac{(1-\lambda_{r,a})^2 r}{Jv}\Big\{[rF_{z_r}\mu(\lambda_{r,a}) - T] \\ &\quad + \frac{J}{r\,m\,(1-\lambda_{r,a})}(F_{z_r}\mu(\lambda_{r,a}) + F_{z_f}\mu(\lambda_f))\Big\}. \end{aligned} \tag{3.13}$$

In what follows, the SOSM traction controller will be designed taking into account the absolute wheel slip dynamics. However, its intrinsic robustness properties will allow to employ the same controller also when the relative wheel slip is used as controlled variable, as will be shown in the simulation tests. Even though the SOSM controller is designed based on the nonlinear wheel slip dynamics, in order to be able to validate the analytical model against the frequency response estimates obtained on the basis of experimental data collected on an instrumented vehicle (see Section 3.3.1), a linearization of the slip dynamics to obtain a transfer function description has to be performed. To this aim, the absolute slip dynamics in (3.13) are considered and, as a first step, the system equilibria need to be computed.

Thus, let $\dot{\lambda}_{r,a} = 0$ and look for the equilibrium points characterized by a constant longitudinal slip value $\lambda_{r,a} = \bar{\lambda}_{r,a}$ (note that the equilibrium characterized by $\mu(\lambda) = 0$ and $T = 0$ is meaningless for traction control purposes as it corresponds to the coasting-down condition with no torque applied). From (3.13) it is easy to find that the equilibrium values for the driving torque T are given by

$$\bar{T} = rF_{z_r}\mu(\bar{\lambda}_{r,a}) + \frac{J}{rm(1-\lambda_{r,a})}(F_{z_r}\mu(\lambda_{r,a}) + F_{z_f}\mu(\lambda_f)). \tag{3.14}$$

According to the assumption of regarding v as a slowly varying parameter, the model can be linearized around an equilibrium point defined by $\delta T = T - \bar{T}$ and $\delta\lambda_{r,a} = \lambda_{r,a} - \bar{\lambda}_{r,a}$. Defining the slope of the $\mu(\lambda)$ curve around an equilibrium point as

$$\mu_1(\bar{\lambda}) := \left.\frac{\partial\mu(\lambda)}{\partial\lambda}\right|_{\lambda=\bar{\lambda}},$$

the linearized absolute wheel slip dynamics have the form

$$\begin{aligned}\delta\dot{\lambda}_{r,a} = &\left\{\frac{F_{z_r}}{\bar{v}}\left[\frac{2(1-\bar{\lambda}_{r,a})}{J}r^2 + \frac{1}{m}\right]\mu(\bar{\lambda}_{r,a}) + \left[\frac{(1-\bar{\lambda}_{r,a})^2F_{z_r}}{\bar{v}}\left(\frac{r^2}{J}+\frac{1}{m}\right)\right]\mu_1(\bar{\lambda}_{r,a})\right.\\ &\left.+\frac{\bar{\lambda}_{r,a}}{m\bar{v}}F_{z_f}\mu(\bar{\lambda}_f) - \frac{2(1-\bar{\lambda}_{r,a})r}{J\bar{v}}\bar{T}\right\}\delta\lambda_{r,a} + \frac{(1-\bar{\lambda}_{r,a})^2r^2}{\bar{v}}\delta T.\end{aligned} \tag{3.15}$$

From the linearized dynamics (3.15) it is immediate to derive the expression of the first-order transfer function $G_{\lambda_{r,a}}(s)$, which will be employed in Section 3.3.1 for model validation against experimental data.

To better evaluate the suitability for TC design of the discussed dynamical model, this is compared to data collected on a hypersport motorbike, which has been used to perform experiments tailored to the identification of actual rear wheel slip dynamics. The considered motorbike is propelled by a 1,000 cc four-stroke engine; it weighs about 160 kg (without rider) and can deliver more than 200 HP. For confidentiality reasons other details of the motorbike are kept undisclosed. The vehicle is equipped with

- an Electronic Throttle Body (ETB) which allows to electronically control the position of the throttle valve independently of the rider's request;
- an Electronic Control Unit (ECU) that allows to control the throttle. The clock frequency of the ECU is 1 kHz;
- two wheel encoders to measure the wheels angular velocity;
- a one-dimensional optical velocity sensor. This sensor measures the true longitudinal velocity and it will be used to compute the instantaneous absolute rear wheel slip.

To experimentally identify the slip dynamics, a frequency sweep response has been employed. The test was carried out on a 3.5 km straight dry asphalt patch; the rider is asked to bring the motorcycle to a given constant engine speed in a given gear. After steady-state conditions are reached, the rider presses a button which starts

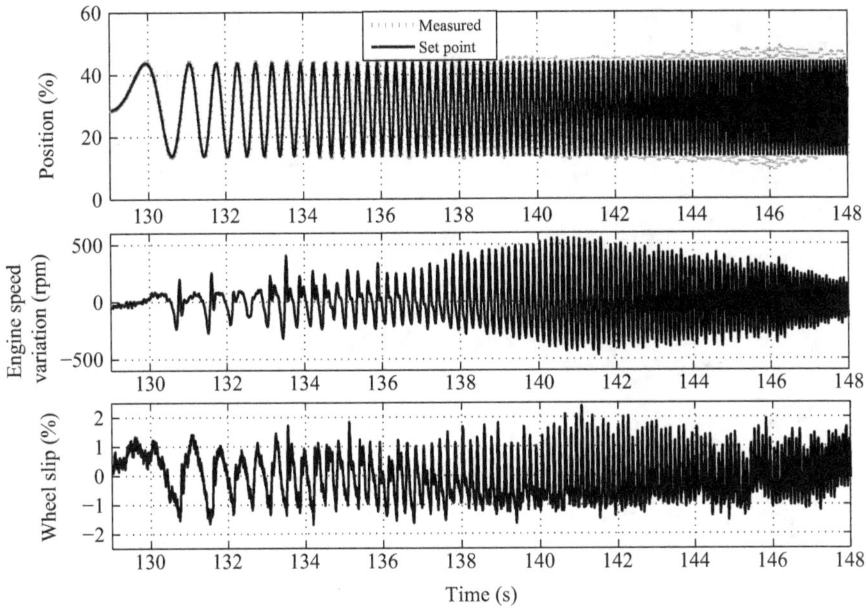

Figure 3.6 Plot of (top): throttle position set point (solid line) and throttle output position (dotted line); (middle) engine speed variation; and (bottom) absolute rear wheel slip

the test. The throttle control is taken over by the ECU and the excitation signal (a frequency sweep) is applied around the neighborhood of the initial condition. In the following, reference is made to the absolute rear wheel slip $\lambda_{r,a}$, which is defined as in (3.7). Figure 3.6 shows the set-point and output throttle position, the engine speed variation and the rear wheel slip measured during a frequency sweep test. For confidentiality reasons, the time scale is omitted.

From the experiments, the throttle set point position θ^o and the absolute rear wheel slip $\lambda_{r,a}$ are recorded to estimate the frequency response $\hat{G}_{\lambda_{r,a}}(j\omega)$. Such a non-parametric estimate of the frequency response is obtained by windowed spectral analysis of the input/output cross-spectral densities, [52], and is shown with the dashed line in Figure 3.7. For confidentiality reasons, the frequencies in Figure 3.7 are shown normalized with respect to the closed loop frequency of the servo-loop throttle control (ω_c).

From Figure 3.7 it can be observed that the measured slip dynamics has a resonance of around 0.7–0.8 ω_c. The fact that this resonance is also visible on the engine speed (see middle plot in Figure 3.6) suggests that it is due to the transmission elasticity.

As such, to complement the analytical wheel slip model (3.15) so that it accounts for the additional dynamic elements which emerge in the measured data, the actuator dynamics and the transmission elasticity need to be considered.

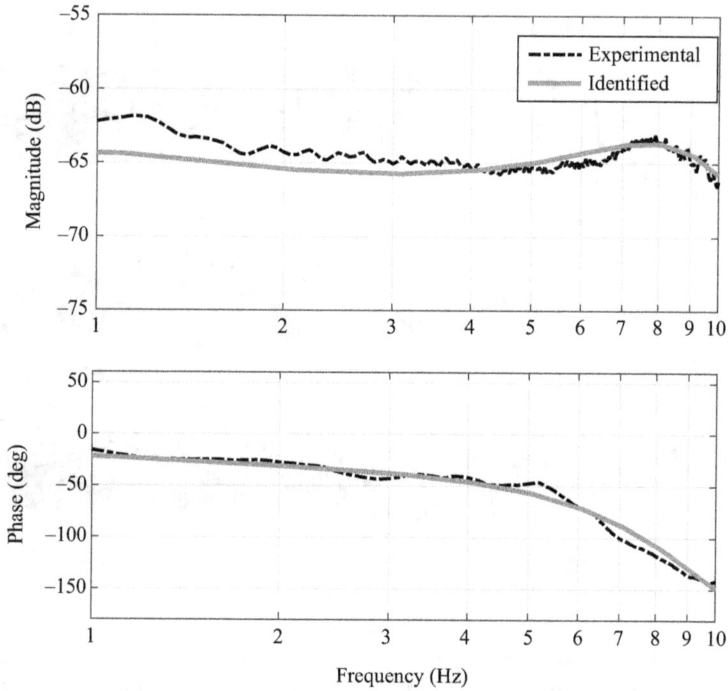

Figure 3.7 Experimental frequency response (dashed line) and analytical transfer function (solid line) from throttle set-point to rear wheel slip

3.3.1.1 Actuator model

The considered actuator is an Electronic Throttle Body (ETB), which is comprised of butterfly valves actuated by an electrical motor through a reduction system. Figure 3.8 shows the system-theoretic representation of the throttle control system. As can be seen, it comprises the electrical part of the DC motor dynamics $G_{el}(s)$, the mechanical DC motor component due to the electromotive force F_e and the outer ETB system made by the planetary gear, the return spring and the LTI throttle dynamics $G_{th}(s)$. The system is completed by the position control loop which regulates the throttle position θ to a desired set-point θ^o.

This mechanical system is rendered complex by packaging, cost and reliability constraints. These constraints give rise to dominant friction and backlash behavior in the transmission, making the control of the valve difficult, [53]. The packaging constraints are even more strict when the system is being designed for racing motorcycles. For traction control purposes, the dynamics of the controlled ETB need to be modeled. These, as discussed in [35], can be described in terms of the transfer function (see also Figure 3.8):

$$G_{ETB}(s) = \nu(\omega_e)\frac{1}{\tau s + 1}e^{-ds}, \tag{3.16}$$

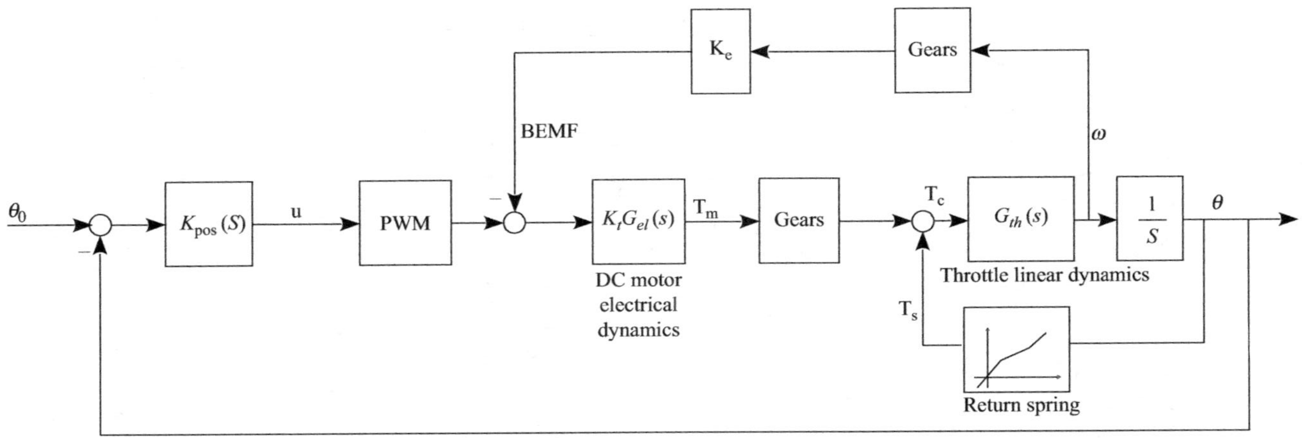

Figure 3.8 Schematic representation of the electronic throttle architecture

that is a first-order low-pass filter with time-varying gain $\nu(\omega_e)$, where ω_e is the engine speed, and a pure delay d.

3.3.1.2 Transmission model

For modeling the transmission elasticity, a mass-spring-damper description has been chosen, which can be therefore represented by means of the following second-order transfer function:

$$G_{transm}(s) = \frac{\omega_n^2}{s^2 + 2\xi\omega_n s + \omega_n^2}, \tag{3.17}$$

where the natural frequency $\omega_n = \sqrt{k/m}$ and $\xi = c/2\sqrt{k\,m}$ and m, c and k are the mass, damping coefficient and spring stiffness of the transmission, respectively.

Thus, the overall analytical model of the motorbike dynamics $G_{m-bike,TC}(s)$ (see also Figure 3.4) is

$$G_{m-bike,TC}(s) = G_{th}(s)\, G_{\lambda_{r,a}}(s)\, G_{transm}(s),$$

and it is given by the cascade of the controlled ETB dynamics (3.16), the transmission (3.17) and the analytical transfer function $G_{\lambda_{r,a}}(s)$ derived from the linearized model (3.15). The overall transfer function is shown with the solid line in Figure 3.7. As can be appreciated, the fitting between measured data and the analytical model can be regarded as quite satisfactory. Note that the fitting is better for frequencies above $0.3\omega_c$, whereas it worsens at lower frequencies. This is due to the fact that only a few periods of the periodic excitation can be completed at low frequency because of track length constraints, and this results in a poor quality of the frequency response estimate because of the very limited number of data points available.

Note that, as both the ETB and the transmission dynamics are characterized by 0 dB DC gain and roll-off frequencies higher than that of the vehicle dynamics, the approach of designing the SOSM controller based on the wheel slip dynamics only is sensible. In fact, the boundedness constraints on which it relies are still valid also in presence of these additional dynamics. It is worth pointing out that the obtained results, which allowed to experimentally validate the dynamical model on experimental data, are quite significant for the application at hand, where the definition of analytically tractable yet reliable control-oriented models is in general quite hard to achieve.

3.3.2 The complete motorcycle braking dynamics

For the design of braking controllers in motorcycles, one usually deals with front-braking only, so that the second-order model (3.5) can be reduced to the so-called single-corner model, which is typically used also in four-wheeled vehicles braking system design, that is

$$\begin{cases} J\dot{\omega} = rF_x - T_b \\ m\dot{v} = -F_x, \end{cases} \tag{3.18}$$

where

- $\omega = \omega_f$ [rad/s] is the angular speed of the front wheel;
- v [m/s] is the longitudinal speed of the vehicle body. In case front-wheel braking is assumed, the vehicle speed can be estimated using the speed of the rear wheel, which is basically in free rolling during the braking maneuver, unless a stoppie occurs (i.e., the braking is so strong that the whole vehicle load is transferred to the front wheel and the rear wheels leaves the ground);
- $T_b = T_f$ [Nm] is the front braking torque;
- $F_x = F_z\mu(\lambda)$ [N] is the longitudinal tire–road contact force.

All the other parameters are as in (3.5).

For the analysis of the braking dynamics, it is useful to introduce the normalized linear wheel deceleration, defined as

$$\eta := \frac{\dot{\omega} r}{g} \tag{3.19}$$

Observe that η is the linear deceleration of the contact point of the tire, normalized with respect to the gravitational acceleration g. It is particularly useful as it can be easily compared with the longitudinal body deceleration.

It is worth noting that in system (3.18) the state variables are v and ω. As λ, v and ω are linked by the algebraic relation that defines the wheel slip during braking, i.e.,

$$\lambda = (v - \omega r)/v,$$

it is possible to replace ω with λ as state variables in the single corner model. Employing the expression of $F_x = F_z\mu(\lambda)$ introduced above and substituting $\dot{\lambda} = -\frac{r}{v}\dot{\omega} + \frac{r\omega}{v^2}\dot{v}$, $\omega = \frac{v}{r}(1-\lambda)$ into (3.18), it yields:

$$\dot{\lambda} = -\frac{1}{v}\left[\frac{(1-\lambda)}{m} + \frac{r^2}{J}\right]F_z\mu(\lambda) + \frac{r}{vJ}T_b, \quad m\dot{v} = -F_z\mu(\lambda). \tag{3.20}$$

To better understand the criticalities of braking control design, let us derive the linearization of the model around its equilibria.

First notice that, by setting $\dot{v} = 0$ and $\dot{\omega} = 0$ in system (3.20), the corresponding equilibrium is given by $\lambda = 0$ and $T_b = 0$. This corresponds to a constant-speed condition without braking; apparently, this equilibrium condition is trivial and meaningless for the design of a braking controller.

The equilibrium points we are interested in – during braking – are characterized by $\dot{\lambda} = 0$ (constant longitudinal slip $\lambda = \overline{\lambda}$) and $\eta = \overline{\eta}$. Starting from these conditions, it is possible to find the set of admissible equilibrium points. Consider the expression of $\lambda = 1 - r\omega/v$. By letting $\dot{\lambda} = 0$ we obtain

$$\frac{(\dot{\omega}v - \dot{v}\omega)}{v^2} = 0,$$

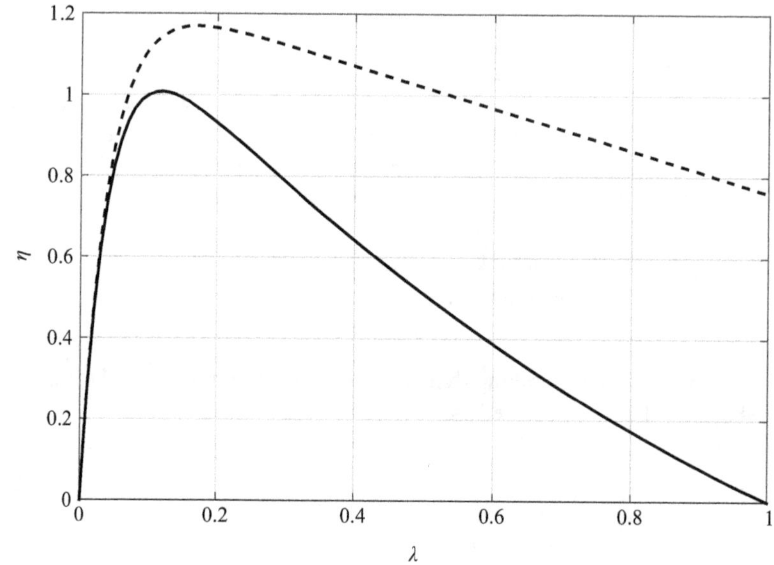

Figure 3.9 Equilibrium manifold $\overline{\eta}(\overline{\lambda})$ in the case of $F_z = mg$ and dry asphalt

namely $\dot{\omega} = \omega\dot{v}/v$. By plugging $\omega = v(1-\lambda)/r$, the second equation of system (3.20) and the definition of η given in (3.19) into this expression, we finally obtain the analytic expression of the set of admissible equilibrium points, which has the form:

$$\overline{\eta}(\overline{\lambda}) = \frac{F_z}{mg}(1-\overline{\lambda})\mu(\overline{\lambda}). \tag{3.21}$$

In Figure 3.9 the equilibrium manifold (3.21) is displayed in the (λ, η)-domain in the case of $F_z = mg$ and dry asphalt and compared with the $\mu(\lambda)$ curve. It is interesting to observe that the peak of the curve $\overline{\eta}(\overline{\lambda})$ is located slightly before the peak of $\mu(\lambda)$. Moreover, the equilibrium normalized wheel deceleration $\overline{\eta}$ goes to zero as $\overline{\lambda}$ approaches 1. Also notice that (since $\overline{\eta}(\overline{\lambda})$ is a non-monotone function) for each admissible value $\overline{\eta}$ there are two admissible slip equilibrium points.

From (3.21) it is easy to find the corresponding values for the braking torque, namely:

$$\overline{T}_b = F_z\mu(\overline{\lambda})\left(r + \frac{J}{rm}(1-\overline{\lambda})\right). \tag{3.22}$$

In practice, since $r \gg J(1-\overline{\lambda})/rm$, the equilibrium braking force is given by $\overline{T}_b = rF_z\mu(\overline{\lambda})$. Consider now the following variables, defined around an equilibrium point (characterized by $\overline{T}_b, \overline{\lambda}, \overline{\omega}, \overline{\eta}$)

$$\delta T_b = T_b - \overline{T}_b; \quad \delta\lambda = \lambda - \overline{\lambda}; \quad \delta\eta = \eta - \overline{\eta}.$$

The management of the state variable v is tricky. Usually (see e.g., [48]) a simple quasi-static assumption is made: v is assumed a slowly varying parameter since it is assumed that the longitudinal dynamics of the vehicle are much slower than the rotational dynamics of the wheel. To enlarge the range of validity of the linearized model, we use an intermediate approach: the linearization is done by explicitly considering the variations of v, locally around the non-equilibrium value $\overline{v}$ (namely, $\delta v = v - \overline{v}$); $\overline{v}$ is then considered a slowly varying parameter in the linearized model.

Using these variables and the definition

$$\mu_1(\overline{\lambda}) := \left.\frac{\partial \mu}{\partial \lambda}\right|_{\lambda=\overline{\lambda}},$$

which represents the slope of the $\mu(\lambda)$ curve around an equilibrium point, the transfer function $G_\lambda(s)$ from δT_b to $\delta\lambda$ is given by

$$G_\lambda(s) = \frac{\frac{r}{J\overline{v}}}{s + \left[\frac{\mu_1(\overline{\lambda})F_z}{m\overline{v}}\left((1-\overline{\lambda}) + \frac{mr^2}{J}\right)\right]}. \tag{3.23}$$

Similarly, the transfer function $G_\eta(s)$ from δT_b to $\delta\eta$ is given by

$$G_\eta(s) = \frac{\frac{r}{Jg}\left(s + \left[\frac{\mu_1(\overline{\lambda})F_z}{m\overline{v}}(1-\overline{\lambda})\right]\right)}{s + \left[\frac{\mu_1(\overline{\lambda})F_z}{m\overline{v}}\left((1-\overline{\lambda}) + \frac{mr^2}{J}\right)\right]}. \tag{3.24}$$

$G_\lambda(s)$ and $G_\eta(s)$ are both first-order transfer functions, characterized by the same pole; however, $G_\lambda(s)$ is a strictly proper transfer function, whereas $G_\eta(s)$ is characterized by a zero.

The stability and minimum-phase properties of the system around a steady-state condition can now be easily analyzed:

Stability condition for $G_\lambda(s)$ and $G_\eta(s)$
The two transfer functions are stable if and only if

$$\frac{\mu_1(\overline{\lambda})F_z}{m\overline{v}}\left((1-\overline{\lambda}) + \frac{mr^2}{J}\right) > 0,$$

which can be reduced to $\mu_1(\overline{\lambda}) > 0$.

This means that $G_\lambda(s)$ and $G_\eta(s)$ are open-loop unstable if the equilibrium $\overline{\lambda}$ occurs beyond the peak of the curve $\mu(\lambda)$.

Minimum phase condition for $G_\eta(s)$
$G_\eta(s)$ is minimum phase if and only if

$$\frac{\mu_1(\overline{\lambda})F_z}{m\overline{v}}(1-\overline{\lambda}) > 0,$$

which can be reduced to $\mu_1(\overline{\lambda}) > 0$.

This means that $G_\eta(s)$ is non-minimum phase if the equilibrium $\overline{\lambda}$ occurs beyond the peak of the curve $\mu(\lambda)$.

The analysis showed that the equilibrium point $\overline{\lambda}$ strongly affects the stability and minimum-phase properties of $G_\lambda(s)$ and $G_\eta(s)$. It also shows the well-known fact

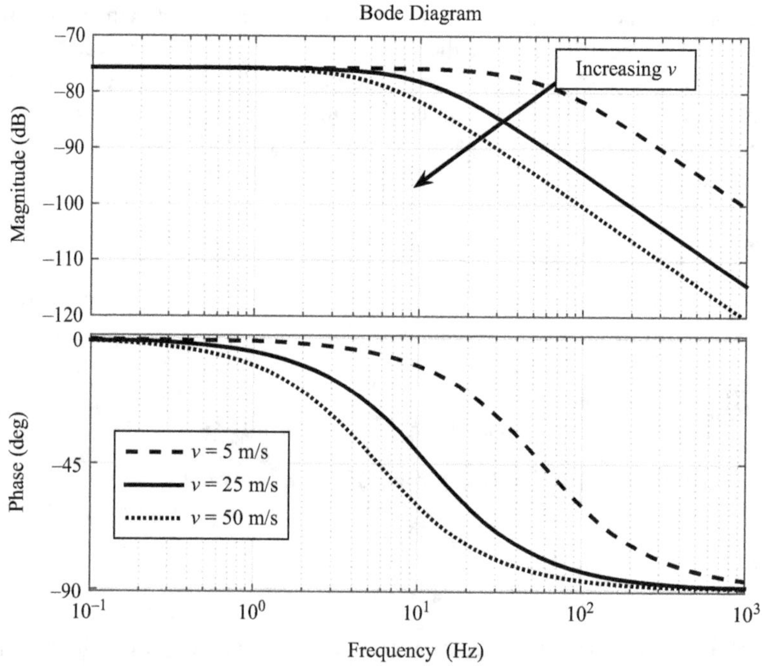

Figure 3.10 Magnitude and phase Bode diagrams of $G_\lambda(j\omega)$, at different longitudinal speeds (dry asphalt; $\overline{\lambda} = 0.05$)

that, for constant values of the braking torque T_b, the system equilibria (specifically those associated with slip values beyond the peak of the tire–road friction curve) are unstable and non-minimum phase. This instability can hardly be handled by non-professional drivers and it represents the main motivation for the design of automatic Anti-lock Braking Systems.

It is interesting to observe that the longitudinal vehicle speed v (considered as a slowly varying parameter) does not affect the stability and minimum-phase properties; however, it has the very obnoxious effect of scaling the wheel dynamics. As a matter of fact, notice that both the pole and the zero are characterized by the multiplying factor $1/\overline{v}$. In Figure 3.10 the magnitude Bode plots of $G_\lambda(s)$ (in the case of dry asphalt, at $\overline{\lambda} = 0.05$) are displayed for three different values of $\overline{v}$. Note that the wheel dynamics at $v = 3$ m/s are a decade faster than those at $v = 30$ m/s. Clearly, this scaling effect must be somehow taken into account in the design of a braking-controller, motivating the efforts of speed-dependent controller designs both for four and two-wheeled vehicles, which was proposed by several authors, see e.g., [21,48] and references therein.

3.3.2.1 Actuator model

For braking control design, two different families of actuation systems have been conceived. The traditional ones, developed for cars and still present on nearly all passengers cars are the so-called hydraulically actuated brakes (HAB), in which pressure exerted by the driver on the pedal is transmitted to the hydraulic system *via* a valve (called *build* valve) which communicates with the brake cylinder. The hydraulic system has a second valve (called *dump* valve) which can discharge the pressure and which is connected to a low pressure accumulator. A pump completes the overall system. The braking force acts on the wheel cylinder, which transmits it to the pads and, finally, to the brake discs (see also [54] for more details).

According to its functioning principles, the HAB actuator is only capable of providing three different control actions:

- *Increase* the brake pressure: in this case the build valve is open and the dump closed.
- *Hold* the brake pressure: in this case both valves are closed, and
- *Decrease* the brake pressure: in this case the build valve is closed and the dump open.

Even if they allow a discrete modulation of the braking pressure, yet HABs are characterized by a long life-cycle and high reliability, and this is the main motivation which has up to now prevented the new generation of braking systems (electro-hydraulic and electro-mechanical) to enter the mass production.

However, if one is interested in having an actuator that permits to finely control the braking torque, a Brake-by-Wire (BBW) solution, either electro-hydraulic or electro-mechanical represents an interesting scenario.

Different technological solutions have been proposed: the electro-hydraulic one (EHB) is based on a hydraulic system which is activated by an electric motor or pump that is controlled by an electronic unit, see e.g., [55–58]. The electro-mechanical brake (EMB) solution does not have the hydraulic part but there is an electric motor as actuator that provides the braking torque, see e.g., [59,60]. For motorcycle applications, a dedicated EHB solution has been developed, which was presented in [61–63]: an electro-mechanical actuator, that pushes backward and forward the piston of a master cylinder connected to a traditional hydraulic brake. Compared to well-known EMB and EHB solutions, this one has the advantage to keep the usual hydraulic layout adding just the actuator. This saves weight, space and cost. With this architecture, the actuator control problem consists in tracking a desired pressure reference. The functioning principles of such a braking actuator are depicted in Figure 3.11.

To properly control the braking torque in a braking maneuver, the BBW actuator must have on board a dedicated servo-control system. The one developed for the motorcycle version of the BBW system described above was presented in [63], and consists of a pressure control system designed with an adaptive cascade architecture.

As can be seen by inspecting Figure 3.12, the controller design phase ensures a linear closed-loop behavior, a bandwidth of approximately 15 Hz and the capability of continuously modulating the braking torque. For the purposes of this chapter, the closed-loop actuator dynamics is modeled as

$$G_{\text{bbw}}(s) = \frac{\omega_{\text{act}}}{s + \omega_{\text{act}}} e^{-s\tau}, \tag{3.25}$$

with $\omega_{\text{act}} = 95$ rad/s and $\tau = 5$ ms.

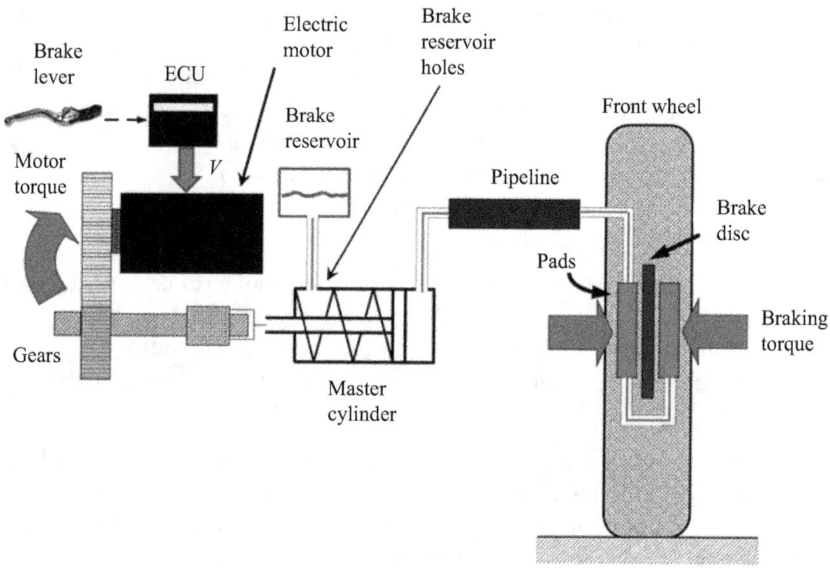

Figure 3.11 Functional diagram of the motorcycle braking system

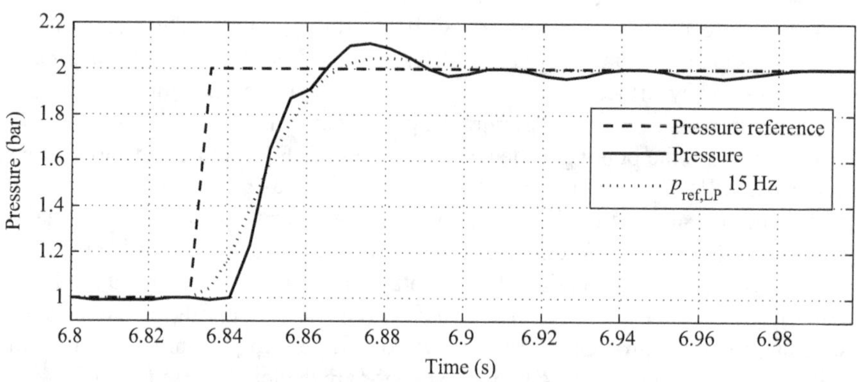

Figure 3.12 Time histories of the closed-loop braking pressure in the motorcycle BBW actuator: reference signal (dashed line), measured pressure (solid line) and pressure given by a low-pass filter with a bandwidth of 15 Hz (dotted line)

3.4 The SOSM control approach

It is well known, see e.g., [5], that in the formulation of the SOSM algorithm starting from the *n*th order nonlinear system known as the perturbed chain of integrators with bounded uncertainties, it is possible to define an appropriate sliding manifold which allows to rewrite the system dynamics as composed of an asymptotically stable linear part controlled by the sliding variable and of the second-order dynamics made by the sliding variable itself and its first time derivative. Such a second-order dynamical system, usually referred to as the *auxiliary* system, has again the form of a perturbed chain of integrators of order two, and it is the system of interest for the design of the SOSM controller and for the study of its closed-loop stability properties.

In this section, we review the traditional SOSM control strategy based on the *auxiliary* system.

3.4.1 Preliminaries

For the discussion on the S-SOSM algorithm, we need to introduce the following definition.

Definition 3.1. *A continuous function* $\alpha(\cdot) : \mathbb{R}^+ \to \mathbb{R}^+$ *is a class* $\mathcal{K}$ *function if* $\alpha(0) = 0$, $\alpha(s) > 0$ *for all* $s > 0$ *and it is strictly increasing.*

Further, it is worth recalling the structure and the basic features of the SOSM controller. For simplicity, we work on the *auxiliary* system, which has the form:

$$\begin{aligned} \dot{z}_1 &= z_2 \\ \dot{z}_2 &= f(z(t), v(t)) + g(z(t))v(t), \end{aligned} \tag{3.26}$$

where $z(t) \in \mathbb{R}^2$ is the system state, $v(t)$ is the auxiliary control signal (defined as the time derivative of the control signal acting on the original system) and $f(z(t), v(t))$ and $g(z(t))$ are uncertain, sufficiently smooth functions, satisfying all the conditions ensuring existence and uniqueness of the solution [64], together with the following bounds:

$$\begin{aligned} 0 < G_1 \leq g(z(t)) &\leq G_2 \\ |f(z(t), v(t))| &< F. \end{aligned} \tag{3.27}$$

The SOSM control problem is formulated as follows: given system (3.26), where $g(z(t))$ and $f(x(t), v(t))$ satisfy (3.27), design the auxiliary control signal $v(t)$ so as to steer $z_1(t)$ and $z_2(t)$ to zero in finite time.

The SOSM controller is such that, under the assumption of being capable of detecting the extremal values z_{Max} of the signal z_1, the following result can be proved:

Theorem 3.1. *Given system* (3.26), *where* $g(z(t))$ *and* $f(x(t), v(t))$ *satisfy* (3.27), *the auxiliary control law:*

$$\begin{aligned} v(t) &= -\alpha V \operatorname{sign}\left(z_1 - \frac{1}{2} z_{Max}\right) \\ \alpha &= \begin{cases} \alpha^* & \text{if} \quad [z_1 - \frac{1}{2} z_{Max}][z_{Max} - z_1] > 0 \\ 1 & \text{else}, \end{cases} \end{aligned} \tag{3.28}$$

where V *is the control gain,* α *is the so-called modulation factor, and* z_{Max} *is a piecewise constant function representing the value of the last singular point of* $z_1(t)$

(i.e., the most recent value z_{1M} such that $z_2(t_M) = 0$), causes the convergence of the system trajectory on the sliding manifold $z_1 = z_2 = 0$ in finite time provided that the control parameters α^* and V are chosen so as to satisfy the following constraints:

$$\alpha^* \in (0,1] \cap \left(0, \frac{3G_1}{G_2}\right)$$
$$V > \max\left\{\frac{F}{\alpha^* G_1}, \frac{4F}{3G_1 - \alpha^* G_2}\right\}. \tag{3.29}$$

The control law (3.28) is such that, by following the theoretical development provided in [5], it can be proved that the trajectories on the (z_1, z_2) plane are confined within limit parabolic arcs including the origin. The absolute values of the coordinates of the trajectory intersections with the z_1 and z_2 axis decrease in time. As shown in [5], under conditions (3.29) the following relationships hold:

$$|z_1| \leq |z_{Max}|, \quad |z_2| \leq \sqrt{|z_{Max}|} \tag{3.30}$$

and the convergence of $z_{Max}(t)$ to zero takes place in finite time. As a consequence, the origin of the state space, i.e., $z_1 = z_2 = 0$, is reached in finite time since z_1 and z_2 are both bounded by $\max(|z_{Max}|, \sqrt{|z_{Max}|})$. This, in turn, implies that the control objective is attained.

3.5 Traction and braking control design

3.5.1 Traction controller design

The main advantage of SOSM control [3,4], with respect to the first-order case [2], is that, in the case of relative degree one systems, it can generate continuous control actions, while keeping the same robustness properties with respect to matched uncertainties [65], and a comparable design simplicity.

In this setting the SOSM controller will be designed based on the nonlinear absolute rear slip dynamics only, disregarding the actuator dynamics (see Section 3.3.1). This allows us to work on a plant model with relative degree 1, thus exploiting the possibility of designing a continuous control law. The effect of the actuator dynamics will be taken into account in the simulations, and its impact on the closed-loop system analyzed in Section 3.6. The traction controller is designed to make the rear wheel slip $\lambda_{r,a}$ track the desired value λ_r^*. The error between the current slip and the desired slip is chosen as the sliding variable, i.e.,

$$s_{r,a} = \lambda_{r,a} - \lambda_r^*, \tag{3.31}$$

and the control objective is to design a continuous control law $\mathscr{T}$ capable of steering this error to zero in finite time. Then, the chosen sliding manifold is given by

$$s_{r,a} = 0 \tag{3.32}$$

The first and second derivatives of the sliding variable $s_{r,a}$ are

$$\begin{cases} \dot{s}_{r,a} = \dot{\lambda}_{r,a} - \dot{\lambda}^*_{r,a} \\ \ddot{s}_{r,a} = \varphi_{r,a} + h_{r,a}\dot{\mathcal{T}}_r \end{cases} \tag{3.33}$$

where $\dot{\lambda}_{r,a}$ is given by (3.13), and h_r and φ_r are defined as

$$h_{r,a} := \frac{v}{J\omega_r^2 r} = \frac{(1-\lambda_{r,a})^2 r}{Jv} \tag{3.34}$$

$$\begin{aligned} \varphi_{r,a} &:= -\frac{\ddot{v}}{r\omega_r} + 2\frac{\dot{v}\dot{\omega}_r}{r\omega_r^2} - 2\frac{v\dot{\omega}_r^2}{r\omega_r^3} - \ddot{\lambda}^*_{r,a} - \frac{v\dot{F}_{x_r}}{J\omega_r^2} \\ &= \frac{r(1-\lambda_{r,a})^2}{Jv}\left\{\frac{2(-rF_{x_r}+T)}{v}[m(F_{x_f}+F_{x_r}) \right. \\ &\quad \left. -r(1-\lambda_{r,a})(-rF_{x_r}+T)] - r\dot{F}_{x_r} - \frac{J(\dot{F}_{x_r}+\dot{F}_{x_f})}{rm(1-\lambda_{r,a})}\right\} - \ddot{\lambda}^*_{r,a}. \end{aligned} \tag{3.35}$$

Combining (3.5) with (3.9) yields:

$$|\dot{v}| \le \frac{2\Psi}{m} = f_1. \tag{3.36}$$

Further, taking into account the first time derivative of (3.5), (3.10) and (3.36), one has that

$$|\ddot{v}| \le \frac{2\Gamma}{m} = f_2. \tag{3.37}$$

Finally, from (3.3) and (3.9), it results:

$$|\dot{\omega}_r| \le \frac{-r\Psi + T}{J} = f_3(T). \tag{3.38}$$

Relying on (3.36)–(3.38), and assuming, as it is the case in traction maneuvers, $v > 0, \omega_r > 0$ and $\lambda_{r,a} \in [0,1)$ one has that $\varphi_{r,a}$ is bounded. From a physical viewpoint, this means that, when a constant driving torque T is applied, the second time derivative of the rear wheel slip is bounded.

Note that in order to design a SOSM controller it is not necessary that a precise evaluation of $\varphi_{r,a}$ is available. In the sequel of the paper it will be only assumed that a suitable bound of $\varphi_{r,a}$, i.e., $\Phi_r(v, \omega_r, T)$ such that

$$|\varphi_{r,a}| \le \Phi_r(v, \omega_r, T) \tag{3.39}$$

is known. Similar considerations can be made for $h_{r,a}$ which can be regarded as an unknown bounded function with the following known bounds:

$$0 < \Gamma_{r1}(v, \omega_r) \le h_{r,a} \le \Gamma_{r2}(v, \omega_r). \tag{3.40}$$

In order to design a second-order sliding mode control law, introduce the auxiliary variables $y_1 = s_{r,a}$ and $y_2 = \dot{s}_{r,a}$. Then, system (3.33) can be rewritten as

$$\begin{cases} \dot{y}_1 = y_2 \\ \dot{y}_2 = \varphi_{r,a} + h_{r,a}\dot{\mathcal{T}}, \end{cases} \tag{3.41}$$

where $\dot{\mathcal{T}}$ can be regarded as an auxiliary control input [3]. As a consequence, the control problem can be reformulated as follows: given system (3.41), where $\varphi_{r,a}$ and $h_{r,a}$ satisfy (3.39) and (3.40), respectively, and y_2 is unavailable for measurement, design the auxiliary control signal $\dot{\mathcal{T}}$ so as to steer y_1, y_2 to zero in finite time.

The SOSM controller herein proposed is of sub-optimal type [3]. This implies that, under the assumption of being capable of detecting the extremal values s_{r_M} of the signal $y_1 = s_{r,a}$, the following result can be proved:

Theorem 3.2. *Given system (3.41), where $\varphi_{r,a}$ and $h_{r,a}$ satisfy (3.39) and (3.40), respectively, and y_2 is not measurable, the auxiliary control law:*

$$\dot{\mathcal{T}} = -\eta V_r \operatorname{sign}\left(s_{r,a} - \frac{1}{2}s_{rM}\right) \tag{3.42}$$

$$\eta = \begin{cases} \eta^* & \text{if } [s_{r,a} - s_{rM}/2]s_{rM} > 0 \\ 1 & \text{if } [s_{r,a} - s_{rM}/2]s_{rM} \le 0 \end{cases}$$

where V_r is the control gain, η is the so-called modulation factor, and s_{rM} is a piece-wise constant function representing the value of the last singular point of $s_r(t)$ (i.e., the most recent value s_{rM} such that $\dot{s}_{r,a}(t_M) = 0$), causes the convergence of the system trajectory to the sliding manifold $s_{r,a} = \dot{s}_{r,a} = 0$ in finite time, provided that the control parameters η^* and V_r are chosen so as to satisfy the following constraints:

$$\eta^* \in (0, 1] \cap \left(0, \frac{3\Gamma_{r1}}{\Gamma_{r2}}\right) \tag{3.43}$$

$$V_r > \max\left\{\frac{\Phi_r}{\eta_r^*\Gamma_{r1}}, \frac{4\Phi_r}{3\Gamma_{r1} - \eta^*\Gamma_{r2}}\right\} \tag{3.44}$$

Proof. The control law (3.28) is a sub-optimal second-order sliding mode control law. So, by following a theoretical development as that provided in [5] for the general case, it can be proved that the trajectories on the $(s_{r,a}, \dot{s}_{r,a})$ plane are confined within limit parabolic arcs including the origin. The absolute values of the coordinates of the trajectory intersections with the $s_{r,a}$, and $\dot{s}_{r,a}$ axis decrease in time. As shown in [5], under conditions (3.43) and (3.44) the following relationships hold:

$$|s_{r,a}| \le |s_{rM}|, \quad |\dot{s}_{r,a}| \le \sqrt{|s_{rM}|} \tag{3.45}$$

and the convergence of $s_{rM}(t)$ to zero takes place in finite time [5]. As a consequence, the origin of the plane, i.e., $s_{r,a} = \dot{s}_{r,a} = 0$, is reached in finite since $s_{r,a}$ and $\dot{s}_{r,a}$ are both bounded by max $(|s_{rM}|, \sqrt{|s_{rM}|})$. This, in turn, implies that the traction control objective is attained. □

3.5.2 Braking controller design

Also in the braking control context, the controller is designed to steer the wheel slip λ_f to its desired value λ_f^*.

Again, the error between the current slip and the desired slip is chosen as the sliding variable, i.e.:

$$s = \lambda_f - \lambda_f^*, \tag{3.46}$$

and the control objective is to design a continuous control law $\mathcal{T}_i$ capable of steering this error to zero in finite time. Then, the chosen sliding manifold is given by $s = 0$.

The first and second time derivatives of the sliding variable s_i are

$$\begin{cases} \dot{s}_f &= \dot{\lambda}_f - \dot{\lambda}_f^* \\ \ddot{s}_f &= \varphi_f + h_f \dot{\mathcal{T}}_f, \end{cases} \tag{3.47}$$

where $\dot{\lambda}_f$ for braking control can be computed from the first and the second of (3.20).

Carrying on the needed computations, for the braking control problem one obtains that h_f and φ_f are given by

$$h_f := \frac{r}{Jv} \tag{3.48}$$

$$\varphi_f := \frac{r\omega_i}{v^2}\ddot{v} + 2\frac{r\dot{\omega}_i\dot{v}}{v^2} - 2\frac{r\omega_i\dot{v}^2}{v^3} - \frac{r^2}{Jv}\dot{F}_{xi} - \ddot{\lambda}_i^*. \tag{3.49}$$

Combining the third of (3.3) with (3.9), it yields:

$$|\dot{v}| \leq \frac{2\Psi}{m} = f_1. \tag{3.50}$$

Further, taking into account the first time derivative of the third of (3.3), (3.10) and (3.50), one has that

$$|\ddot{v}| \leq \frac{2\Gamma}{m} = f_2. \tag{3.51}$$

Finally, from the first and second of (3.3) and (3.9), one gets that, for braking control:

$$|\dot{\omega}_f| \leq \frac{r\Psi_f - T_f}{J} = f_3(T_f). \tag{3.52}$$

Relying on (3.50)–(3.52), and assuming $v > 0$, $\omega_f > 0$, hence $\lambda_f \in (0, 1)$ one has that φ_f is bounded.

From a physical viewpoint, this means that, when a constant torque is applied, the second time derivative of the rear wheel slip is bounded. Note that, also to design a SOSM braking controller, we only need to assume that suitable bounds of φ_f are known.

Note that, in principle, different SOSM algorithms are available in the literature. In particular, the most widely used are the super-twisting and the sub-optimal algorithms, [3]. As a matter of fact, the difference between these two algorithms lies in *shape* of the closed-loop trajectories. In this respect, as discussed in [3], the commutation of the sign of $\dot{\mathscr{T}}$ in (3.28) is anticipated when using the sub-optimal algorithm with respect to the twisting controller case. This makes the typical closed-loop trajectories in the sliding variables space different from those of the twisting and super-twisting algorithms. In fact, depending on the control parameters, the sup-optimal algorithm allows the closed-loop trajectories both to twist around the origin and to leap, so that the sliding variable converges monotonically to zero with consequent elimination of undesired transient oscillations. Most importantly, sub-optimal control algorithm features reduced convergence time and control effort as compared to both twisting and super-twisting controller. This is due to the fact that the sub-optimal control algorithm, as its name suggests, is derived from the time-optimal control philosophy, i.e., it mimics the *bang-bang* optimal control. Hence, the choice of the sub-optimal algorithm is motivated by the fact that it allows a shorter transient of the sliding variable, which is a desirable feature in the considered safety-oriented application.

3.6 Analysis of the closed-loop performance

This section is devoted to assess the performance of the SOSM controller via a simulation study. For the traction control system, an extensive simulation analysis is shown, which at first considers a relatively simple Simulink®-based in-plane motorcycle model (i.e., a model which describes the vehicle longitudinal and vertical dynamics in straight-running conditions), which takes into account tire elasticity and tire relaxation dynamics and models the ETB dynamics.

Further, to validate the proposed SOSM controller in a setting as close as possible to real on-bike experiments, some simulation results obtained on a full-fledged commercial motorcycle simulator are presented. The considered simulation environment is the Mechanical Simulation Corp. MSC BikeSim® simulation environment, based on the AutoSim symbolic multi-body software, [66], which also models transmission and engine dynamics and provides a very accurate description of the road–tire interaction forces, [38]. As for the motorcycle model, the BikeSim Hypersport motorcycle has been used, whose relevant geometric parameter values can be found in [10].

For the braking control system, again, results obtained with the full motorcycle model are reported.

All the simulations have been carried out with fixed step integration and a sampling frequency of 1 kHz, which is that available on vehicle ECUs.

3.6.1 Traction control

To get a first view of the wheel slip tracking capabilities of the designed SOSM controller, let us refer to Figure 3.13, which shows the time histories of relative and absolute slip and of the braking torque in a traction maneuver where the slip set-point was set to 0.1.

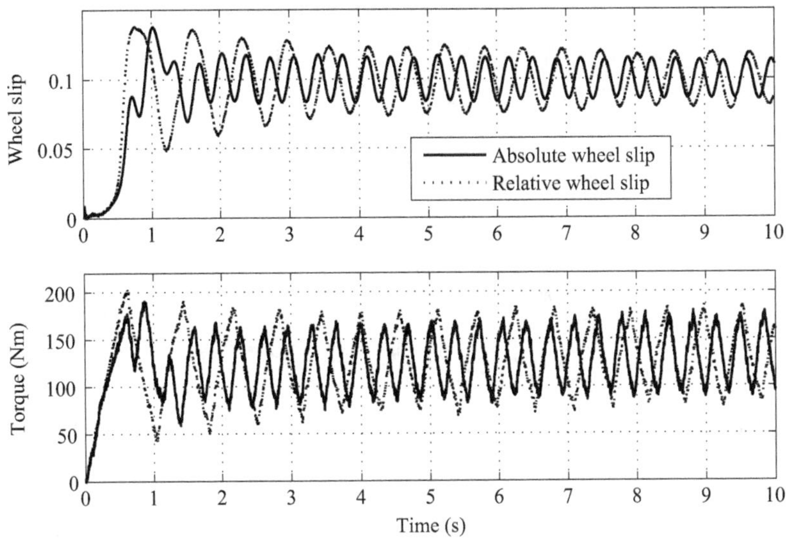

Figure 3.13 Time histories of relative (solid line) and absolute (dotted line) wheel slip – top plot; time histories of the corresponding driving torque – bottom plot, obtained in a traction maneuver where the slip set-point was set to $\lambda_r^* = 0.1$

As can be seen, the difference in using the relative wheel slip is not significant, also in a realistic simulation setting, confirming the suitability of the choice dictated by practical needs.

To consider challenging yet realistic situations, note that in traction control applications it is crucial that the control algorithm can correctly manage sudden changes in the road conditions, which possibly occur during strong accelerations. Such a situation is often referred to as a μ-jump. Figure 3.14 shows the time histories of the relative rear wheel slip, vehicle and wheel speeds and driving torque in a traction maneuver on the full multibody simulator where a μ-jump from $\mu = 0.2$ to $\mu = 0.4$ with relative rear slip $\lambda_{r,r}$ as a controlled variable.

Inspecting Figure 3.14, note also that the wheel slip exhibits small oscillations: analyzing the period of such oscillations one finds that it corresponds to the actuator bandwidth. Such oscillations are due to the fact that the presence of the unmodeled ETB dynamic increases the relative degree of the system (note that the pure delay in (3.16) has been modeled via a second-order Padé approximation, hence with no additional increase in the relative degree). As a consequence, the transient process converge to a periodic motion [67].

However, the amplitude of such oscillations is such that it can be well tolerated in the specific application. Instead, the use of a higher order SM controller, which would be needed in principle to formally deal with a plant with relative degree higher

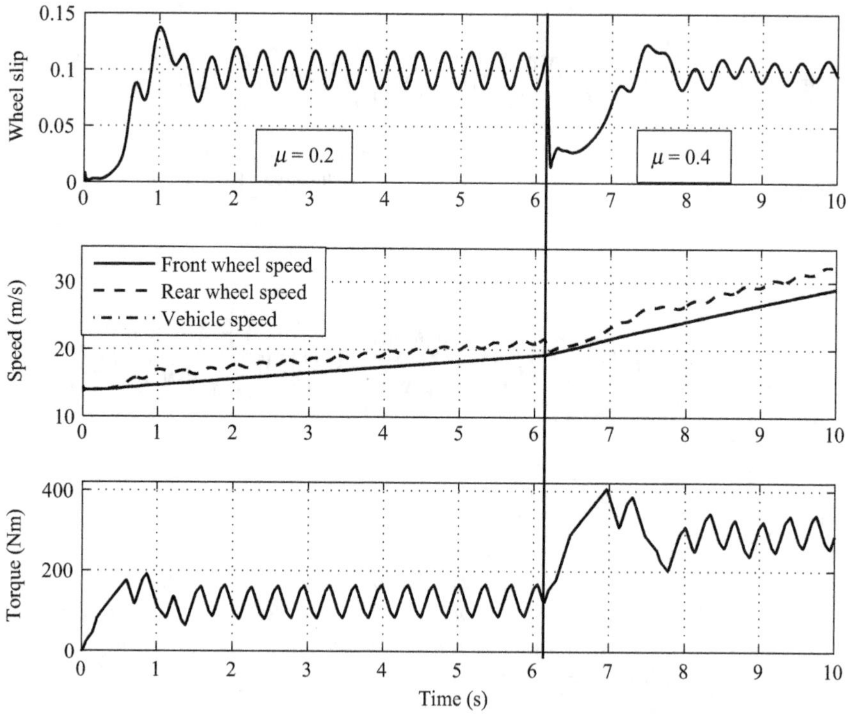

Figure 3.14 Plot of (top): relative rear wheel slip; (middle): front wheel (solid line) rear wheel (dashed line) and vehicle (dash-dotted line) speed; (bottom): driving torque in a traction maneuver on the full multibody simulator with a μ-jump from $\mu = 0.2$ to $\mu = 0.4$ with relative rear slip $\lambda_{r,r}$ as controlled variable

than one by means of a continuous control law, is not advisable in automotive control, as higher order derivatives of the controlled variable need to be computed and this cannot be done reliably due to measurements noise.

Further, Figure 3.14 shows that the proposed controller can guarantee safety also in very critical maneuvers. As can be seen, however, tracking the wheel slip in face of the more realistic vehicle model offered by the full multibody simulator (where also engine and transmission are modeled) is more critical, and this results in slightly increased oscillations amplitude than encountered with the simpler model. However, the results are still more than acceptable for a real vehicle implementation. This is confirmed by experimental data analysis (kept undisclosed for confidentiality reasons), which showed that, in acceleration maneuvers which were considered safe and comfortable by a professional driver, the relative rear wheel slip exhibited oscillations which were much larger than those obtained in the simulation results shown in Figure 3.14.

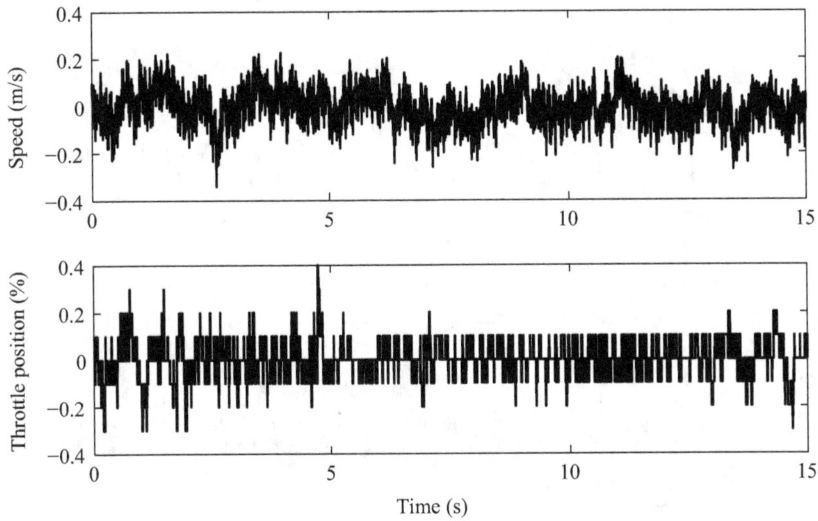

Figure 3.15 Plot of (top): measured noises on rear wheel speed; (bottom) measured noise on the throttle position

Finally, disturbances need to be taken into account. As previously mentioned, to handle both matched and unmatched disturbances a SOSM controller scheme complemented with error prefiltering has been designed. Further, in order to validate the controller in a realistic setting, the employed noises used in these simulations are wheel speed and throttle position noise measurement errors which have been recorded on the instrumented vehicle and are shown in Figure 3.15. Note that the wheel speed errors magnitude (for readability, Figure 3.15 shows the rear wheel speed noise only: a comparable noise magnitude is present on the front wheel speed) is such that the induced oscillations on the relative wheel slip are approximately of ± 0.02.

The results of these simulations are shown in Figure 3.16, where the behavior of the relative wheel slip and the driving torque is shown both with and without error prefiltering. As can be seen, with error prefiltering the closed-loop behavior of the rear wheel slip exhibits oscillations of much smaller magnitude than without error prefiltering, so that the traction control systems can guarantee safety even in the presence of real-life disturbances, thereby confirming the suitability of the proposed SOSM controller for motorcycle traction control applications.

3.6.2 Braking control

Before testing the SOSM algorithm, similarly to what was done for traction control, the BikeSim simulation model was complemented with the actuator dynamics For testing the performance of the braking SOSM controller, a μ-jump test on the BikeSim simulation model was considered. A jump from $\mu = 0.85$ to $\mu = 0.6$ is shown.

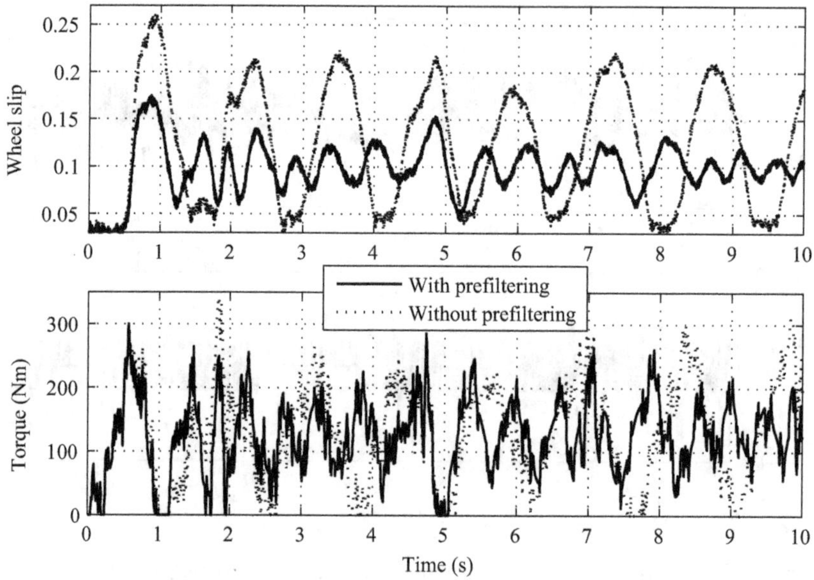

Figure 3.16 Plot of (top): relative rear wheel slip with (solid line) and without (dotted line) error pre-filtering; (bottom): driving torque with (solid line) and without (dotted line) error pre-filtering in a traction maneuver on the full multibody simulator fed with measured noises

Figure 3.17 shows the time histories of the front wheel slip and of the braking torque during such a maneuver. As can be seen, the jump in the friction coefficient causes an overshoot in the wheel slip, which is then recovered. Again, oscillations are present in the wheel slip due to the larger relative degree of the simulation model with respect to that assumed at design time. The amplitude of such oscillations grows larger as the braking maneuver proceeds, mainly due to the fact that the vehicle speed gets lower, and the wheel dynamics faster.

To better appreciate the performance of the braking controller, Figure 3.18 shows the time histories of both front wheel speed and vehicle body speed in the same maneuver. Of course, the wheel speed exhibits a variation consistent with that present in the wheel slip at the moment of time in which the friction changes.

To further test the braking control performance and directly compare them to those of the traction control system, a second μ-jump braking maneuver was considered, with the same grip variation from 0.2 to 0.4 tested with the traction controller (see Figure 3.14). The results are plotted in Figure 3.19, showing very similar performance between traction and braking control systems.

A final test for the braking control system was that of analyzing its performance in the presence of measurement noises. The noise on the wheel speed was the same used

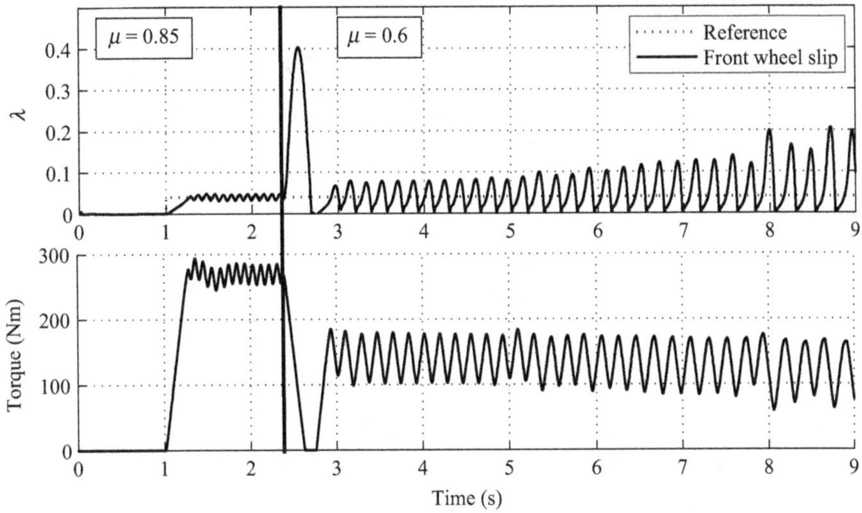

Figure 3.17 Top plot: time histories of the front wheel slip (solid line) and slip reference (dotted line); bottom plot: time histories of the braking torque, in a μ-jump braking maneuver on the full multibody simulator

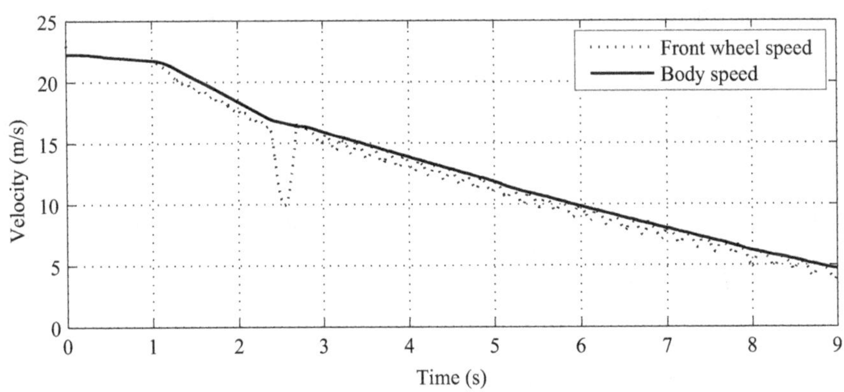

Figure 3.18 Time histories of the front wheel speed (solid line) and of the vehicle speed (dotted line), in a μ-jump braking maneuver on the full multibody simulator

for the traction control system, while for the braking torque a properly scaled version of the noise used for the ETB in traction control was used (see Figure 3.15). To test the system in a very challenging setting, the braking maneuver, with measurement noises added to the simulator, was carried out on a very low grip road ($\mu = 0.2$).

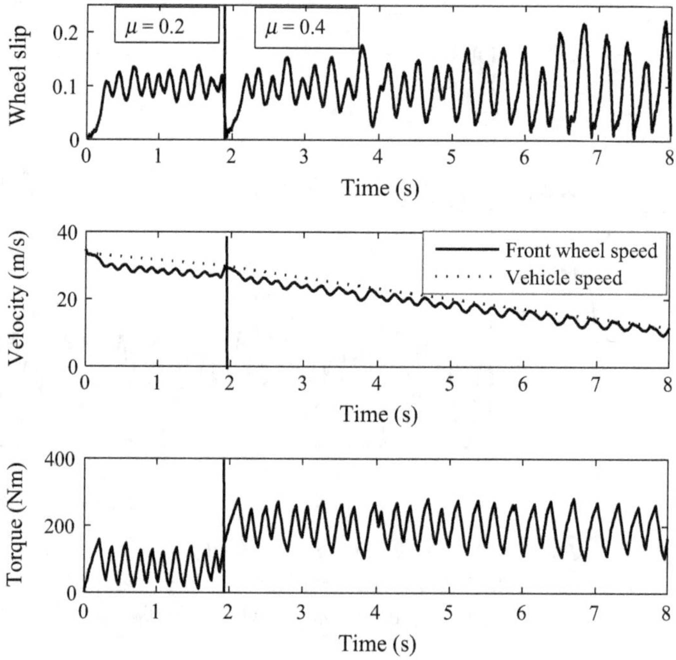

Figure 3.19 Plot of (top): front wheel slip; (middle): front wheel (solid line) and vehicle (dotted line) speed; (bottom): braking torque in a traction maneuver on the full multibody simulator with a μ-jump from $\mu = 0.2$ to $\mu = 0.4$

The results, in terms of time histories of both wheel slip and braking torque, can be seen in Figure 3.20. To perform the maneuver, the error prefiltering scheme used also for traction control was implemented.

3.7 Concluding remarks

In this chapter, the problem of longitudinal control of two-wheeled vehicles was considered. The main controllers to be designed are traction and braking controller. Their main features and challenges were extensively discussed, and the main solutions available in the state of the art were reviewed. Further, a control-oriented modeling of the two dynamics of interest was proposed, which served as a basis for the SOSM controller design. Finally, the chapter offered a validation setting which can be considered rather close to real bike experiments, based on the test of the proposed algorithms on a full-vehicle commercial simulator. The chapter also considered the technological issues related to the actuation systems, discussing their features and their impact on the controller design.

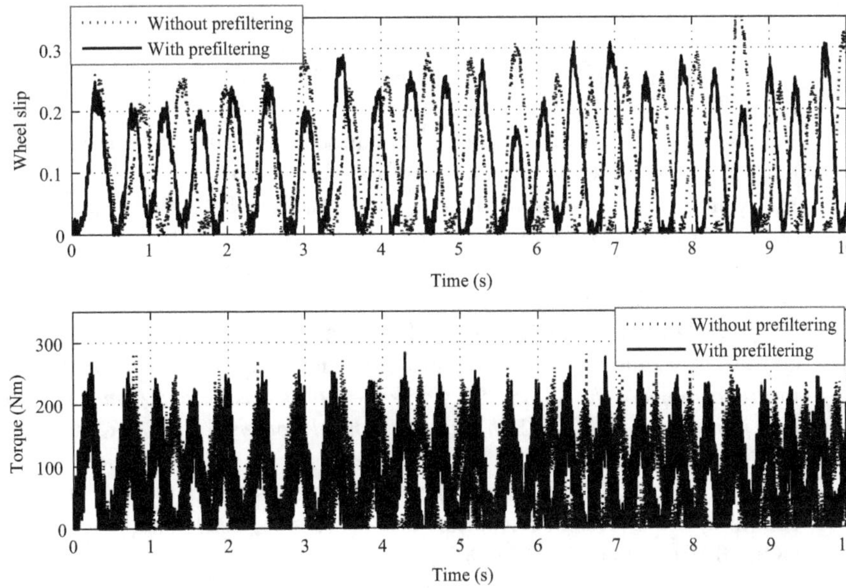

Figure 3.20 Plot of (top): front wheel slip with (solid line) and without (dotted line) error pre-filtering; (bottom): braking torque with (solid line) and without (dotted line) error pre-filtering in a traction maneuver on the full multibody simulator fed with measured noises and low grip ($\mu = 0.2$)

References

[1] S. Savaresi, M. Tanelli, and M. Corno. *Modelling, Simulation and Control of Two-Wheeled Vehicles*. Chichester, West Sussex, UK: Wiley, 2014.

[2] V. Utkin, J. Guldner, and J. Shi. *Sliding-Mode Control in Electromechanical Systems*. Taylor & Francis, London, UK, 1999.

[3] G. Bartolini, A. Ferrara, F. Levant, and E. Usai. "On second order sliding mode controllers," in K. D. Young and Ü. Özgüner, editors, *Lecture Notes in Control and Information Sciences*, vol. 247, pp. 329–350, 1999. Springer-Verlag, Berlin.

[4] L. Fridman and F. Levant. "Higher order sliding modes," in W. Perruquetti and J. P. Barbot, editors, *Sliding Mode Control in Engineering*, vol. 217, pp. 53–101, 2002. Marcel Dekker, New York.

[5] G. Bartolini, A. Ferrara, and E. Usai. "Chattering avoidance by second-order sliding mode control," *IEEE Transactions on Automatic Control*, 43(2): 241–246, 1998.

[6] M. Tanelli, R. Sartori, and S. M. Savaresi. "Combining slip and deceleration control for brake-by-wire control systems: a sliding-mode approach," *European Journal of Control*, 6(6):593–611, 2007.

[7] H.B. Pacejka. *Tyre and Vehicle Dynamics*. Butterworth Heinemann, Oxford, 2002.

[8] S.M. Savaresi and M. Tanelli. *Active Braking Control Design for Road Vehicles*. London, UK: Springer, 2010.

[9] V. Cossalter, A. Doria, and R. Lot. "Optimum suspension design for motorcycle braking," *Vehicle System Dynamics*, 34(3):175–198, 2000.

[10] M. Corno, S. M. Savaresi, M. Tanelli, and L. Fabbri. "On optimal motorcycle braking," *Control Engineering Practice*, 16(6):644–657, 2008.

[11] R. S. Sharp. "Limit braking of a high-performance motorcycle," *Vehicle System Dynamics*, 47(5):613–625, 2009.

[12] C. Spelta, D. Delvecchio, R. Cantoni, R. Lazzari, and S. M. Savaresi. "Analysis and design of handling-oriented control strategies for semi-active suspensions," in *Proceedings of the 48th IEEE Conference on Decision and Control, 2009 held jointly with the 2009 28th Chinese Control Conference. CDC/CCC 2009*, pp. 7633–7638, Dec. 2009.

[13] P. Seiniger, K. Schroeter, and J. Gail. "Perspectives for motorcycle stability control systems," *Accident Analysis and Prevention*, 2011. Available online, DOI: 10.1016/j.aap.2010.11.018.

[14] G. Panzani, M. Corno, and S. M. Savaresi. "On adaptive electronic throttle control for sport motorcycles," *Control Engineering Practice*, 21(1): 42–53, 2013.

[15] A Keawtubtimthong, D. Koolpiruck, S. Wongsa, Y. Laoonual, and A Kaewpunya. "Development of engine control technique for flex-fuel motorcycle," in *Electrical Engineering/Electronics Computer Telecommunications and Information Technology (ECTI-CON), 2010 International Conference on*, pp. 159–162, May 2010.

[16] P. Cardinale, C. D'Angelo, and M. Conti. "A traction control system for motocross and supermotard" in *2008 International Workshop on Intelligent Solutions in Embedded Systems*, pp. 1–15, Jul. 2008.

[17] X. Song and Z. Sun. "Pressure-based clutch control for automotive transmissions using a sliding-mode controller," *IEEE/ASME Transactions on Mechatronics*, 17(3):534–546, Jun. 2012.

[18] F. Todeschini, M. Corno, G. Panzani, and S. M. Savaresi. "Adaptive position-pressure control of a brake by wire actuator for sport motorcycles," *European Journal of Control*, 20(2):79–86, 2014.

[19] F. Todeschini, M. Corno, G. Panzani, S. Fiorenti, and S. M. Savaresi. "Adaptive cascade control of a brake-by-wire actuator for sport motorcycles," *IEEE/ASME Transactions on Mechatronics*, 20(3):1310–1319, 2015.

[20] D. Green. "A comparison of stopping distance performance for motorcycles equipped with ABS, CBS and conventional hydraulic brake systems," in *International Motorcycle Safety Conference*, pp. 26–30, 2006.

[21] M. Corno, S. M. Savaresi, and G. J. Balas. "On linear-parameter-varying (LPV) slip-controller design for two-wheeled vehicles," *International Journal of Robust and Nonlinear Control*, 19(12):1313–1336, 2009.

[22] C.-Y. Lu and M.-C. Shih. "Application of the Pacejka magic formula tyre model on a study of a hydraulic anti-lock braking system for a light motorcycle," *Vehicle System Dynamics*, 41(6):431–448, 2004.

[23] S. Formentin, P. De Filippi, M. Tanelli, and S. Savaresi. "Model-free control for active braking systems in sport motorcycles," in *Proceedings of the Eighth IFAC Symposium on Nonlinear Control Systems*, pp. 873–878, Bologna, Italy, 2010.

[24] M. Tanelli and A Ferrara. "Active braking control for two-wheeled vehicles via switched second order sliding modes," in *American Control Conference (ACC), 2011*, pp. 3930–3935, Jun. 2011.

[25] C.-Y. Lu and M.-C. Shih. "Design of a hydraulic anti-lock braking modulator and an intelligent brake pressure controller for a light motorcycle," *Vehicle System Dynamics*, 43(3):217–232, 2005.

[26] C. K. Huang and M. C. Shih. "Design of a hydraulic anti-lock braking system (ABS) for a motorcycle," *Journal of Mechanical Science and Technology*, 24(5):1141–1149, 2010.

[27] S. Formentin, P. De Filippi, M. Tanelli, and S. Savaresi. "Model-free control for active braking systems in sport motorcycles," *Nonlinear Control Systems*, 43(14): 873–878, 2010.

[28] G. Panzani, S. Formentin, and S. M. Savaresi. "Active motorcycle braking via direct data-driven load transfer scheduling," in *IFAC Proceedings Volumes*, 45(16):1257–1262, 2012.

[29] P. Cardinale, C. D'Angelo, and M. Conti. "Traction control system for motorcycles," Hindawi Publishing Corporation, *EURASIP Journal on Embedded Systems*, 2009, Article ID 161373, 10pp, 2009. doi:10.1155/2009/161373.

[30] M. Corno and S. M. Savaresi. "Experimental identification of engine-to-slip dynamics for traction control applications in a sport motorbike," *European Journal of Control*, 16(1):88–108, 2010.

[31] S. Formentin, P. De Filippi, and S. M. Savaresi. "Optimal design of experiment for the identification of throttle-to-slip dynamics in two-wheeled vehicles," in *2010 IEEE International Conference on Control Applications (CCA)*, pp. 142–147, Sep. 2010.

[32] M. Massaro, R. Sartori, and R. Lot. "Numerical investigation of engine-to-slip dynamics for motorcycle traction control applications," *Vehicle System Dynamics*, 49(3):419–432, 2011.

[33] M. Tanelli, C. Vecchio, M. Corno, A. Ferrara, and S. M. Savaresi. "Traction control for ride-by-wire sport motorcycles: a second-order sliding mode approach," *IEEE Transactions on Industrial Electronics*, 56(9): 3347–3356, 2009.

[34] M. Corno, G. Panzani, and S. M. Savaresi. "Traction-control-oriented state estimation for motorcycles," *IEEE Transactions on Control Systems Technology*, 21(6):2400–2407, Nov. 2013.

[35] M. Corno, M. Tanelli, S. M. Savaresi, L. Fabbri, and L. Nardo. "Electronic throttle control for ride-by-wire in sport motorcycles," in *IEEE*

Multi-conference on Systems and Control, San Antonio, Texas, USA, pp. 233–238, 2008.

[36] A. Beghi, L. Nardo, and M. Stevanato. "Observer-based discrete-time sliding mode throttle control for drive-by-wire operation of a racing motorcycle engine," *IEEE Transactions on Control Systems Technology*, 14:767–775, 2006.

[37] H. Lee and M. Tomizuka. "Adaptive vehicle traction force control for intelligent vehicle highway systems (IVHSS)," *IEEE Transactions on Industrial Electronics*, 50(1):37–47, 2003.

[38] R. S. Sharp, S. Evangelou, and D. J. N. Limebeer. "Advances in the modelling of motorcycle dynamics," *Multibody System Dynamics*, 12:251–283, 2004.

[39] I. Boniolo, S. M. Savaresi, and M. Tanelli. "Roll angle estimation in two-wheeled vehicles," *IET Control Theory & Applications*, 3(1):20–32, 2009.

[40] U. Kiencke and L. Nielsen. *Automotive Control Systems*. Springer-Verlag, Berlin, 2000.

[41] V. Cossalter, A. Doria, and R. Lot. "Steady turning of two-wheeled vehicles," *Vehicle System Dynamics*, 31(3):157–181, 1999.

[42] V. Cossalter, R. Lot, and F. Maggio. "On the stability of motorcycle during braking," in *SAE Small Engine Technology Conference and Exhibition*, Graz, Austria, Sep. 2004. SAE Paper number: 2004-32-0018/20044305.

[43] D. J. N. Limebeer, R. S. Sharp, and S. Evangelou. "The stability of motorcycles under acceleration and braking," *Proceedings of the Institute of Mechanical Engineers, Part C, Journal of Mechanical Engineering Science*, 215(9):1095–1109, 2001.

[44] R. S. Sharp. "Stability, control and steering responses of motorcycles," *Vehicle System Dynamics*, 35:291–318, 2001.

[45] R. S. Sharp and D. J. N. Limebeer. "A motorcycle model for stability and control analysis," *Multibody System Dynamics*, 6:123–142, 2001.

[46] R. S. Sharp. "The stability and control of motorcycles," *Journal of Mechanical Engineering Science*, 13(5):316–329, 1971.

[47] S. M. Savaresi and M. Tanelli. *Active Braking Control Systems Design for Vehicles*. Springer, London, UK, 2010.

[48] T. A. Johansen, I. Petersen, J. Kalkkuhl, and J. Lüdemann. "Gain-scheduled wheel slip control in automotive brake systems," *IEEE Transactions on Control Systems Technology*, 11(6):799–811, Nov. 2003.

[49] M. Tanelli, C. Vecchio, M. Corno, A. Ferrara, and S. M. Savaresi. "Traction control for ride-by-wire sport motorcycles: a second order sliding mode approach," *IEEE Transactions on Industrial Electronics*, 56(9):3347–3356, 2009.

[50] M. Tanelli, M. Prandini, F. Codecà, A. Moia, and S. M. Savaresi. "Analysing the interaction between braking control and speed estimation: the case of two-wheeled vehicles," in *47th IEEE Conference on Decision and Control*, Cancún, Mexico, 2008.

[51] M. Tanelli, A. Astolfi, and S. M. Savaresi. "Robust nonlinear output feedback control for brake-by-wire control systems," *Automatica*, 44(4):1078–1087, 2008.

[52] R. Pintelon and J. Schoukens. *System Identification: A Frequency Domain Approach*. IEEE Press, New York, NY, 2001.

[53] J. Deur, D. Pavkovic, P. Nedjeljko, M. Jansz, and D. Hrovat. "An electronic throttle control strategy including compensation of friction and limp-home effects," *IEEE Transactions on Industrial Electronics*, 40:821–834, 2004.

[54] S. M. Savaresi and M. Tanelli. *Active Braking Control Systems Design for Vehicles*. Springer-Verlag, London, UK, 2010.

[55] D. Reuter, E. Lloyd, J. Zehnder, and J. Elliott. "Hydraulic design considerations for EHB systems," *SAE World Congress*, 2003.

[56] N. D'Elfio, A. Morgando, and A. Sorniotti. "Electro-hydraulic brake systems: design and test through hardware-in-the-loop simulation," *Vehicle System Dynamics: International Journal of Vehicle Mechanics and Mobility*, 44(1), 378–392, 2006.

[57] A. Sorniotti and G. Repici. "Hardware in the loop with electro-hydraulic brake systems," in *Ninth WSEAS International Conference on Systems*, 2005.

[58] V. Milanès, C. Gonzàlez, E. Naranjo, E. Onieva, and T. De Pedro. "Electro-hydraulic braking system for autonomous vehicles," *International Journal of Automotive Technology*, 11(1):89–95, 2010.

[59] J. Kwak, B. Yao, and A. Bajaj. "Analytical model development and model reduction for electromechanical brake system," ASME 2004 *International Mechanical Engineering Congress and Exposition, Dynamic Systems and Control, Parts A and B*, Anaheim, CA, USA, November 13–19, 2004 Conference Sponsors: Dynamic Systems and Control Division, 2004. ISBN: 0-7918-4706-3.

[60] C. Line. "Modelling and control of an automotive electromechanical brake," Ph.D. Thesis, University of Melbourne, 2007.

[61] A. Dardanelli, G. Alli, and S. Savaresi. "Modeling and control of an electro-mechanical brake-by-wire actuator for a sport motorbike," in *Mechatronic Systems*, pp. 524–531, 2010.

[62] G. Panzani, M. Corno, F. Todeschini, S. Fiorenti, and S. Savaresi. "Analysis and control of a brake by wire actuator for sport motorcycles," in *13th Mechatronics Forum International Conference*, Linz, Austria, Sep. 17–19, 2012.

[63] F. Todeschini, M. Corno, G. Panzani, S. Fiorenti, and S.M. Savaresi. "Adaptive cascade control of a brake-by-wire actuator for sport motorcycles," *IEEE/ASME Transactions on Mechatronics*, 20(3):1310–1319, 2015.

[64] H.K. Khalil. *Nonlinear Systems*, 2nd edn. Prentice-Hall, Upper Saddle River, NJ, 1996.

[65] C. Edwards and K. S. Spurgeon. *Sliding Mode Control: Theory and Applications*. Taylor & Francis, London, UK, 1998.

[66] R. S. Sharp, S. Evangelou, and D. J. N. Limebeer. "Multibody aspects of motorcycle modelling with special reference to Autosim," in J. A. C. Ambrosio, editor, *Advances in Computational Multibody Systems*, pp. 45–68. Houten, The Netherlands: Springer-Verlag, 2005.

[67] I. Boiko, L. Fridman, A. Pisano, and E. Usai. "Performance analysis of second-order sliding-mode control systems with fast actuators," *IEEE Transactions on Automatic Control*, 52(6):1053–1059, 2007.

Chapter 4

Lateral vehicle dynamics control via sliding modes generation

Enrico Regolin[1], Dzmitry Savitski[2], Valentin Ivanov[2], Klaus Augsburg[2], and Antonella Ferrara[1]

This chapter introduces the application of Sliding Mode Control techniques by solving the lateral Vehicle Stability Control (VSC) problem. In Section 4.1 the linearized single track model is presented, which is typically used for the design of lateral stability controllers. The formulation of the yaw rate control problem is also explained, starting from the equations of motion. In Section 4.2, we propose a survey of the different methods for the control of the vehicle lateral dynamics, which are known from the research literature. The control structure and the specific sliding mode controllers which are presented in this chapter are illustrated in Section 4.3 and assessed in Section 4.4, where the results of the simulations are reported and discussed.

4.1 Vehicle modeling and problem formulation

The driver control on the steering wheel while the vehicle is moving at a certain speed is a dynamic process: the turning of the wheels causes the side slip and the generation of lateral forces. The lateral forces change the overall attitude of the vehicle. As a consequence, the vehicle course angle is being changed and lateral forces are also generated on the non-steering wheels. The resulting forces determine the vehicle trajectory. However, when the vehicle speed is moderate and the cornering stiffness is high, the linearity of the vehicle response gives the driver the feeling of a kinematic motion, which is entirely determined by the trajectory of the mid-planes of the wheels.

The vehicle stability belongs to the most complex attributes that should be handled by active safety systems. A comprehensive analysis of vehicle stability parameters has been introduced in fundamental works of Karnopp [1], Mitschke [2], and other

[1]Dipartimento di Ingegneria Industriale e dell'Informazione, University of Pavia, Via Ferrata 5, 27100 Pavia, Italy
[2]Automotive Engineering Group, Department of Mechanical Engineering, Technische Universität Ilmenau, Ehrenbergstr. 15, 98693 Ilmenau, Germany

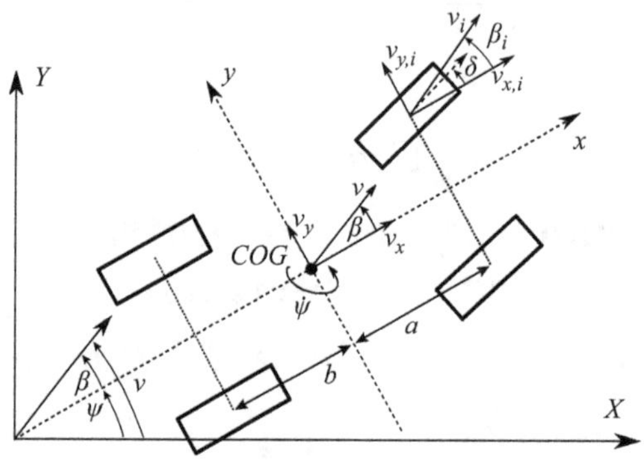

Figure 4.1 Full vehicle model

authors. The characteristics of directional vehicle stability can be explained making reference to Figure 4.1.

The directional stability is a characteristic of the vehicle (i) to keep the course demanded by the driver control and (ii) to stabilize the direction of motion against external disturbances. In the case of vehicle maneuvers with lateral dynamics (turning, obstacle avoidance, etc.) the vector of vehicle velocity v is deviated from the longitudinal X-axis of earth-fixed coordinate system XY on a certain course angle (Figure 4.1). The course angle is the sum of the yaw angle ψ and the vehicle side slip angle β:

$$\nu = \psi + \beta \tag{4.1}$$

ψ is the angle between longitudinal axes of earth-fixed XY and vehicle-fixed xy coordinate systems. Due to deformation processes in tire–road contact and tire side slip, the vector v is deviated from the x-axis on the additional angle β.

For the design of a control system for the lateral stability of the vehicle, a proper dynamic model is required. Such a model should represent the physical phenomenon introduced above. In the following subsection, it will be further illustrated how the linear single track model (or "bicycle" model) is derived. For this purpose, the lifting, rolling, and pitching motion are neglected, which means we are working under the hypothesis that the vehicle consists of a rigid body moving on a flat surface.

4.1.1 Dynamic equations

Given the hypothesis of rigid body, the motion of the body is described by a 3 degree-of-freedom model. For the determination of a set of equations expressed in the reference frame of the center of gravity (COG) of the vehicle (it is assumed that the vehicle mass is concentrated at the COG), different procedures can be adopted: in [3] a method based on the Lagrange equations is used, while in [4] the equations are obtained by calculating explicitly the acceleration of the COG. For a vehicle with mass m and inertia moment around the normal axis J_z, having defined the longitudinal and lateral velocities in the vehicle fix coordinate system as v_x, v_y, and the yaw angle $\dot{\psi}$, the equations of the dynamics can be written as

$$\begin{cases} m(\dot{v}_x - \dot{\psi} v_y) = F_{x,tot} \\ m(\dot{v}_y + \dot{\psi} v_x) = F_{y,tot} \\ J_z \ddot{\psi} = M_{z,tot} \end{cases} \tag{4.2}$$

where $F_{x,tot}$, $F_{y,tot}$ are the overall longitudinal and lateral forces applied at the COG, and $M_{z,tot}$ is the overall yaw moment around the normal axis.

Instead of the representation v_x and v_y, it is a common a practice to use the velocity vector v and the side slip angle β. For this purpose, since angle β is in general small, the following approximations can be considered:

$$v_x = v \cos(\beta) \approx v \tag{4.3}$$

$$v_y = v \sin(\beta) \approx v\beta \tag{4.4}$$

Equation (4.2) is reduced to

$$\begin{cases} m(\dot{v} - \dot{\psi} v \beta) = F_{x,tot} \\ mv(\dot{\beta} + \dot{\psi}) + m\beta\dot{v} = F_{y,tot} \\ J_z \ddot{\psi} = M_{z,tot} \end{cases} \tag{4.5}$$

The term $m\beta\dot{v}$ is often neglected, as it is typically much smaller than the other terms. The system complexity can be reduced further, by assuming the vehicle velocity is a pseudo constant during the steering maneuver: the effect of the longitudinal forces is neglected, so that only the second two equations in (4.5) are considered:

$$\begin{cases} mv(\dot{\beta} + \dot{\psi}) = F_{y,tot} \\ J_z \ddot{\psi} = M_{z,tot} \end{cases} \tag{4.6}$$

The equations from models (4.2), (4.5), and (4.6) represent the generic dynamic evolution of a rigid body, with external influences from forces $F_{x,tot}$, $F_{y,tot}$, and moment $M_{z,tot}$. They do not consider the wheels disposition or, in a more general way, the interaction between the wheels and the surface. Further detailing of the model (4.6) requires the explicit formulation of the force $F_{y,tot}$ and the moment $M_{z,tot}$.

4.1.2 Cornering forces

A schematic representation of the cornering forces acting on the wheels is shown in Figure 4.2. The forces determining the $F_{y,tot}$ term on the right-hand side of (4.2), (4.5), and (4.6) are composed of several components:

- cornering forces generated on all wheels due to the tire–road interactions determined by the steering maneuver;
- projection of longitudinal forces F_x (on steering wheels) on the y-axis, which is generally really small due to the small angle δ that can be handled during stable maneuvers at non-vanishing vehicle speed;
- aerodynamic effects due to the vehicle body shape, which are in general much smaller than the aerodynamic effects on the longitudinal dynamics;

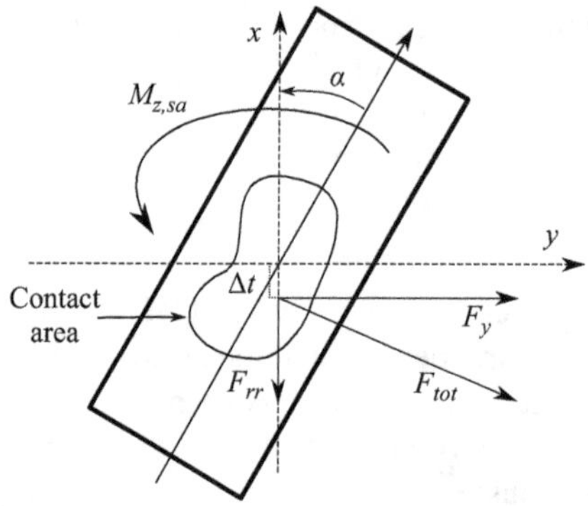

Figure 4.2 Wheel cornering forces

- external influences due to e.g., lateral wind, road inclination along the y-axis, etc., which are often unpredictable.

Due to limited impact or non-predictability of other components on the lateral dynamics, we can consider the cornering forces alone, indicated as F_y in Figure 4.2.

An extensive explanation of the tire–road interaction, which determines the lateral dynamics, can be found in [3]. Here follows a short overview of such phenomena, which can be useful to get a basic understanding of how the models for the lateral dynamics are obtained.

4.1.2.1 Tire–road interaction

We recall here, how the longitudinal forces in the pneumatic are exerted thanks to the tire deformation, which determines the wheel-slip. Similarly, the lateral (or cornering) forces are originated due to the lateral deformation, which affects the tire and its side slip angle α. Also in this case, the fact that the wheel has a side slip angle does not necessarily mean that the wheel is slipping on the road in the contact zone. The portion of contact zone, which is actually slipping, increases with the side slip angle. The lateral wheel force F_y is not applied to the center of the contact zone, but at a point behind it, located at a distance Δt. For this reason, a moment $M_{z,sa} = F_y \Delta t$ is generated, which tends to align the wheel towards its velocity v. In Figure 4.3, F_y and $M_{z,sa}$ are plotted against the side slip angle α. Note that, similarly to the longitudinal case, a side force coefficient $\mu_y = F_y/F_z$ can be defined. While in the case of longitudinal wheel-slip control there is often a direct interest in controlling the slip range towards the maximum of the $\mu(\lambda)$ curve, when controlling the lateral dynamics it is assumed that the control action keeps the side slip angle low, therefore the knowledge of the behavior of μ_y is not strictly necessary for the control algorithm, therefore the curve does not have the same importance in this chapter.

Both the lateral force F_y and the aligning torque $M_{z,sa}$ grow for low values of the side slip angle α, almost linearly in case of the force. For higher values of α, the force F_y tends to stabilize or slightly decrease, while $M_{z,sa}$ shows a sharp decrease, leading to a change of sign. Several factors influence the behavior of the curve $F_y(\alpha)$ at higher

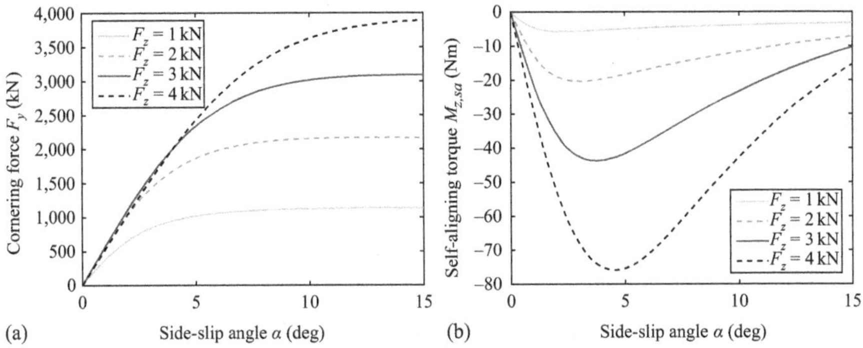

Figure 4.3 Example of dependencies $F_y(\alpha)$ and $M_{z,sa}(\alpha)$: (a) cornering forces and (b) self-aligning torques

slips, such as tire design, road conditions, camber angle, and others. The impact of these factors is in general limited for low values of α: this region is the stable area, in which the tires should normally be operated, as the lateral forces applied to the wheel or proportional to the steer request by the driver. The stability control systems are designed to keep the vehicle operation in this region, and therefore a linearized local representation of the curve $F_y(\alpha)$ can be adopted. This allows one to define a "cornering stiffness" coefficient C_y as the slope of $F_y(\alpha)$ in the origin, i.e.:

$$C_y = \left. \frac{\partial F_y}{\partial \alpha} \right|_{\alpha=0}, \tag{4.7}$$

so that the following approximation takes place in the stable area:

$$F_y(\alpha) = C_y \alpha \tag{4.8}$$

4.1.2.2 Forces linearization

Based on the representation in Figure 4.1, the ith wheel side slip angle is defined as the difference between β_i and the steering angle δ_i, where β_i is the angle between the velocity of the ith wheel and the direction of the vehicle (x-axis). It follows that:

$$\alpha_i = \beta_i - \delta_i = \arctan\left(\frac{v_{yi}}{v_{xi}}\right) - \delta_i = \arctan\left(\frac{v_y + \dot{\psi} x_i}{v_x - \dot{\psi} y_i}\right) - \delta_i \tag{4.9}$$

where x_i and y_i are the coordinates of the wheel in the xy frame (the time dependence of the variables has been omitted for the sake of simplicity). In particular, for the front wheels $x_i = a$ and for the rear wheels $x_i = -b$. Equation (4.9) can be linearized, by noting that $\dot{\psi} y_i \ll v$. Therefore:

$$\alpha_i = \beta_i - \delta_i \approx \frac{v_y + \dot{\psi} x_i}{v} - \delta_i = \beta + \frac{x_i}{v}\dot{\psi} - \delta_i \tag{4.10}$$

Note that the coordinate y_i of the center of the contact area of the wheel does not appear in the expression for the side slip angle α_i. Furthermore, if the differences between the steering angles δ_i of the wheels on the same axle are neglected, the values

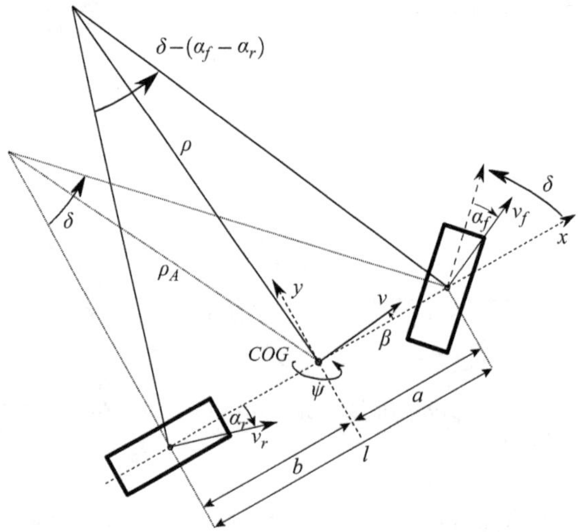

Figure 4.4 Bicycle model and cornering dynamics

of their side slip angles are then equal. These considerations allow one to work in terms of axles instead of single wheels, and therefore to use the so-called "single track model" instead of a four-wheeled one (see Figure 4.4).

The explicit expressions of the side slip angles of the front and rear axles of a vehicle with front steering, assuming the steering angle δ is equal for both front wheels, are then:

$$\begin{cases} \alpha_f = \beta + \frac{a}{v}\dot{\psi} - \delta \\ \alpha_r = \beta - \frac{b}{v}\dot{\psi} \end{cases} \tag{4.11}$$

With all considerations expressed above, we can re-write the lateral forces acting on the front and rear axle. The dynamic generation mechanism of tire forces is also modeled by introducing the tire lateral relaxation lengths l_f and l_r:

$$\begin{cases} F_{y,f} + \frac{l_f}{v}\dot{F}_{y,f} = -C_f\alpha_f = -C_f(\beta + \frac{a}{v}\dot{\psi} - \delta) \\ F_{y,r} + \frac{l_r}{v}\dot{F}_{y,r} = -C_r\alpha_r = -C_r(\beta - \frac{b}{v}\dot{\psi}) \end{cases} \tag{4.12}$$

Note that, in the new expression, the cornering stiffness C_i, $i = f, r$ is now referred to the entire axle, and not to the single wheel.

4.1.3 Yaw moment generation

Similarly to the lateral force, the total yaw moment $M_{z,tot}$ around the normal axis is composed of several components, which are affecting the system behavior with different impact:

- the resulting moment due to the effect of the longitudinal forces F_x on each tire. For the front axle, due to the small steering angle ($\cos(\delta) \approx 1$), the projection can be neglected and the forces considered in their entirety, therefore the component can be expressed as $\sum_i F_{x_i} y_i$, with y_i being the coordinate of the ith wheel;

- the resulting moment due to the effect of the lateral forces F_y. Same considerations expressed above for the small steering angle hold in this case, which means that the resulting moment is $\sum_i F_{y_i} x_i$. When dealing with a single track model, the expression of the moment simplifies to $aF_{y,f} - bF_{y,r}$, where $F_{y,f}$ and $F_{y,r}$ are now the lateral forces applied to the entire axle, and not to the single wheels;
- in case of the front axle only, there is an additional component, resulting from the longitudinal and lateral forces projected on the opposite axes, during steering maneuvers. Also in this case, the yaw moment generated can be neglected due to the small steering angles considered;
- self-aligning torques $M_{z,sa}$ generated during the steering maneuver, due to the cornering forces not being applied exactly at the center of the tire contact zone (see Section 4.1.2). The magnitude of such moment (see Figure 4.3(b)) is in general considerably smaller, approximately two orders of magnitude, compared to the one exerted due to the action of longitudinal and lateral forces on the vehicle frame, therefore these can also be neglected;
- aerodynamic and other externally generated moments, which have minor impact and are not considered for our vehicle model (see [3] for more details).
- an externally generated torque $M_{\dot{\psi}}$ used for the control of the system.

Due to the considerations above, one can use the first two components for the generation of the overall yaw moment considered in the model:

$$M_{z,tot} = \sum_i F_{y_i} x_i - \sum_i F_{x_i} y_i \tag{4.13}$$

In case of the single track model, operating under the hypothesis of zero (or small) longitudinal forces applied to the wheels, (4.13) reduces to:

$$M_{z,tot} = aF_{y,f} - bF_{y,r} \tag{4.14}$$

In addition to the components described above, a Vehicle Stability Control (VSC) system which ensures a desired vehicle behavior is obtained by generating an additional yaw moment $M_{\dot{\psi}}$. In most stability control applications, the yaw moment requested by the control system is generated by producing a braking torque on selected wheels, by using active differentials on the axle (typically the rear one), or by producing desired torque levels, both positive and negative, on each wheel.

Both yaw rate $\dot{\psi}$ and vehicle side slip β control can be generally realized through different approaches to the architecture of active safety systems. Although in recent years there have been great technological advancements in this field, which have translated in several mixed configurations, the most important devices for the generation of $M_{\dot{\psi}}$ are summarized as follows:

- *Selective braking*. The lateral braking control applies different braking forces to the four wheels independently, so as to produce a difference in braking force between the left and right wheels, which generates the yaw moment. As this control uses braking forces, it appears to the driver like deceleration, but the control is effective because it can generate yaw moment under a wide range of conditions of vehicle operation.

- *Active differential*. The lateral torque vectoring control transfers the torque from the left wheel to the right wheel, and vice versa, to generate an amount of braking torque on one wheel while generating the same amount of driving torque on the other wheel. The control of this type, therefore, can generate the yaw moment at any time regardless of the engine torque. Another advantage is that it does not affect the total driving and braking forces acting on the vehicle: that is no conflict with acceleration and deceleration operations is observed. Although this control affects the steering reaction force when applied to the front wheels, it does not produce any adverse effects when applied to the rear wheels.
- *Torque vectoring*. In case of electric vehicles with multiple motors, the torque of each wheel can be individually modulated, in order to satisfy the requests of the electronic control unit. One can already see, the system being an over-actuated one, how a control algorithm which generates a yaw moment demand:

$$M_{\dot{\psi}} = -\sum_i F_{x_i} y_i \tag{4.15}$$

requires a further step, which determines the torque allocation on the wheels. This optimization problem has already been investigated in several works, such as [5].

Active differential and *torque vectoring* can be considered both as torque-based control systems (redistribution of driving torques between front/rear or left/right wheels to achieve required yaw motion and maintain the vehicle side slip in safe limits), as opposed to *Selective Braking*, which is a brake-based control. The latter is realized in such systems like ESC or ESP. Some examples of brake- and torque-based VSC will be introduced in Sections 4.2.2 and 4.2.3.

Independently from the technique adopted for its generation, the control signal $M_{\dot{\psi}}$ can be included in the model in Figure 4.4 as a unique contribution, so that the final expression of the single track model is obtained as

$$\begin{cases} mv(\dot{\beta} + \dot{\psi}) = F_{y,f} + F_{y,r} \\ J_z \ddot{\psi} = aF_{y,f} - bF_{y,r} + M_{\dot{\psi}} \\ F_{y,f}(t) + \frac{l_f}{v}\dot{F}_{y,f}(t) = -C_f\alpha_f = -C_f(\beta + \frac{a}{v}\dot{\psi} - \delta) \\ F_{y,r}(t) + \frac{l_r}{v}\dot{F}_{y,r}(t) = -C_r\alpha_r = -C_r(\beta - \frac{b}{v}\dot{\psi}) \end{cases} \tag{4.16}$$

4.1.4 Vehicle steering properties

It is now clear, from the formulation of the vehicle model through (4.16), how an active safety system, which generates a controlled yaw moment around the vehicle normal axis, can be employed to change the steady state and dynamic behavior of the car, improving its handling properties.

Ideally, during the steering maneuver, the vehicle behavior should follow as accurately as possible the kinematic motion that theoretically occurs at low velocities, i.e., when the steering angle of the wheels determines the curvature radius of the vehicle via a fixed ratio. Such steering angle is generally referred to as Ackermann angle δ_A. Starting from the representation of the kinematic cornering in Figure 4.4, having defined as $l = a + b$ the distance between the front and rear axles, the relation between δ_A and the curvature radius ρ can be written as

$$\tan\delta_A = \frac{l}{\rho^2 - b^2} \xrightarrow{|\delta_A| \ll 1, b \ll \rho} \delta_A \approx \frac{l}{\rho} \tag{4.17}$$

One can see, considering the previously introduced model (4.16), that the vehicle inputs are the steering angle δ, commanded by the driver, and the external forces and moments applied to the vehicle COG. More in general, the most significant variables describing the behavior of the vehicle are its speed $v(t)$, lateral acceleration $a_y(t)$, yaw rate $\dot{\psi}(t)$, and side slip angle $\beta(t)$. Regarding the vehicle as a rigid body moving at constant speed v, the following relationship between such quantities takes place:

$$a_y(t) = v \cdot (\dot{\psi}(t) + \dot{\beta}(t)) \tag{4.18}$$

In case of a purely kinematic maneuver with constant vehicle velocity, the side slip angle can be assumed constant ($\dot{\beta} = 0$) and the overall acceleration is equal to its lateral component, which in case of uniform circular motion is equal to the constant $a_y = v^2/\rho$. Thus, from (4.18), the following relation between yaw rate and curvature radius/Ackermann steering angle is found:

$$\dot{\psi} = \frac{v}{\rho} = \frac{v \cdot \delta_A}{l} \tag{4.19}$$

From (4.19) it appears how, at constant velocity, a good handling property of the vehicle is characterized by a linear relation between lateral acceleration/yaw rate and steering angle. In practical situations, road conditions and the dynamics which determine the generation of the cornering forces cause that such linear relation is lost for increasing values of the steering angle. The hypothesis of kinematic steering implies that steering angle δ, side slip angle β, and yaw rate $\dot{\psi}$ are all constant (and therefore their derivatives are equal to zero). By operating the proper substitutions ($\dot{\beta} = 0$, $\ddot{\psi} = 0$, $M_{\dot{\psi}} = 0$) in the first two equations in (4.16), the following explicit forms for the cornering forces are obtained:

$$F_{y,f} = \frac{b}{l}\frac{mv^2}{\rho} \tag{4.20}$$

$$F_{y,r} = \frac{a}{l}\frac{mv^2}{\rho} \tag{4.21}$$

By substituting (4.20), (4.21) in (4.8), the difference between side slip angles of front (α_f) and rear (α_r) wheels can be computed as

$$\alpha_f - \alpha_r = \underbrace{-\frac{m}{l}\frac{bC_r - aC_f}{C_f C_r}}_{SSG}\underbrace{\frac{v^2}{\rho}}_{a_y} = SSG \cdot a_y \tag{4.22}$$

where the *self-steering gradient SSG* characterizes the typical driving behavior of a given vehicle for a given steering angle. When *SSG* has a positive value, the vehicle

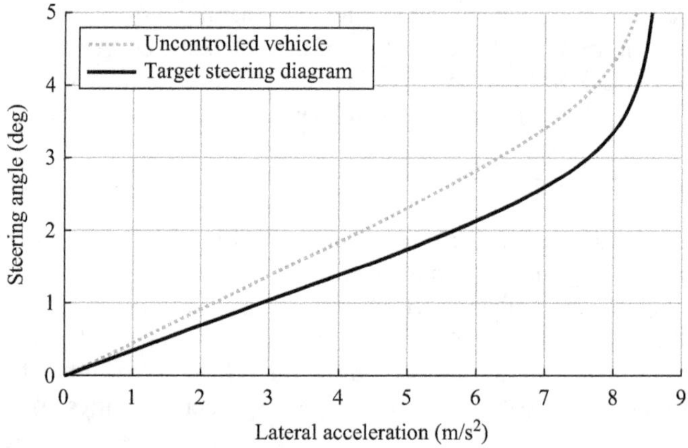

Figure 4.5 Uncontrolled vehicle (dotted), and target (solid) steering diagrams. Vehicle speed: 100 km/h

is characterized by an *understeering* behavior, conversely, for negative values of *SSG* it shows an *oversteering* one, while for vanishing values of *SSG* the vehicle has a *neutral steering* characteristic.

The difference $\alpha_f - \alpha_r$ is a key quantity for the description of the steering behavior, as it is a direct indication of the deviation of the required steering angle δ from the ideal Ackermann angle δ_A. By merging the equations in (4.11) and using the relation (4.19), the following expression for the steering angle is found:

$$\begin{aligned} \delta &= a\frac{\dot{\psi}}{v} - \alpha_f + \beta = \delta_A - \alpha_f + \alpha_r \\ &= \delta_A + SSG \cdot a_y \end{aligned} \tag{4.23}$$

Equation (4.23) shows clearly how the linearity of the relation $\delta - a_y$, typical of the kinematic motion at the Ackermann angle δ_A, tends to be lost for increasing values of a_y. In Figure 4.5, the so-called *steering diagram* is shown for an understeering vehicle, in the controlled and non-controlled cases.

A further indication of this behavior can be seen by substituting (4.23) in (4.19), thus obtaining:

$$\frac{\dot{\psi}}{\delta} = \frac{v}{l + SSG \cdot v^2} \tag{4.24}$$

The *yaw amplification factor* over velocity $\frac{\dot{\psi}}{\delta}$ is small for large values of *SSG* (understeering vehicle) and large for small values of *SSG* (oversteering vehicle).

In steady-state motion $\dot{\beta}(t) = 0$ the lateral acceleration is proportional to yaw rate through the vehicle speed. In this situation, let us consider the uncontrolled car behavior: for each constant speed value, by means of standard steering pad maneuvers it is possible to obtain the steady-state lateral acceleration a_y corresponding to different values of the steering angle δ.

Such curves are mostly influenced by road friction and depend on the tire lateral force–slip characteristics. At low acceleration the shape of the steering diagram is linear and its slope is a measure of the readiness of the car: the lower this value, the higher the lateral acceleration reached by the vehicle with the same steering angle, the better the maneuverability and handling quality perceived by the driver [6]. At high lateral acceleration the behavior becomes nonlinear showing a saturation value, that is the highest lateral acceleration the vehicle can reach. The intervention of an external yaw moment source can be considered as a yaw moment $M_{\dot{\psi}}$ acting on the car center of gravity: such a moment is capable of changing, under the same steering conditions, the behavior of a_y, modifying the steering diagram according to some desired requirements. Thus, a target steering diagram (as shown in Figure 4.5, solid line) can be introduced to take into account the performance improvements to be obtained by the control system. More details about the generation of such target steering diagrams are reported in Section 4.2.1.1.

Therefore, the choice of yaw rate $\dot{\psi}$ as the controlled variable is fully justified, also considering its reliability and ease of measurement on the car. A reference generator will provide the desired values $\dot{\psi}_{ref}$ for the yaw rate $\dot{\psi}$ needed to achieve the desired performances by means of a suitably designed feedback control law.

4.1.5 Control problem formulation

The vehicle side slip angle β and the derivative $\dot{\psi}$ (yaw rate) are used as the main control indicators of active safety systems responsible for the VSC.

The β-variable is generally a less sensitive indicator and can be limited during the control process in relation to reference values depending on vehicle velocity, road conditions, and other factors. For example, the work [7] proposes the following empirical equation for maximum allowable β-value during a turning maneuver, having defined a nominal velocity value $v_{\beta,nom} = 40$ m/s:

$$\beta_{\max} = 10^\circ - 7^\circ \cdot \frac{v^2}{v^2_{\beta,nom}} \tag{4.25}$$

The yaw rate $\dot{\psi}$ is widely used in algorithms of active safety control systems because this variable can be measured by a corresponding on-board sensor. However the definition of the reference yaw rate dynamics is a complex task requiring consideration of tire cornering properties and response characteristics of the vehicle, as it will be explained in Section 4.2.1.1. Figure 4.6 shows typical response characteristics used in analysis of yaw rate dynamics. It can be seen, from the analysis of the response to a step input of lateral force, that tires with higher cornering stiffness can provide lower growth of yaw rate during the maneuver (Figure 4.6(a)), while the understeer or oversteer behavior of the vehicle can be also evaluated through the yaw amplification factor (4.24) expressed as a function of velocity. A detailed discussion of yaw dynamics can be found in handbooks of Milliken and Milliken [8] and Wong [9]. Advanced variants of VSC systems handle more complex quantities, thus allowing to consider the full variation of vehicle side slip angle and yaw rate both in steady-state and transient vehicle maneuvers. Among the first examples of such approach, one can

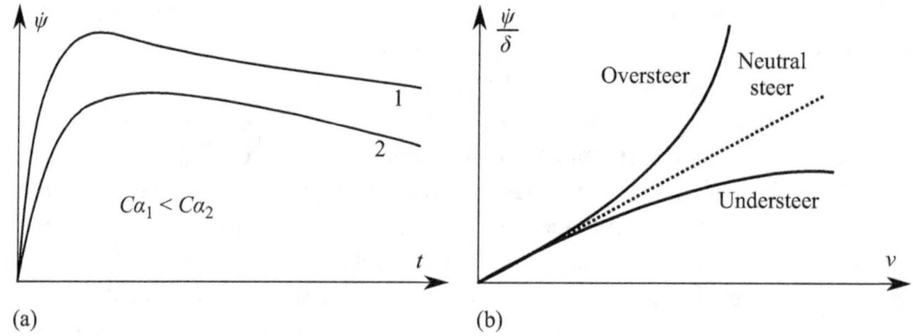

Figure 4.6 Generic yaw rate dynamics: (a) yaw rate evolution comparison and (b) yaw amplification factor

consider the work presented by Shibahata, Shimada, and Tomari in [10,11], where dependencies between the stabilizing yaw moment $M_{\dot{\psi}}$, the vehicle side slip angle and difference between the front and rear tire slip angles for various driving modes are investigated, in order to obtain static maps for the determination of $M_{\dot{\psi}}$. In this context, as the stabilizing yaw moment is meant the moment to be applied to the vehicles center of gravity to compensate redundant yawing motion. The use of such extended characteristics contributes to more precise consideration of non-linear and transient lateral dynamics.

4.2 Yaw rate control survey

4.2.1 Vehicle stability control: yaw rate and side slip angle

A basic process of the VSC can be explained with the help of Figure 4.7. In a general case, the traction/brake demand and the steering demand from the driver are firstly processed by the Vehicle Control Unit (VCU). The VCU determines required individual brake pressure P_{br} and driveline torques T_d to be realized by the brake system and/or the driveline (it is assumed in the considered case that the vehicle has individual wheel drive). Then the actual brake T_{br} and/or traction T_{tr} torque from the actuators of the braking system and driveline is transferred to the corresponding wheel. As a result, the wheel torques T_w and the wheel velocities are being changed. The feedback to the VCU, of the actual values of the variables influencing the vehicle dynamics, is usually realized through the wheel rotational velocity sensors ω_w and the sensors of the lateral vehicle acceleration a_y and yaw rate $\dot{\psi}$.

The availability of the yaw rate sensor is beneficial for the vehicle dynamics control algorithms, and makes it a natural choice for the control variable, whereas the adoption of the vehicle side slip as the control parameter is more difficult due to lack of corresponding on-board measurement procedures in serial vehicles. For this reason the vehicle side slip is in general calculated in the vehicle dynamics control algorithms by means of estimators/observers. Most of the known side slip estimators

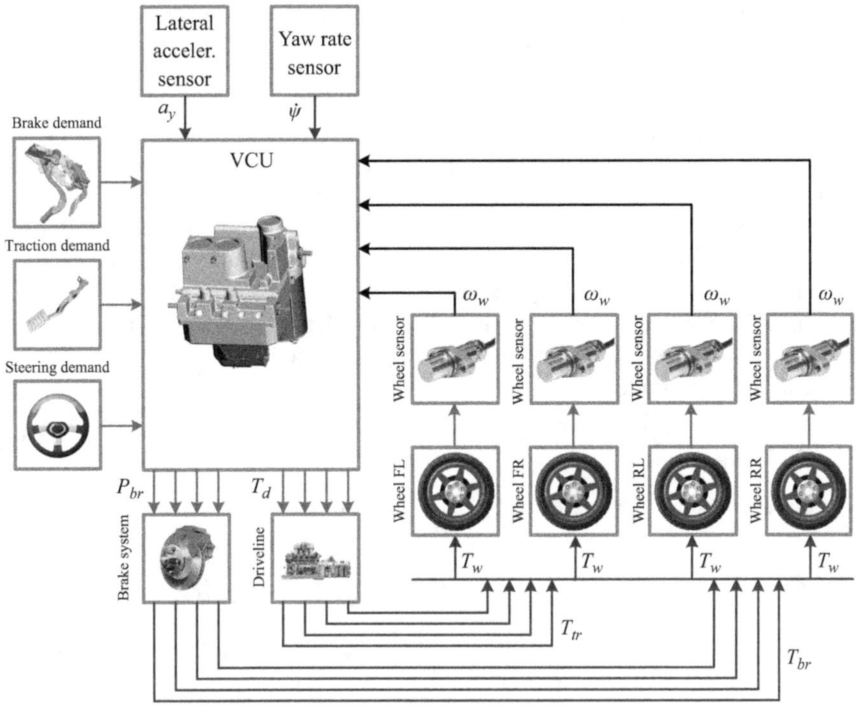

Figure 4.7 General VSC circuit

use the relationship, which can be derived from the equations describing the lateral vehicle motion (see Figure 4.1) in the following simplified form:

$$\beta = \arctan\left(\frac{\hat{v}_y}{v_x}\right) \tag{4.26}$$

where $\hat{v}_y$ is the estimated value of vehicle lateral velocity. Various methods have been proposed for the observation of $\hat{v}_y$, including nonlinear observers. The study [12] presents diverse approaches for the side slip angle estimation based on least squares estimation (extended and unscented Kalman filter) and Bayesian estimation (particle filter).

Instead of the nonlinear observation, some alternative methodologies propose to use an on-board tire model. For example, the work [13] describes the calculation of the lateral vehicle acceleration for the β-estimation based on the following description of the v_y dynamics, derived from the second equation in (4.2):

$$\dot{v}_y = \frac{F_{y,f} + F_{y,f}}{m} - v_x \cdot \dot{\psi} \tag{4.27}$$

where lateral front and rear forces $F_{y,f}$, $F_{y,r}$ are obtained from the tire model stored in the control system. The authors of [13] explain that this approach is more advantageous for real-time vehicle control systems because usual estimators (like Kalman filters and others) require the pseudo-integration of yaw rate and lateral acceleration, which can lead to considerable integration errors because of nonlinearity of vehicle dynamics.

4.2.1.1 Reference and actual values of yaw rate and side slip angle

There are several approaches for the definition of reference values of yaw rate and side slip angle in the corresponding vehicle controllers. A short overview is given below for the most widespread approaches in this regard.

The reference yaw rate $\dot{\psi}^*$ can be derived from a look-up-table given for the vehicle as a family of $\delta - \dot{\psi}$ dependencies composed for variable longitudinal accelerations. Figure 4.8 introduces an example of a tree of corresponding curves that were computed from the steady-state circle test for a generic car model. Having defined the maximum allowed lateral acceleration $a_{y,\max}$, the maximum value of yaw rate [7] can be in addition limited to:

$$\dot{\psi}_{\max} = \frac{a_{y,\max} - a_x \sin \beta^*}{v_x \cos \beta^*} \tag{4.28}$$

The reference side slip angle β^* in (4.28) can be calculated as

$$\beta^* = \begin{cases} \beta_{ss,\max} \frac{v^2}{v_{ss}^2}, & v \leq v_{ss} \\ \beta_{ss,\max}, & \text{otherwise} \end{cases} \tag{4.29}$$

where v is the actual absolute vehicle velocity, $\beta_{ss,\max}$ and v_{ss} are correspondingly the maximal side slip angle and absolute vehicle velocity given for the point where the influence of velocity on yaw rate becomes negligible. Equation (4.29) refers to the source [7], where $\beta_{ss,\max} = 3°$ and $v_{ss} = 40$ m/s have been recommended on a statistical basis, from the experimental results for different types of vehicles.

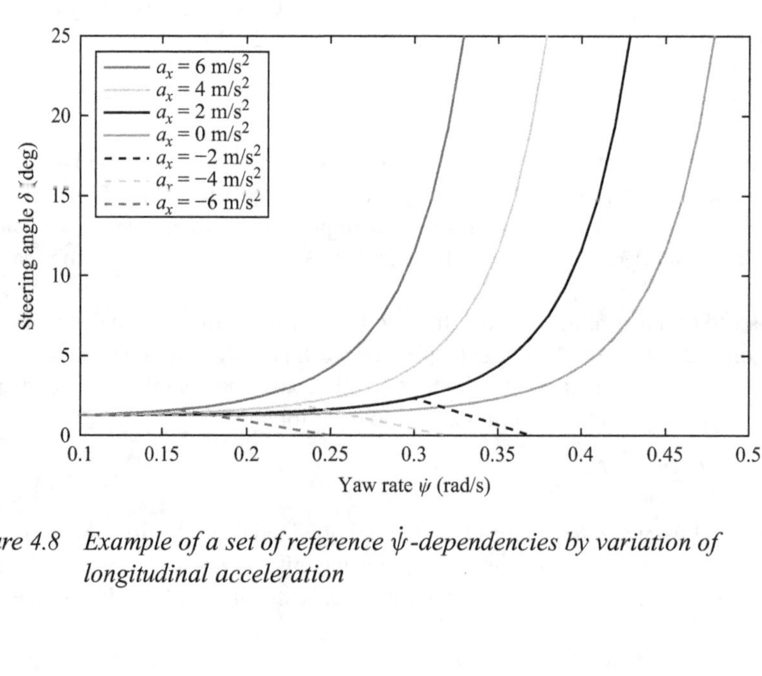

Figure 4.8 Example of a set of reference $\dot{\psi}$-dependencies by variation of longitudinal acceleration

Note that, unlike the reference yaw rate, the reference side slip angle should be considered as a maximum allowed value during the maneuver performed. Additionally, in several situations, it is necessary to estimate the dynamic situations on surfaces with the low friction, where the vehicle stability can be critical already at small driving velocities with low side slip angles.

The value of actual vehicle side slip angle β can be computed as in (4.26) where the actual longitudinal velocity is estimated or made available by a vehicle model, e.g.:

$$v_x = \int (a_x + v_y \cdot \dot{\psi})dt \tag{4.30}$$

and the actual lateral velocity can be found as

$$v_y = \int (a_y - v_x \cdot \dot{\psi})dt \tag{4.31}$$

The values of vehicle accelerations a_x and a_y as well as yaw rate $\dot{\psi}$ are usually obtained from corresponding vehicle sensors.

4.2.1.2 Parallel utilization of side slip angle and yaw rate by the stability control

The parallel utilization of the yaw rate and side slip angle in a VSC system can be realized with different approaches. In [14] a decoupling control approach is presented (see Figure 4.9), under the hypothesis that the steering configuration allows for an actuator to correct the driver request (typically a servo-steering device should allow this). At a given velocity, the linearized transfer functions from inputs δ, $M_{\dot{\psi}}$ to states β, $\dot{\psi}$ are defined: $G_n(s)$, $n = I, II, III, IV$. In such a case the control parameters and $\dot{\psi}$ can be presented as follows:

$$\begin{bmatrix} \beta \\ \dot{\psi} \end{bmatrix} = \begin{bmatrix} G_I(s) & G_{II}(s) \\ G_{III}(s) & G_{IV}(s) \end{bmatrix} \begin{bmatrix} \delta \\ M_{\dot{\psi}} \end{bmatrix} \tag{4.32}$$

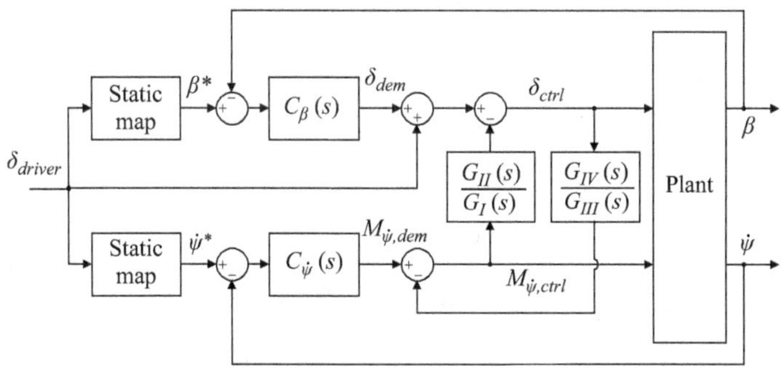

Figure 4.9 Decoupling of yaw rate and side slip angle control

and the control law which decouples the two outputs can be written as

$$\delta = \delta^{dem} - \frac{G_{II}(s)}{G_I(s)} \cdot M_{\dot{\psi}} \tag{4.33}$$

$$M_{\dot{\psi}} = M_{\dot{\psi}}^{dem} - \frac{G_{IV}(s)}{G_{III}(s)} \cdot \delta \tag{4.34}$$

where δ^* is the demanded value of the steering angle, $M_{\dot{\psi}}^*$ is the demanded value of the yaw moment of the vehicle, values obtained from controllers $C_\beta(s)$, $C_{\dot{\psi}}(s)$.

4.2.2 Brake-based stability control: example

Up-to-date ESC systems have usually a very complex architecture covering formulation of control signals for actuators, estimation of vehicle dynamics parameters, fail-safe functions, etc. However a key element of the system is in fact the controller defining the correcting yaw moment $M_{\dot{\psi}}$ to keep the yaw rate or the side slip angle in required limits. In general, the braking of a selected wheel takes place when the actual yaw rate is above the reference yaw rate. As a result, the lateral force on the selected wheel, which concurs to the generation of the overall yaw moment $M_{z,tot}$, is being reduced. The simplified procedure of the relevant brake-based yaw control can be explained with the control scheme in Figure 4.10.

The reference yaw rate for actual driving conditions $\dot{\psi}_{ref}$ is being generated using the estimated value of the tire–road friction coefficient μ_{est}, the vehicle velocity v_x, and the front wheel steering angle δ. The vehicle models introduced in Section 4.1 can be used for this purpose. Then, the reference yaw rate is compared with the actual yaw rate value $\dot{\psi}$, which can be obtained directly from the yaw rate sensor installed on the vehicle. Assuming only the front wheels are used for the generation of the correcting yaw moment, for the selected wheel the reference slip generator computes the value of λ^* corresponding to the demanded correcting yaw moment of the vehicle. For example, the study [15] proposes a procedure with a PID controller for the wheel slip. The reference slip value λ^* can be set up also by the tire–road friction estimator, in the case of straight-line motion of the vehicle, since there is no need for a corrective

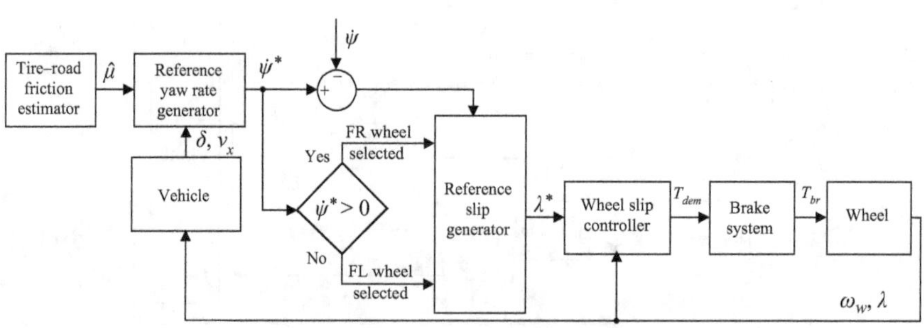

Figure 4.10 Example of brake-based yaw moment control layout

yaw moment $M_{\dot{\psi}}$. The reference and actual wheel slip values are used by the wheel slip controller, which is actually part of anti-lock brake system of the vehicle. The wheel slip controller allocates the demanded brake torque T_{dem} to be realized by the brake system on the corresponding wheel, in order to generate the correcting yaw moment of the vehicle. The actual brake torque T_{br} is then applied to the selected wheel.

In more complex situations, several wheels can be simultaneously controlled with individually tuned values and rates of corresponding brake torques. It should be noted that the motor braking can also be used in brake-based stability control systems in certain control situations.

The variant of another architecture of the brake-based stability control system is depicted in Figure 4.11. This system uses both the yaw rate and the side slip angle as the controlled variables and can be realized through diverse control methods. In particular, the paper [16] proposes to generate the reference yaw rate in accordance with a PD formulation. The reference yaw rate can be computed based on (4.24):

$$\dot{\psi}^* = \frac{v_x \delta}{l + \hat{SSG} \cdot v_x^2} \tag{4.35}$$

where $\hat{SSG}$ is the estimate of the *self-steering gradient* defined in (4.22), obtained from the observation of the linearized cornering stiffness coefficients C_f and C_r. Examples of online estimation methods for the cornering stiffness coefficient can be found in [17].

The reference value of side slip angle from Figure 4.11 can be calculated with different methods, as discussed in Section 4.1.5, while the reference value of slip ratio is used in the wheel slip controller to set the brake pressure demand. The works [16,18] propose the formulation of the brake pressure for each i-wheel using a sliding mode approach for this purpose.

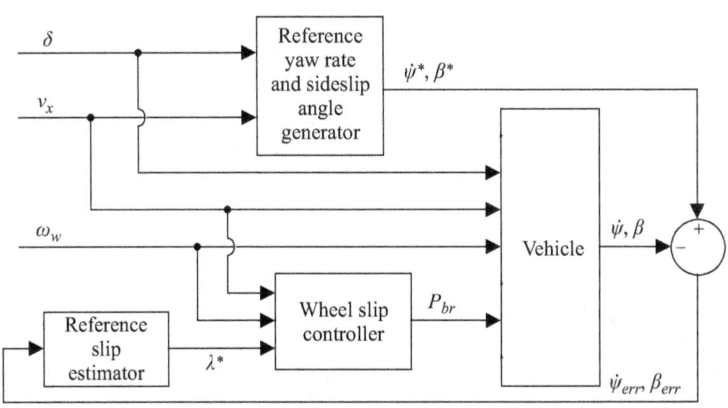

Figure 4.11 Example of stability control system architecture involving yaw rate and side slip control (adapted from [16])

There are many other control methods, which are known from published studies and can be applied for the brake-based stability control systems. Their analysis lies outside of the targets of the presented work.

4.2.3 Torque-based stability control: example

As mentioned in Section 4.1.3, stability control systems based on torque vectoring are mainly relevant to all-wheel drive vehicles with active inter-axle and inter-wheel differentials or to electric vehicles with individual in-wheel or in-board motors. The related torque vectoring systems are appropriately discussed in particular in works of Sawase, Piyanbongkarn, Cheli, and other authors [19,20] for vehicles with active differentials and by De Novellis, Kaiser, and other authors [21,22] for electric vehicles.

The operating principle of a TV system in a general case can be explained as follows. Considering a left cornering maneuver without TV control, a situation can arise where one or more tires reach the friction limit. To avoid unstable vehicle behavior under these conditions, the distribution of the individual wheel torques should be changed. For the case of left–right TV control, the torque distribution should be such that the overall lateral force F_y acting on the COG remains unchanged. The main idea is to distribute torques with the aim of keeping the vehicle stable and also increasing traction efficiency. Different strategies are being proposed for this purpose. For instance, the following methods are known from [11,23]:

1. *Constant torque distribution* – traction and braking forces of AWD vehicle are distributed between the right and left rear wheels while their distribution between the front and rear wheels is kept constant.
2. *Slip-based torque distribution* – traction and braking forces of the vehicle are distributed between the wheels in accordance with the appointed left/right or front/rear wheel slip difference.
3. *Torque distribution in proportion to the vertical load* – optimization of the ratio of the vertical road on a wheel to the total load on all wheels.

A variant of wheel torque distribution during the TV control can be implemented by rewriting the total yaw moment $M_{\dot{\psi}}$ as the sum of a component $M_{\dot{\psi},f}$ originated by the torque distribution on the front axle, and $M_{\dot{\psi},r}$ on the rear one:

$$M_{\dot{\psi}} \cdot (k_f + k_r) = M_{\psi,f} + M_{\psi,r} \tag{4.36}$$

where $k_{f,r} \in [0, 1]$ are the TV distribution ratios defining the torque distribution between the vehicle front and rear axles. Note that the condition $k_f + k_r = 1$ shall be satisfied. Having defined I_w the inertia moment of the wheel, r its radius, ω_w, T_w, and F_x its rotational speed, torque applied and longitudinal tire–road contact force, one can take into account a simplified torque balance on the wheel from the 1-DOF vehicle model:

$$I_w \dot{\omega}_w = -rF_x + T_w \tag{4.37}$$

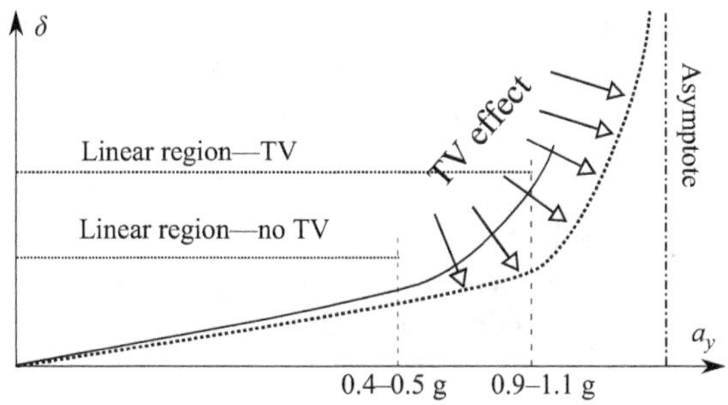

Figure 4.12 Modifications of the vehicle understeer characteristic achievable through torque vectoring

so that the shares of the driving torque to be redistributed between the left and right wheels are

$$T_{w,ij} = \frac{k_i \cdot M_{\dot{\psi}} \cdot r}{t_i} + I_w \dot{\omega}_{ij} \tag{4.38}$$

where the indexing $i = f, r$ and $j = l, r$ is adopted for front/rear and left/right, respectively, and t_i is the axle track width. The TV control law can be formulated, for instance, as in [24], where a PD controller is proposed for the definition of the required TV distribution ratios k_f and k_r.

Besides the PD and model-based controllers, other methods like sliding mode or H^∞ have found applications in TV systems [25]. It should be mentioned that the torque vectoring control allows not only to correct the vehicle yaw dynamics but also influences the vehicle agility by changing understeer/oversteer characteristics. As indicated by Figure 4.12, the vehicle response is linear within a certain lateral acceleration threshold, which is usually about 0.4–0.5 g at constant vehicle velocity. Beyond this threshold value, the response becomes and remains non-linear until the maximum lateral acceleration of the vehicle, i.e., its steady-state cornering limit, is reached. Implementation of individual wheel torque control can positively influence the understeer behavior as follows: (i) Extension of the linear region and reduction of the understeer gradient. By doing so, vehicle responsiveness is enhanced and the stability limit for steady-state cornering is extended. (ii) Increase of the maximum level of lateral acceleration. In general, this change improves the lateral performance of the vehicle.

4.3 Vehicle stability control via sliding mode control

The control of handling and stability of wheeled vehicles has been actively investigated in recent decades. One of the first relevant solutions was introduced by Bosch in [26], for vehicles with electro-hydraulic braking system, aimed at enhancing vehicle stability during evasive maneuvers. After this development, the VSC has undergone extensive development. Nowadays, with the introduction of electric vehicles with individual electric motors for each wheel, this topic gains renewed attention. Such electric vehicle configuration makes it possible to obtain a significant performance improvement. In general, high-performance actuators like individual electric motors and electro-hydraulic braking systems provide more control agility in this regard. However, simultaneous use of the electric motors and the braking system for the handling and stability control requires a proper consideration of the system response. It also requires to solve a problem of overactuation to generate the control laws. In this respect, this section introduces a solution to the following problems, related to VSC:

- set up of a control layout fitting to the electric vehicle with four individual electric motors and decoupled electro-hydraulic braking system;
- selection of a proper reference value $\dot{\psi}^*$ for the vehicle yaw rate, which should provide the required handling characteristics;
- development of a high-level control law for the yaw rate tracking, which minimizes the error between the reference and the measured yaw rate.

The main attention is given to the application of different sliding mode control strategies in the high-level controller and to the benchmarking of these strategies via simulation. The target control should satisfy requirements in terms of handling, promptness of response and robustness.

4.3.1 Overall control structure

The overall control structure depicted in Figure 4.13 represents a multi-layered controller for the minimization of the tracking error between the desired and actual yaw rate of the vehicle. The desired vehicle motion is based on the nonlinear "bicycle" model, which corresponds to model (4.16) illustrated in Section 4.1, except for the nonlinearity caused by the nonlinear tire model, which is used for the calculation of the lateral forces F_y.

The reference model is adjusted in order to achieve a wider linear area in the steering diagram compared to the baseline vehicle, which is affected by understeer (see

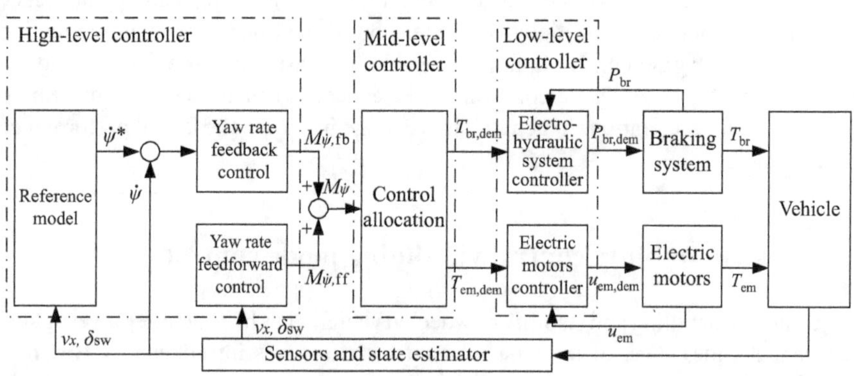

Figure 4.13 Vehicle lateral dynamics control scheme

Figure 4.5 in Section 4.1.4). It uses as inputs the longitudinal vehicle velocity v_x and the steering wheel angle δ_{SW}, which is preferred to the steering angle of the wheel δ, as it is a directly controlled quantity. To minimize deviation between the reference and actual yaw rate, the control law includes feedback and feedforward parts. Therefore, the control input $M_{\dot{\psi}}$ generated by the high-level controller is expressed as:

$$M_{\dot{\psi}} = M_{\dot{\psi},\text{ff}} + M_{\dot{\psi},\text{fb}}, \tag{4.39}$$

where $M_{\dot{\psi},\text{ff}}$ is the feedforward and $M_{\dot{\psi},\text{fb}}$ is the feedback stabilizing torque. The feedforward contribution is based on the measurements of the steering wheel angle δ_{SW} and the estimated vehicle velocity v_x. The feedback part considers the difference between the measured yaw rate $\dot{\psi}$ and the desired value obtained from the reference model $\dot{\psi}^*$.

The control input $M_{\dot{\psi}}$ is distributed between the individual wheel actuators. This distribution is computed by the mid-level controller, which solves the control allocation problem. The derived information about the demanded friction brake torque $T_{\text{br,dem}}$ and electric motor torque $T_{\text{em,dem}}$ is delivered to the low-level controllers, where the physical signals $u_{\text{em,dem}}$ and $P_{\text{br,dem}}$ are generated, which represent electric voltage and friction brake pressure, respectively. They are based on the feedback information about the electric motor voltage u_{em} and the estimated braking pressure in the brake calipers P_{br}. The actual values of torque applied to the wheels, electric T_{em} and friction T_{br}, are a result of the operation of the overall control system.

4.3.2 Reference model

Several approaches are known for the generation of the reference yaw rate. They can use both linear and nonlinear vehicle models. The model selection depends on the type of relation between the wheel slip angle α and the lateral force F_y applied to the wheel. This relation determines the forces which affect the vehicle dynamics described by the single track model. In this section three types of reference model are introduced:

- *linear*, based on the model (4.16) illustrated in Section 4.1. The relation between the lateral tire force and the wheel slip angle, as shown in (4.8), is considered;
- *linear bounded*, where the yaw rate reference $\dot{\psi}^*$ is saturated above a threshold steering wheel angle value;
- *nonlinear*, based on a parametrized Dugoff tire model. The basics of this model can be found in [27].

In its pure form, the linear model neglects the friction limits, fact that can lead to an unpredictable vehicle behavior when the vehicle reaches high values of lateral acceleration. In case the linear bounded model is used, such misbehavior is avoided by applying the following law:

$$\dot{\psi}^* = \begin{cases} \dot{\psi}^* & \text{if } \ |\dot{\psi}| \leqslant |\frac{\mu g}{v_x}| \\ \pm\frac{\mu g}{v_x} & \text{otherwise} \end{cases} \tag{4.40}$$

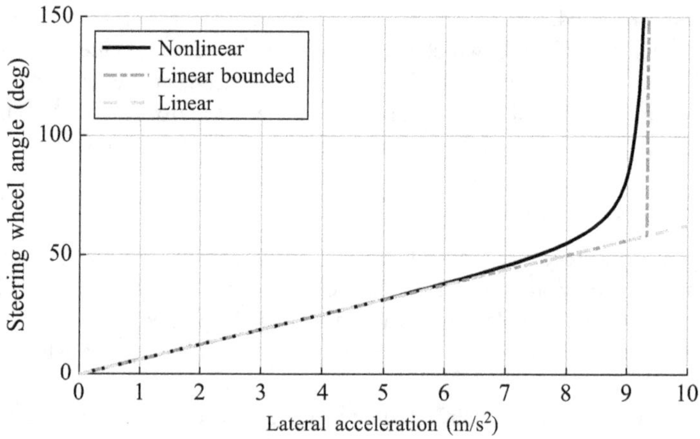

Figure 4.14 Comparison of reference vehicle models on the steering diagram

where g is the gravity acceleration, μ is the coefficient of friction between the tire and road, and v_x is the vehicle longitudinal velocity.

Figure 4.14 compares the outputs of the three introduced reference models. As can be seen, the linear range of the steering diagram never ends in the case of a pure linear model. It may lead to an excessive torque demand and a consequent loss of feeling for the driver, in terms of recognition of the approaching friction limits. Hence, this variant is not applicable in terms of handling and vehicle safety. Bounding the linear range by considering the friction limits solves the issue of vehicle safety. Nevertheless, in terms of handling, such bounded variant is characterized by a sudden change of the vehicle steering characteristics, which is uncomfortable for the driver. Therefore, using the nonlinear model for the reference signal generation can be considered as the optimal solution. It limits the steering gradient according to friction conditions, and provides at the same time a smooth transition between linear and nonlinear areas. Such approach will be utilized in the examples illustrated in this chapter as the reference model to derive the desired yaw rate of the vehicle.

4.3.3 Feedforward control

According to the control theory, a feedforward control may be applied, in addition to the feedback control, in order to improve the dynamic response of the system. This requires a well-modeled plant or the presence of known experimental results. As for the vehicle yaw rate control, the application of a feedforward control allows more agile vehicle reaction to the driver's input. One example can be found in [28], where the vehicle is considered as a linear "bicycle" model with known parametrization and cornering stiffness of the tires. With such an approach, the feedforward stabilizing torque is calculated from the transfer function applying its contribution during the transient vehicle behavior. Other approaches (see e.g., [29]) are based on the use

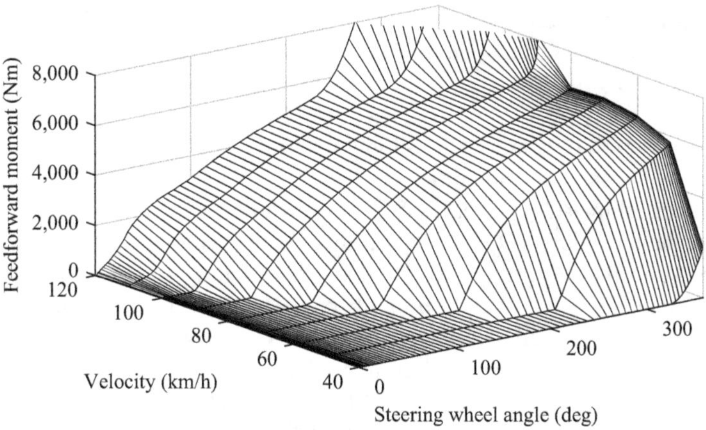

Figure 4.15 Feedforward stabilizing torque

of quasi-static vehicle models. A static map can be obtained via simulation, which provides a feedforward torque from the steady-state vehicle behavior. This approach is efficient if, for example, the desired steering diagram is known. In this case, the demanded stabilizing torque can be derived for a wide range of steering wheel angles.

The feedforward mapping illustrated in Figure 4.15 is obtained by performing a slowly increasing steer test at a constant vehicle velocity. During this procedure, the steering wheel angle δ_{SW} is increased from 0° up to 360° with a rate of 13.5°/s. In this testing procedure the steering wheel angle instead of wheel angle is used (as in Figure 4.5 and more in general Section 4.1), because it can be measured directly by the standard sensor on the vehicle. The ratio between steering wheel angle and steer angle of the wheels is specific for each type of vehicle. Several tests with variation of the initial vehicle velocity were performed in order to obtain the values of feedforward demand in relation to the vehicle velocity. These tests were performed using a 14 DoF vehicle model. During the simulation, the corrective torque around the vertical vehicle axis $M_{\dot{\psi}}$, which allows to reach the reference vehicle behavior, was measured. The illustrated feedforward control will be applied in combination with the sliding mode feedback controllers introduced in the next subsections.

4.3.4 Second-order sliding mode control design

A control law based on a second-order sliding mode (SOSM) approach, as proposed in [28], is now illustrated for the yaw rate control.

SOSM control generalizes the basic sliding mode control idea, with the difference that it acts on the second order time derivative of the system deviation from the sliding manifold, instead of the first derivative, as it happens in first order sliding mode control design [30]. The sliding manifold is a subspace of the system state space, in which the evolution of the controlled dynamic system is confined by the control action.

Having defined the sliding variable $\sigma(x)$ as a smooth function of the state x of the considered dynamical system, the second order sliding mode is determined by

$$\sigma(x) = \dot{\sigma}(x) = 0. \tag{4.41}$$

The surface $\sigma(x) = 0$ identifies the sliding manifold, which according to the theory is reached by the state in finite time.

It has been shown [31,32] that SOSM may provide higher accuracy than first order sliding mode algorithms (FOSM), while keeping the same robustness with respect to matched uncertainties. In addition to this, when dealing with a system of relative degree equal to one, SOSM has the property of producing a continuous control signal, which allows to obtain reduced chattering effect, when compared to FOSM [31]. In the considered problem, the actual system relative degree is in practice unknown, although it is certainly greater than one, due to the high complexity of the vehicle, and to the simplifications operated for its modeling as a dynamic system.

As the control objective is to make the error between the actual yaw rate and the reference yaw vanish, the sliding variable is chosen as

$$\sigma(t) = \dot{\psi}(t) - \dot{\psi}^*(t) \tag{4.42}$$

To design the proposed controller, it is useful to observe that the first and second time derivative of the sliding variable are, respectively:

$$\dot{\sigma}(t) = (aF_{y,f}(t) - bF_{y,r}(t) + M_{\dot{\psi}}(t))/J_z - \ddot{\psi}^*(t) \tag{4.43}$$

$$\ddot{\sigma}(t) = (a\dot{F}_{y,f}(t) - b\dot{F}_{y,r}(t) + \dot{M}_{\dot{\psi}}(t))/J_z - \overset{\dots}{\dot{\psi}}{}^*(t) \tag{4.44}$$

The following auxiliary system is then derived from (4.43) and (4.44), by defining the auxiliary variables $y_1(t) = \sigma(t)$ and $y_2(t) = \dot{\sigma}(t)$:

$$\begin{cases} \dot{y}_1(t) = \dot{\sigma}(t) = y_2(t) \\ \dot{y}_2(t) = \ddot{\sigma}(t) = \lambda(t) + \tau(t) \end{cases} \tag{4.45}$$

where $\tau(t) = M_{\dot{\psi}}(t)/J_z$ is the auxiliary control variable and

$$\lambda(t) = \frac{(a\dot{F}_{y,f}(t) - b\dot{F}_{y,r}(t))}{J_z} - \overset{\dots}{\dot{\psi}}{}^*(t) \tag{4.46}$$

From considerations on the physical quantities in play, one can assume that $\lambda(t)$ is bounded, i.e.:

$$|\lambda(t)| \leq \Lambda \tag{4.47}$$

with the value $\Lambda > 0$ depending on the operating condition of the vehicle. Equation (4.16) and the tire characteristic can be used for the estimation of Λ. The quantity y_2 is considered an unmeasurable quantity, it being the first derivative of y_1 which depends on $\lambda(t)$ and $\tau(t)$.

After these preliminary considerations, the control problem can be reformulated as follows: given system (4.45), where $\lambda(t)$ satisfies (4.47), and y_2 is unavailable for measurement, design the auxiliary control signal $\tau(t)$, such that y_1 and y_2 are both brought to zero in finite time.

The SOSM controller proposed in [28], which we consider here, is of sub-optimal type (see [33]). It works under the assumption that it is possible to detect the extremal values of the signal y_1. In practical terms this requires the use of a peak detector or, in alternative, an estimation of the derivative of the tracking error $\dot{y}_1(t)$, and the identification of the time instances $\bar{t}$ such that $\dot{y}_1(\bar{t}) = 0$. Under the hypothesis that the extremal values are detectable, having defined $y_{1M}(t)$ as a piece-wise constant function representing the value of the last singular point of $y_1(t)$ (i.e., the most recent value $y_{1M}(\bar{t})$), it can be proved [28] that the following control law for the auxiliary system fulfills the requirement of bringing both y_1, y_2 to zero in finite time:

$$\tau(t) = \frac{\dot{M}_{\dot{\psi}}(t)}{J_z} = -K_{SL}\ \text{sign}\left\{y_1(t) - \frac{1}{2}y_{1M}(t)\right\} \tag{4.48}$$

where the control gain K_{SL} is chosen such that

$$K_{SL} > 2\Lambda \tag{4.49}$$

The final yaw moment request is then derived from (4.48) via integration.

Note that, the saturation of the control action was already considered in the yaw rate reference generation (see Subsection 4.3.3) and is not taken into account in the control design. It is possible, in alternative, to consider the saturation explicitly in the design, proceeding as in [34].

4.3.5 Integral sliding mode control design

The ISM approach presented in this section is based on the method proposed in [25]. The yaw rate tracking is implemented by means of two separate controllers, a classic PID and a sliding mode one, which are set up as shown in Figure 4.16.

According to the theory, the adoption of an ISM controller allows one to control a nonlinear affine system of the form:

$$y = f(x) + b(x)u + d(x,t) \tag{4.50}$$

and to obtain the rejection of the bounded matched disturbances $|d(x,t)| < d_{\text{MAX}}$ starting from the initial time instant t_0. Such a result can be achieved thanks to a modification operated to the sliding surface $\sigma(t)$, which ensures the system initial

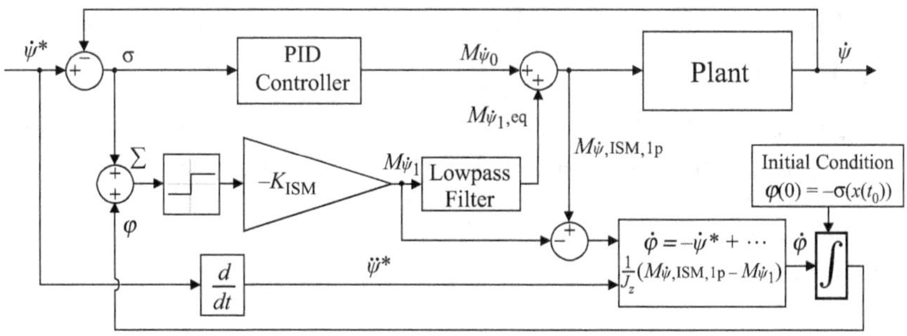

Figure 4.16 Schematization of the control scheme of the ISM control algorithm

condition lies on the sliding surface itself. For this reason, a new sliding surface is defined as

$$\Sigma(t) = \sigma(t) + \varphi(t) \tag{4.51}$$

where the integral term $\varphi(t)$ has an evolution determined by

$$\dot{\varphi} = -\frac{\partial \sigma}{\partial x}(f(x) + b(x)u_0) \tag{4.52}$$

and the *nominal input* u_0 is the control input generated by any suitable controller, which we will refer to as the nominal controller. In our case, we consider a standard PID as the nominal controller, assuming that a suitable calibration of its parameters has been previously obtained.

In order to have the new sliding manifold $\Sigma(t) = 0$ passing through the initial conditions $x(t_0)$ of the controlled system, it is necessary to impose the following initial condition for (4.52):

$$\varphi(t_0) = -\sigma(x(t_0)) \tag{4.53}$$

Then the ISM control law is given by the sum of the nominal control input and the sliding mode component u_1, i.e., $u_{ISM} = u_0 + u_1$. The latter is given by

$$u_1 = -K_{\text{ISM}} \cdot \text{sign}(\Sigma), \quad \text{K}_{\text{ISM}} > \text{d}_{\text{MAX}} \tag{4.54}$$

In the yaw rate control case, system (4.50) can be derived from the second equation in (4.16), with the sliding variable equal to the controlled state, which is the yaw rate tracking error, therefore $\sigma = e_{\dot{\psi}} = \dot{\psi}^* - \dot{\psi}$. The control input is the external control yaw moment, indicated as $M_{\dot{\psi}}$ in (4.16). The disturbance component $d(x, t)$ includes all the contributions to the overall yaw moment which have been neglected in the 2DOF model (see Subsection 4.1.3).

Given the discontinuous nature of signal (4.54), chattering phenomena can occur. A different formulation [35] of the integral component φ, alternative to (4.52), has been adopted in order to alleviate chattering. It is based on the so-called "equivalent control" concept [36], which can be approximated as the output of a first order linear filter, provided the time constant of the filter τ_{lp} has been properly selected, in accordance with the actuators bandwidth. In our application, the equivalent control $u_{1,\text{eq}} = M_{\dot{\psi}1,\text{eq}}$ can be approximated as follows:

$$\tau_{\text{lp}}\dot{M}_{\dot{\psi}1,\text{eq}} + M_{\dot{\psi}1,\text{eq}} = M_{\dot{\psi}1} \tag{4.55}$$

where τ_{lp} is the time constant of the filter suitably chosen in order not to distort the slow component of the control action. Since the equivalent control computed as in (4.55) is an approximation, the rejection of the disturbance terms is guaranteed by suitably selecting the integral term of the sliding manifold. Having defined the overall control signal in the filtered case as $u_{\text{ISM,lp}} = u_0 + u_{1,\text{eq}}$, the evolution of φ should be defined as follows:

$$\dot{\varphi} = -\frac{\partial \sigma}{\partial x}(f(x) + b(x)(u_{\text{ISM,lp}} - u_1)) \tag{4.56}$$

Formulating the control law making reference to the specific yaw rate tracking problem, φ can be expressed as:

$$\dot{\varphi} = -(1)\left(\dot{\psi}^* - \frac{1}{J_z}(M_{\dot{\psi},\mathrm{ISM,lp}} - M_{\dot{\psi}_1})\right) \tag{4.57}$$

$$= -\dot{\psi}^* + \frac{1}{J_z}(M_{\dot{\psi},\mathrm{ISM,lp}} - M_{\dot{\psi}_1}) \tag{4.58}$$

4.4 Controllers assessment

The control methods introduced in the previous section (SOSM and ISM in combination with a static feedforward component) are now assessed via simulation with the software IPG Carmaker on a previously validated vehicle model.

4.4.1 Vehicle specifications and model validation

The sport utility vehicle (SUV), which is used for the case study in this chapter, has an electric powertrain with four individual electric in-wheel motors (IWM). Torque and efficiency maps of electric motors are shown in Figures 4.17(a) and 4.17(b), respectively. Such configuration has several advantages, compared to vehicles with active differentials, thanks to its flexible wheel torque distribution (see Section 4.1.3). In this case a stabilizing torque can be realized by a proper assignment of torque demand to each wheel individually.

The vehicle is also equipped with an electro-hydraulic braking system, which is required to provide fault-tolerant vehicle operation in braking mode. Its presence is required in case the electric powertrain fails, or to fulfill the torque demand when it cannot be fully realized by the electric motors. The most important physical attributes of the investigated SUV for the purpose of lateral stability control are its overall mass, $m = 1{,}963$ kg, the inertia moment around the normal axis, $J_z = 2{,}525$ kg m^2, and the

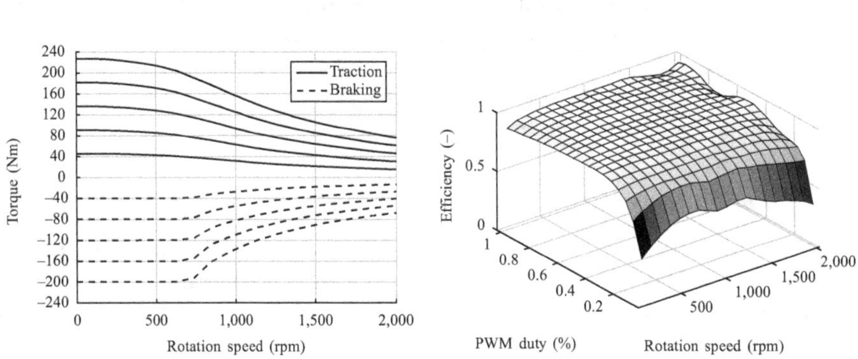

(a) (b)

Figure 4.17 Characteristics of the in-wheel electric motors: (a) motor torque in traction and braking mode and (b) motor efficiency map

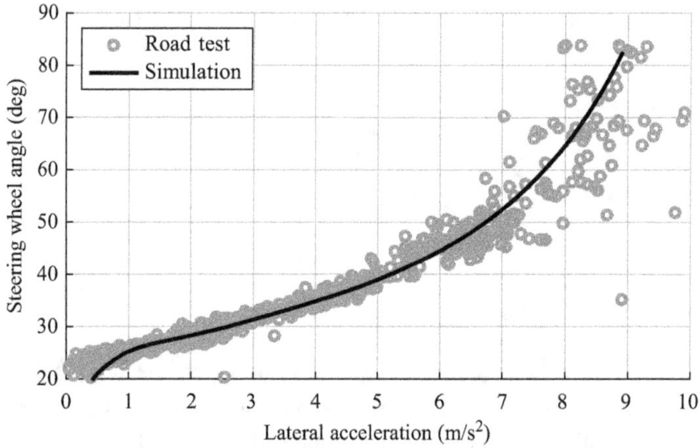

Figure 4.18 Vehicle model validation via skid-pad test

distances of the axles from the COG, $a = 1.07$ m and $b = 1.59$ m for front and rear axles, respectively. Tires are modeled via the so-called MF-Tire approach [37] and parametrized accordingly. Such tire model is applicable for the simulation of handling maneuvers on an even road.

The vehicle model, which has already been utilized by the authors in combination with on road experimental activities [38], was validated experimentally by performing the so-called *skid-pad* test with the real prototype of the electric vehicle. This test, similarly to the *slowly increasing steer* test which will be introduced in the next subsection, allows one to derive the steering diagram of the vehicle. It has the advantage of not requiring an extremely wide track, which is in general necessary with large initial steering radius. On the other hand, it does not represent the vehicle behavior at a specific velocity, since it changes during the test. During the *skid-pad* test, the vehicle drives on a circle of a certain radius. The longitudinal velocity is gradually increased, in order to maintain the lateral acceleration of the vehicle. The steering diagram obtained with the *skid-pad* test, compared against the simulation results, is shown in Figure 4.18. As can be seen, the vehicle model is parametrized properly, thus providing a good fitting of the results.

4.4.2 Simulation results

To evaluate the proposed control laws, a set of simulations is performed with the software IPG CarMaker. They include standard testing procedures, which are usually used during vehicle design. These procedures can be classified in open-loop and closed-loop tests. Such classification is based on the involvement of the driver in testing (driver's model in case of simulation).

- *Open-loop* procedures use a predefined profile of the steering wheel angle and the traction/braking torque. The following open-loop tests are considered for the

Table 4.1 Evaluation of developed control laws

Test	Evaluation metric	Baseline	SOSM	ISM	PID
Slowly increasing steer	RMSE$_{\dot{\psi}}$ (°/s)	2.24	0.41	0.21	0.21
Step steer	RMSE$_{\dot{\psi}}$ (°/s)	3.04	0.16	0.27	0.25
	Rise time (s)	0.42	0.27	0.30	0.31
	Overshoot (%)	10.0	3.1	5.2	6.1
	Steady error (°/s)	1.2	0.09	0.21	0.32
Sine with dwell	RMSE$_{\dot{\psi}}$ (°/s)	17.69	3.62	2.39	2.3
Braking in the turn	RMSE$_{\dot{\psi}}$ (°/s)	31.13	0.13	0.52	0.5
Double lane change	RMSE$_{\dot{\psi}}$ (°/s)	22.9	4.1	2.48	2.44
	Final speed (km/h)	58.9	66.2	67.5	67.5

controllers evaluation: *slowly increasing steer*, *step steer*, *sine with dwell* and *braking in the turn*;

- *Closed-loop* maneuvers are performed by professional drivers trying to follow a certain road path or trajectory. In this context the *double lane change* test is performed.

For the benchmarking, a definition of the *baseline* vehicle is also required. The reference vehicle is represented by the same hardware components but does not have any kind of handling or stability control. This basic configuration of the vehicle has been used also for the validation of the simulation model.

In the next subsections we present the results of the sliding mode control algorithms, together with a standard PID, for the sake of comparison. Several evaluation metrics obtained from the tests, which are mentioned in this section, are reported in Table 4.1.

4.4.2.1 Slowly increasing steer

The *slowly increasing steer* test is performed to evaluate the developed controller in steady-state conditions. According to the standard ISO 4138, the vehicle moves with a constant velocity of 100 km/h, while the steering angle is increased from 0° up to 360°, with a rate of 13.5°/s. Recorded results allow analyzing the vehicle behavior on the steering diagram, considering both linear and nonlinear areas. Compared to the baseline vehicle, the parametrization of the reference model has been performed in such a way to achieve a wider linear area of the steering gradient. This allows one to obtain a better vehicle handling by means of yaw rate control. In terms of control quality, SOSM, ISM, and PID generate similar behaviors, and in general guarantee precise tracking of the given reference trajectory. In Figure 4.19 the three lines are overlapping each other representing almost identical results. Using the root-mean square error of the control variable (RMSE$_{\dot{\psi}}$), it can be seen that PID and ISM have the same deviation of 0.21°/s while SOSM produces a slightly higher value of 0.41°/s. This difference is determined by the chattering around the reference value of the yaw rate at steady state, which is still present by using the SOSM algorithm. In fact, the chattering phenomenon is due to the relative degree of the controlled system being

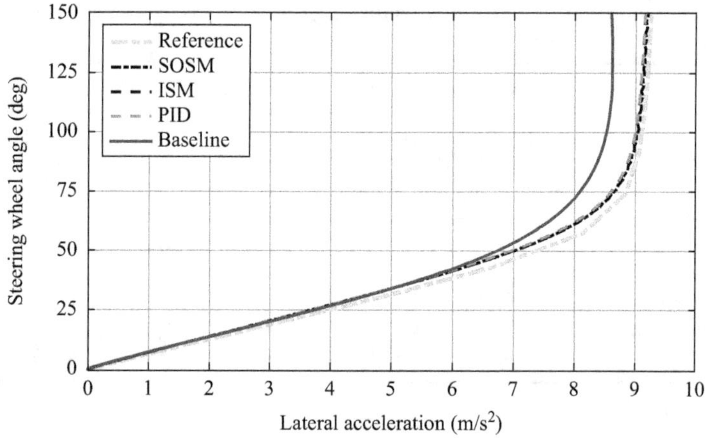

Figure 4.19 Steering diagram obtained with the different controllers

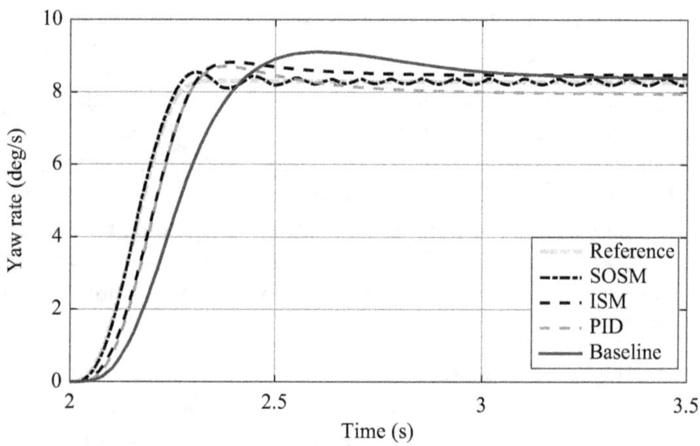

Figure 4.20 Yaw rate step response test

greater than one, as mentioned in Section 4.3.4. From this test, it appears that the different control strategies proposed do not differ substantially in steady maneuvers.

4.4.2.2 Step steer

The *step steer* test is a transient maneuver performed according to the standard ISO 7401. The vehicle moves with the constant velocity of 100 km/h and a step steering input is then applied. It has the rate of 200°/s and the amplitude is defined in such a way to maintain the lateral acceleration of 4 m/s^2. Compared to the baseline vehicle behavior, the vehicle with the proposed high-level controllers has a more agile behavior and is responding to the driver's input much quicker, as it appears in Figure 4.20. In particular, in terms of agility, the SOSM variant shows the quickest response to

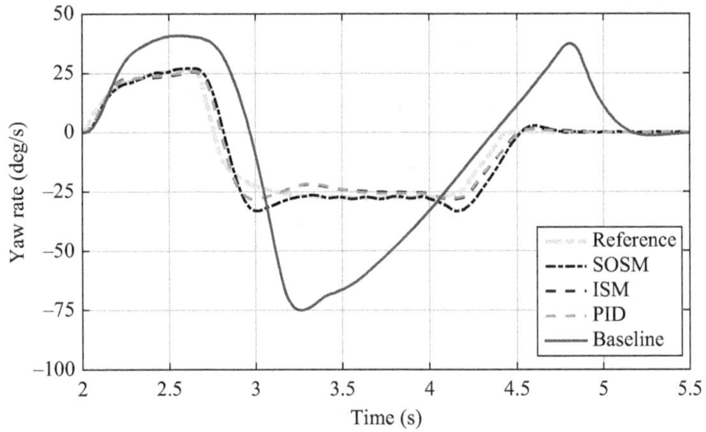

Figure 4.21 Sine with dwell test results for the different controllers: yaw rate tracking

the step input passing from 0% to 100% of the desired value in 0.27 s. In terms of agility, the baseline vehicle provides the slowest response with rise time of 0.42 s, while the PID and ISM response is faster, reaching the desired value in 0.31 and 0.30 s respectively.

Additionally, the lowest overshoot of just 3.1% is provided by the SOSM, while the ISM and PID controllers perform similarly in this regard, both showing a clear improvement over the baseline vehicle performance. Such agile behavior of the SOSM leads to the lowest $RMSE_{\dot{\psi}}$ of the yaw rate, with 0.09°/s. ISM has a higher deviation of 0.21°/s, but it is still a preferable solution in transient tests compared to the PID, which has a $RMSE_{\dot{\psi}}$ value of 0.32°/s. As a conclusion from this test, one could find the SOSM control strategy to be a good solution for sport and racing cars, where agile vehicle response is of high importance. On the other hand, the SOSM variant shows low amplitude oscillations at steady state, which may be felt by the driver as an undesired vibration.

4.4.2.3 Sine with dwell

The *sine with dwell* test evaluates the handling of the vehicle in the presence of counter-steering (by the standard FMVSS 126). In this testing procedure the vehicle has the initial velocity of 80 km/h, and the acceleration pedal is released at the start of the maneuver. After that, a pseudo-sinusoidal steering input with 0.7 Hz frequency is applied. After the third quarter of the sine period is completed, the steering wheel is held at the same position for 500 ms, before the final quarter period is entered, and the maneuver is completed.

It can be observed in Figure 4.21 that, during the counter steering phase, the phase-shift caused by the SOSM variant is slightly higher than those introduced by the other investigated methods. Nevertheless, during the hold phase, the dynamics are faster, and SOSM also allows holding the actual value of the yaw rate at the given

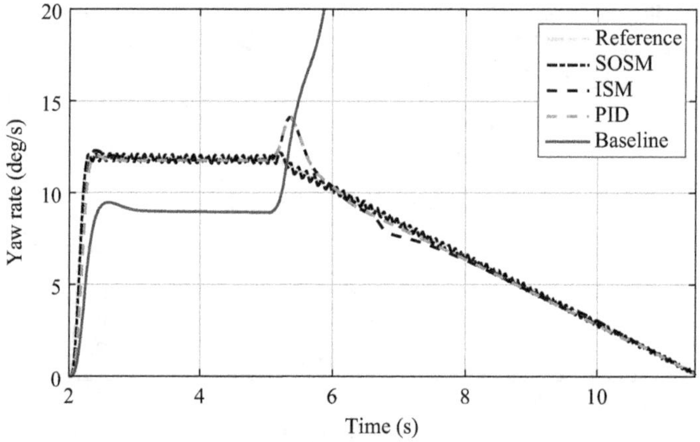

Figure 4.22 Braking in the turn test for the different controllers: yaw rate tracking

reference for almost the entire duration of the hold phase. With less phase shift in terms of yaw rate tracking, PID and ISM show the best performances, with $\mathrm{RMSE}_{\dot{\psi}}$ of 2.3 and 2.39°/s, respectively. The tracking error of the SOSM is comparable to these two cases, although slightly higher with a $\mathrm{RMSE}_{\dot{\psi}}$ of 3.62°/s.

4.4.2.4 Braking in the turn

According to the standard ISO 7975, the *braking in the turn* test is performed with an initial velocity of 90 km/h. This maneuver is done to check how the vehicle responds in presence of reduced vertical loads on the inside wheels. At the beginning, a steering input is applied in order to achieve 0.4 g of lateral acceleration. After 4 s the vehicle decelerates with a reference deceleration of 0.45 g.

In the yaw rate plot in Figure 4.22, an insufficient yaw moment occurs in case of the baseline vehicle. It loses its trajectory and is overturned at the end of maneuver. Its deviation from the possible desired vehicle behavior makes it practically unstable. By applying a stabilizing torque from the powertrain, the loss of vehicle stability is fixed, as can be seen for all proposed control laws in Figure 4.22. The best performance is provided by SOSM, which quickly reaches the reference value of the yaw rate and holding its actual value at the desired one during the whole test. As a result, it provides the best results in terms of yaw rate $\mathrm{RMSE}_{\dot{\psi}}$ with just 0.13°/s. In Figure 4.23 the stable trajectory held by the vehicle with this control type can be observed. In a similar way, although with some higher deviation (≈0.5°/s), the vehicle with ISM and PID controllers copes well with this test. As can be seen, the three proposed controllers remain on the desired trajectory, with the SOSM control reacting faster than the ISM and PID controls. Nevertheless, compared to them, the SOSM control produces more chattering in the yaw rate after reaching the reference value. This can be uncomfortable for the driver if the oscillations of the yaw rate reach a sufficiently high amplitude, so that some filtering should be added with possible slight deterioration of the promptness of the overall control system.

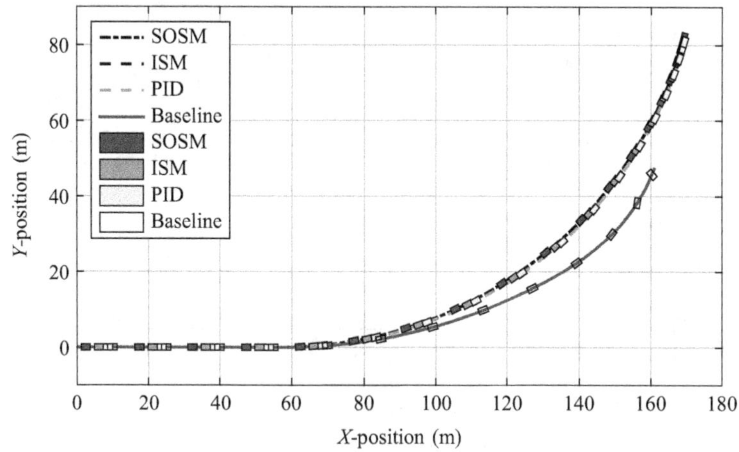

Figure 4.23 Braking in the turn test for the different controllers: vehicle trajectory

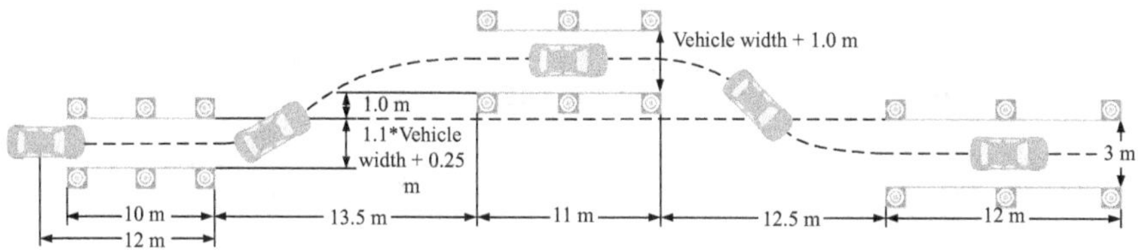

Figure 4.24 Double-lane change test

4.4.2.5 Double lane change

The *double lane change* test, performed in accordance with the standard ISO 3888, is schematically shown in Figure 4.24. It is a closed-loop maneuver, which the vehicle should complete without touching the cones.

This test starts with a 10 m long entry lane, then the driver releases the acceleration pedal while the top gear is selected. From this moment on, the control of the vehicle by the driver is performed simply by turning the steering wheel. The side lane cones are installed with a margin of 1 m and distance of 13.5 m from the entry lane. The resulting boundary width is equal to the width of the vehicle, plus additional 1 m margin. The last lane, which is 3 m wide, begins 12.5 m after the end of the side lane. At the end of the test the vehicle velocity is measured.

The test is performed on a road surface with friction coefficient $\mu = 0.8$ and starting velocity of 80 km/h. Under such conditions, the baseline vehicle loses its trajectory and hits the cones of the last lane, as shown in Figure 4.25. The vehicle, when equipped with ISM and PID controllers, shows quite similar behaviors, completing successfully the whole maneuver, while SOSM shows some slight deviation from the trajectory.

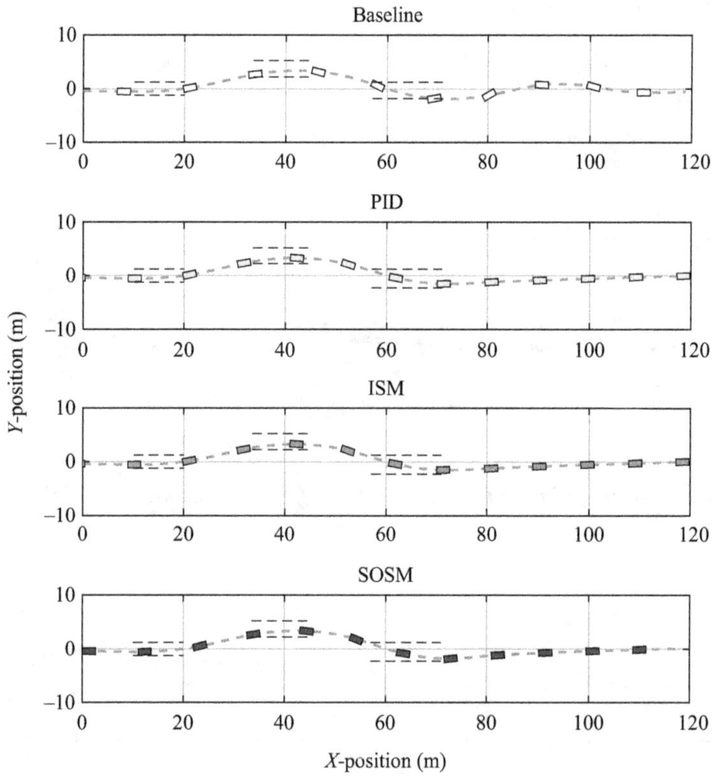

Figure 4.25 Double-lane change test results for the different controllers

With the proposed controllers, the vehicle does not hit the cones on the track, which is the main requirement. In terms of yaw rate deviation from its reference value, SOSM has $RMSE_{\dot{\psi}}$ of 4.1°/s while ISM and PID produce quite similar results of 2.48 and 2.44°/s, respectively (Figure 4.26).

To assess the energy efficiency of the vehicle during this maneuver, the velocity at the end of the test is measured. The baseline vehicle, due to its unstable behavior, loses velocity significantly, its measured value at the end being 58.9 km/h. The adoption of SOSM produces a final velocity equal to 66.2 km/h, while the best results in terms of energy efficiency are guaranteed by ISM and PID, with a final value of velocity of 67.5 km/h.

4.5 Conclusions

In this chapter the control of the vehicle lateral dynamics has been studied. Although the topic was dealt by providing a general overview, the control systems considered in the major discussion were developed specifically for the class of vehicles featuring

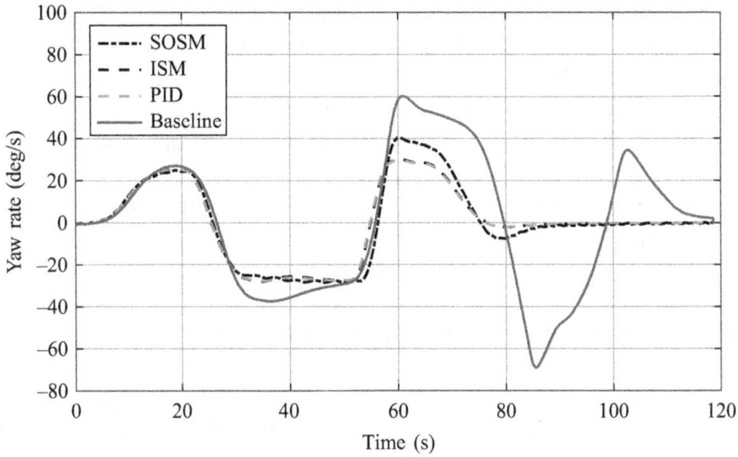

Figure 4.26 Double-lane change test results: yaw rate tracking

four individual electric motors and a decoupled electro-hydraulic braking system. In particular, the vehicle control architecture includes a high-level controller, where the PID, ISM, and SOSM control strategies were implemented as a feedback yaw rate controller. They are used in combination with a feedforward control, in order to obtain agile vehicle response during curvilinear motion. The reference yaw rate is based on a nonlinear vehicle model, and aimed at the achievement of a wider linear area of the steering gradient. The performed simulations with an experimentally validated vehicle model assess the efficacy of the application of the SOSM and ISM methods, with substantial improvements over the baseline vehicle, in terms of handling and stability.

Acknowledgment

The project leading to this application has received funding from the European Union's Horizon 2020 research and innovation program under the Marie Skodowska-Curie grant agreement no. 675999.

References

[1] D. Karnopp, *Dean Karnopp*. London, UK: CRC Press, 2013.
[2] M. Mitschke and H. Wallentowitz, *Dynamik der Kraftfahrzeuge*. Berlin, Heidelberg: Springer-Verlag, 2004.
[3] G. Genta, *Motor Vehicle Dynamics – Modeling and Simulation*, 2nd edition. Singapore: World Scientific, 2003.

[4] D. Schramm, M. Hiller, and R. Bardini, *Vehicle Dynamics, Modeling and Simulation*. Berlin, Heidelberg: Springer-Verlag, 2014.

[5] J. Tjonnaas and T. A. Johansen, "Adaptive optimizing dynamic control allocation algorithm for yaw stabilization of an automotive vehicle using brakes," in *2006 14th Mediterranean Conference on Control and Automation*, 28–30 June, Ancona, Italy, 2006.

[6] S. Data and F. Frigerio, "Objective evaluation of handling quality," *Journal of Automobile Engineering*, vol. 216, no. 4, pp. 297–305, Apr. 2002.

[7] U. Kiencke and L. Nielsen, *Automotive Control Systems – For Engine, Driveline, and Vehicle*. Berlin: Springer-Verlag, 2005.

[8] W. F. Milliken and D. L. Milliken, *Race Car Vehicle Dynamics*. Warrendale: SAE International, 1995.

[9] J. Y. Wong, *Theory of Ground Vehicles*, 4th edition. NJ: Wiley, 2008.

[10] Y. Shibahata, K. Shimada, and T. Tomari, "Improvement of vehicle maneuverability by direct yaw moment control," *Vehicle System Dynamics*, vol. 22, no. 5–6, pp. 465–481, Jul. 1993.

[11] K. Shimada and Y. Shibahata, "Comparison of three active chassis control methods for stabilizing yaw moments," *SAE Technical Paper Series*, Mar. 1994.

[12] Q. Cheng, A. C. Victorino, and A. Charara, "A new nonlinear observer of the sideslip angle and the road bank angle using the particle filter," in *The 11th International Symposium on Advanced Vehicle Control*, Seoul, Korea, 2012.

[13] M. Abe, A. Kato, K. Suzuki, Y. Kano, Y. Furukawa, and Y. Shibahata, "Estimation of vehicle side-slip angle for DYC by using on-board-tire-model," in *The Fourth International Symposium on Advanced Vehicle Control*, AVEC, Nagoya, Japan, 1998.

[14] H. Fujimoto and Y. Yamauchi, "Advanced motion control of electric vehicle based on lateral force observer with active steering," in *IEEE International Symposium on Industrial Electronics*, ISIE, Bari, Italy, Feb. 2010.

[15] M. Bang, S. Lee, C. Han, D. Maciuca, and J. Hedrick, "Performance enhancement of a sliding mode wheel slip controller by the yaw moment control," *Proceedings of the Institution of Mechanical Engineers Part D Journal of Automobile Engineering*, vol. 215, no. 4, pp. 455–468, 2001.

[16] J. Song and W. S. Che, "Comparison and evaluation of brake yaw motion controllers with an antilock brake system," *Proceedings of the IMechE Part D: Journal of Automobile Engineering*, vol. 222, no. 7, pp. 1273–1288, Jul. 2008.

[17] R. Tafner, M. Reichhartinger, and M. Horn, "Estimation of tire parameters via second-order sliding mode observers with unknown inputs," in *13th IEEE Workshop on Variable Structure Systems*, VSS'14, Nantes, France, 29 Jun.–2 Jul. 2014.

[18] J. Song, H. S. Kim, and B. Kim, "Vehicle longitudinal and lateral stability enhancement using TCS and yaw motion controller," *International Journal of Automotive Technology*, vol. 8, no. 1, pp. 49–57, 2007.

[19] K. Sawase and Y. Sano, "Application of active yaw control to vehicle dynamics by utilizing driving/breaking force," *JSAE Review*, vol. 20, no. 2, pp. 289–295, Apr. 1999.

[20] F. Cheli, L. Kakalis, and A. Zorzutti, "A torque vectoring control logic for active high performance vehicle handling improvement," in *The ASME 2007 International Design Engineering Technical Conferences & Computers and Information in Engineering Conference*, Las Vegas, NV, USA, 2007.

[21] L. D. Novellis, A. Sorniotti, and P. Gruber, "Wheel torque distribution criteria for electric vehicles with torque-vectoring differentials," *IEEE Transactions on Vehicular Technology*, vol. 63, no. 2, pp. 1593–1602, May 2013.

[22] G. Kaiser, F. Holzmann, B. Chretien, and M. Korte, "Torque vectoring with a feedback and feed forward controller – applied to a through the road hybrid electric vehicles," in *IEEE Intelligent Vehicles Symposium*, Baden-Baden, Germany, Jun. 2011.

[23] J. Yamakawa and K. Watanabe, "A method of optimal wheel torque determination for independent wheel drive vehicles," *Journal of Terramechanics*, vol. 43, no. 3, pp. 269–285, Jul. 2006.

[24] K. Jalali, T. Uchida, S. Lambert, and J. McPhee, "Development of an advanced torque vectoring control system for an electric vehicle with in-wheel motors using soft computing techniques," *SAE International Journal of Alternative Powertrains*, vol. 2, pp. 261–278, Apr. 2013.

[25] T. Goggia, A. Sorniotti, L. De Novellis, *et al.*, "Integral sliding mode for the torque-vectoring control of fully electric vehicles: theoretical design and experimental assessment," *IEEE Transactions on Vehicular Technology*, vol. 64, no. 5, pp. 1701–1715, May 2014.

[26] A. van Zanten, R. Erhardt, and G. Pfaff, "VDC, the vehicle dynamics control system of Bosch," *SAE Technical Paper 950759*, 1995.

[27] R. He, E. Jimenez, D. Savitski, C. Sandu, and V. Ivanov, "Investigating the parameterization of dugoff tire model using experimental tire-ice data," *SAE International Journal of Passenger Cars-Mechanical Systems*, vol. 10, no. 2016-01-8039, 2016.

[28] M. Canale, L. Fagiano, A. Ferrara, and C. Vecchio, "Vehicle yaw control via second-order sliding-mode technique," *IEEE Transactions on Industrial Electronics*, vol. 55, pp. 3908–3915, Nov. 2008.

[29] L. De Novellis, A. Sorniotti, P. Gruber, *et al.* , "Direct yaw moment control actuated through electric drivetrains and friction brakes: theoretical design and experimental assessment," *Mechatronics*, vol. 26, pp. 1–15, 2015.

[30] C. Edwards and S. Spurgeon, *Sliding Mode Control: Theory and Applications*. Systems and Control, Taylor & Francis, 1998.

[31] G. Bartolini, A. Ferrara, A. Levant, and E. Usai, "On second order sliding mode controllers," in *Variable Structure Systems, Sliding Mode and Nonlinear Control* (K. D. Young and Ü. Özgüner, eds.), Lecture Notes in Control and Information, pp. 329–350, London, UK: Springer-Verlag, 1999.

[32] A. Levant, "Sliding order and sliding accuracy in sliding mode control," *International Journal of Control*, vol. 58, no. 6, pp. 1247–1263, Dec. 1993.

[33] G. Bartolini, A. Ferrara, and E. Usai, "Output tracking control of uncertain nonlinear second-order systems," *Automatica*, vol. 33, pp. 2203–2212, Dec. 1997.

[34] A. Ferrara and M. Rubagotti, “A sub-optimal second order sliding mode controller for systems with saturating actuators,” *IEEE Transactions on Automatic Control*, vol. 54, pp. 1082–1087, Jun. 2009.

[35] V. I. Utkin and J. Shi, “Integral sliding mode in systems operating under uncertainty conditions,” in *Proceedings of the 35th IEEE Conference on Decision and Control*, vol. 4, (Kobe, Japan), pp. 4591–4596, Dec. 1996.

[36] V. I. Utkin, *Sliding Modes in Optimization and Control Problems*. Berlin, Heidelberg: Springer-Verlag, 1992.

[37] H. Pacejka and E. Bakker, “The magic formula tyre model,” *Vehicle Systems Dynamics*, vol. 21, pp. 1–18, 1993.

[38] D. Savitski, K. Hoepping, V. Ivanov, and K. Augsburg, “Influence of the tire inflation pressure variation on braking efficiency and driving comfort of full electric vehicle with continuous anti-lock braking system,” *SAE International Journal of Passenger Cars-Mechanical Systems*, vol. 8, no. 2015-01-0643, pp. 460–467, 2015.

Chapter 5
Stability control of heavy vehicles

Hocine Imine[1] *and Leonid Fridman*[2]

In this chapter the design of active steering assistance systems for heavy vehicles is discussed. These kinds of systems are oriented to avoid the rollover and prevent lane departure of the vehicle. The methodology herein illustrated is based on the super-twisting algorithm. An estimator relying on high order sliding mode observer is developed in order to get information on the vehicle dynamics, such as lateral acceleration limit and the height of the center of gravity. The lateral position and lateral speed are controlled using sliding mode control in order to ensure the stability of the vehicle and avoid accidents. While in standard practical situations the lateral offset and the relative yaw angle are typically measured and the road curvature can be assumed known, the identification of some relevant parameters of the model needs to be carried out in order to increase the robustness of the control system, as discussed in the chapter. Simulation and experimental results are reported, making reference to a tractor model, in order to show the quality of the presented concept.

5.1 Introduction

Accidents involving Heavy Vehicles (HVs) have serious consequences for road users, and incidents induce major congestion or damage to the environment or the infrastructure. Statistics show that accidents involving such vehicles are more serious than light vehicles (LVs). The mortality rate is twice in an accident involving a heavy vehicle (see Figure 5.1). While they constitute only 3% vehicles in traffic, HV represent 10% of the vehicles involved in fatal accidents (13% of HV-related accidents are deadly against 6% for LV, and 85% of deaths are not HV users).

The severity of HV accidents is mainly due to the following reasons:

- HV control is much more difficult than the light vehicle,
- HV dynamics is more complex because of its structure, especially when dealing with an articulated multi-axle vehicle. In effect, the peculiarity with these vehicles

[1]Laboratory for Road Operations, Perception, Simulators and Simulations; French Institute of Science and Technology for Transport, Spatial Planning, Development and Networks, France
[2]UNAM, Departamento de Ingeniería, de Control y Robótica División de Ingeniería Eléctrica, Facultad de Ingeniería UNAM, Mexico

	HGV	LV
Fatalities	13% of accidents	6% of accidents
Serious accidents	36%	29%
% of deaths	15 of HGV users	62% of LV users

Figure 5.1 Distribution of fatal accidents by type of involved vehicles

is that there are several types in traffic. The previous studies notice that the most important type of HV is the tractor/semi-trailer,

- HV dynamics is less stable because of its special characteristics: height of center of gravity with respect to the width of the axles, liquid feed inducing a swinging phenomenon for tank vehicles, the presence of joints induces risk of jackknifing during a brutal downturn,
- Because of its weight, HV has a higher kinetic energy and thus an extension of the stopping distance and an increased risk in a possible impact.

Most often, accidents occur on local roads, and primarily the result of a crash, rollover and jackknifing. Studies show that human error is the cause of 90% of road accidents. Indeed, a short inter-distance between vehicles and driving with speed higher than that allowed in a tight turn or highway exit can lead to dangerous situations. The driver assistance systems seem to be necessary to prevent these risk situations.

However, the good knowledge of driving laws can also constitute the good preventive solutions. As an example, in order to obtain the driving license, the HV driver should know exactly which kind of risks his vehicle can have. He also needs to know how he can anticipate this risk in order to avoid it. The HV load should also be very uniformly distributed inside the vehicle, in order to not create load transfer between left and right side of the vehicle. As we explain later, the load transfer ratio is an indicator for rollover risk. The driver needs to have a good knowledge about how he brakes in order to avoid the jackknifing, control the speed during bad weather conditions, especially to avoid the lane departure, etc.

5.1.1 Main risks of accidents related to HVs

Several types of truck accidents can occur on the road. The most important risks involving HVs are rollover (Roll instability), jackknifing (Yaw instability), lane departure (Transversal instability) and collision (Longitudinal instability). All these accidents are characterized by appropriate risks criteria. These are used in accident analysis and designing appropriate prevention systems (warning systems and active control). In this work, we focus on the criteria of rollover and lane departure, but the developed approaches can be generalized to other criteria and other types of accidents.

In this section, some criteria commonly cited in the literature are described.

5.1.1.1 Jackknifing criteria

When driving with a low wheel grip, and in case of a bad distribution of thrust or tight cornering forces, the loss of control can lead to a jackknifing or a lane departure.

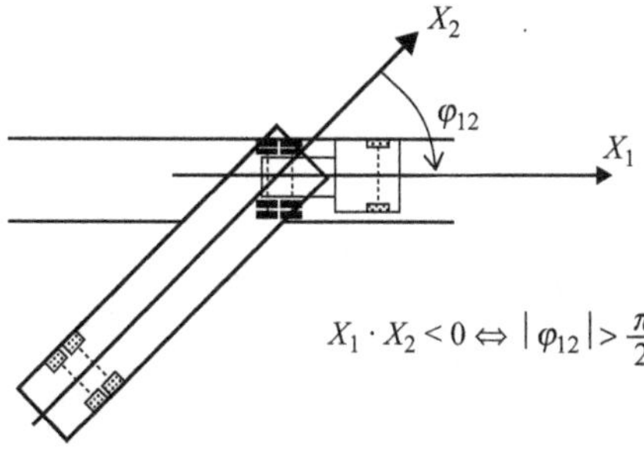

Figure 5.2 Jackknifing criterion

A jackknife is detected when the relative yaw angle of an articulated vehicle exceeds a threshold value, as shown in Figure 5.2.

In other words, the test can be translated by a vector relationship, i.e., *d* by a scalar product between the longitudinal axes of the tractor and the trailer. In [1], the criterion is expressed by the *dot* product of the velocity vector of articulation point and that the longitudinal axis of the trailer. This scalar product should always be positive to stay in the security domain. The authors in [2] developed a jackknifing indicator in a tight corner. This indicator assesses the minimal adhesion $\mu_{\min}$ between tires and the road.

It is expressed by:

$$\mu_{\min} = \sum F_y \Big/ \left(\cos(\Psi_r). \sum F_z\right) \tag{5.1}$$

where F_y and F_z represent, respectively, the lateral and the vertical forces of the drive wheels, and Ψ_r is the joint angle between the tractor and the trailer. In order to avoid the jackknifing, it is necessary that the coefficient $\mu_{\min} < 0.1$.

5.1.1.2 Rollover criteria

The rollover is the most important risk related to HVs. It occurs when the lateral acceleration equals or exceeds the vehicle's rollover limit (which may be assisted by roadway cross-fall or camber). Lateral acceleration in a curve is highly sensitive to speed. The required speed to produce rollover decreases as the radius of curvature reduces. Roll stability is influenced by the center of gravity height (COG), the effective track width, provided by the axles and tires, and the suspensions characteristics. The COG height is affected by the chassis height and the heavy vehicle load. This performance measure is evaluated in terms of the steady-state lateral acceleration at which all wheels on the inside of curvature have lifted off the road surface. This is accomplished by increasing the steering angle of a vehicle until all axles of one side of a given vehicle lift off. The rollover can occur when one wheel of the same axle of the vehicle, lifts off the road surface, as illustrated in Figure 5.3.

The rollover criteria can be divided into two categories: static or dynamic as defined by [3–5]. Following, few criteria commonly used in the literature are presented.

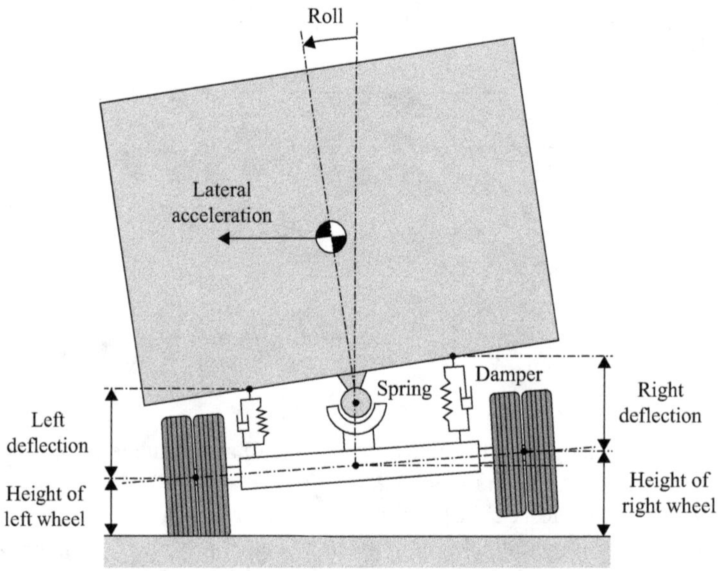

Figure 5.3 Wheel lifts off the road

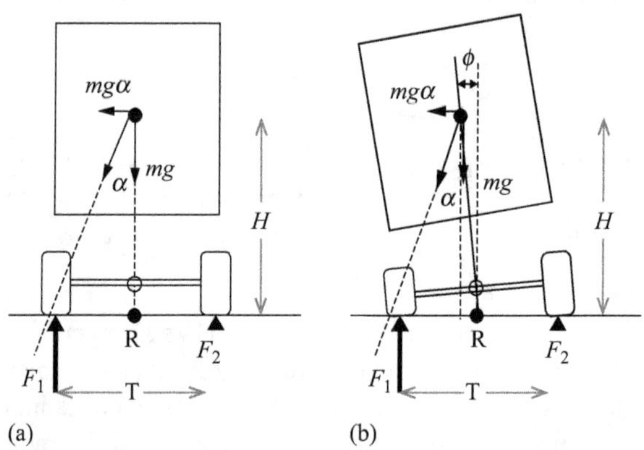

Figure 5.4 Vehicle under the effect of the lateral acceleration: (a) without roll and (b) with roll

SRT: Static rollover threshold

The static rollover threshold represents the lateral acceleration required to produce a rollover. Figure 5.4 shows a vehicle under the effect of the lateral acceleration $ay = \alpha g$, where g is the gravity. Using this model, the rollover occurs when a wheel lifts off the road, i.e., d when the intersection of the acceleration vector with the ground does not belong to the vehicle contact surface.

In this case, the static rollover threshold is calculated by [3]:

$$SRT = \alpha = \frac{T}{2H} - \varphi \tag{5.2}$$

where φ is the tractor roll angle and T is the tractor track width, H is the static center height of gravity related to the road.

LTR: Load transfer ratio

This indicator describes the load transfer between the two sides of the same axle [6–9]. This load transfer report for the vehicle is defined by the difference of loads on the left and right wheels, normalized by the total load:

$$LTR = \frac{F_{zr} - F_{zl}}{F_{zr} + F_{zl}} \tag{5.3}$$

When $F_{zr} = 0$ ($F_{zl} = 0$) the right (left) wheels lift off the road and the rollover coefficient takes on the limit value $LTR = -1$ ($LTR = 1$). For straight driving on a horizontal road for the tire vertical forces, it holds that $F_{nr} = F_{nl}$ which means that $LTR = 0$.

5.1.1.3 Lane departure criteria

The risk of lane departure can be caused by the carelessness of the driver or by driving with an inappropriate speed during a bad weather. On the other hand, roll stability of Heavy Vehicles is affected by the center height of gravity, the track width and the kinematic properties of suspensions. More destabilizing moment arises during the cornering maneuver when the center of gravity of the vehicle shifts laterally. The roll stability of the vehicle can be guaranteed if the sum of the destabilizing moment is compensated during a lateral maneuver. The lane departure depends also on the lateral grip of the tires on the road, the speed, the geometrical configuration of the vehicle, and lane width. The authors in [4,5] proposed a criterion for high speeds and one for low speeds. The high speed lane departure criterion is defined by the lateral distance between the trace of the steered wheels and the wheels of the rear axle. The distance is obtained for a constant curvature bend passage (400 m) and high speed (100 km/h).

The low-speed channel of exit criterion is a measure of distance when the vehicle moves on a highly curved track (intersection, for example). This criterion is based on the calculation of the maximum distance between the midpoint trace of the steering axle and the middle point of the last axle (see Figure 5.5).

5.1.2 State of the art review

In general, two types of preventive strategies can be defined in order to avoid accidents of HV: active and passive prevention. Recently, several solutions have been proposed in order to reduce the number of HV accidents. These techniques are based on the speed steering, braking and suspension control. Several systems have been developed in order to assist the driver to avoid the rollover. Some of them are installed in the infrastructure before a dangerous curvature. In case of overspeed leading to a rollover in the cornering, a warning is sent to the driver in order to decrease the speed [6,7]. Other systems have been installed inboard the truck. They use informative measures

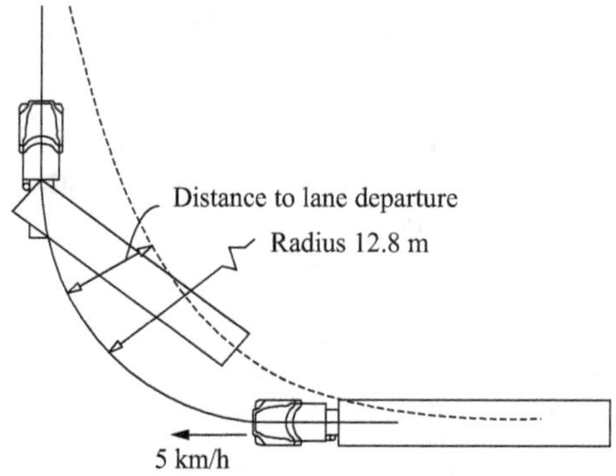

Figure 5.5 Low speed lane departure indicator

coming from sensors, such as vehicle speed and lateral acceleration, in order to send a warning signal to the driver when he goes beyond some risk thresholds [8,9].

Concerning, the lane departure risks, many works and systems exist [10–12]. Some of them which are able to measure the truck speed, are installed in the infrastructure before a dangerous curvature. In case of overspeed leading in the cornering, a warning is sent to the driver in order to decrease the speed [13]. Other systems are installed inboard the truck. In case of active systems, the concept is to minimize the lateral acceleration by steering action [14].

In case of active systems, the concept is to minimize the lateral acceleration by braking action, steering action, suspension action, by anti-roll action or a combination of all [15–17]. However, most of these methods required full information on the state that may limit their practical utility. Indeed, even if all the state measurements are possible, they are typically corrupted by noise. Moreover, the increased number of sensors makes the overall system more complex in implementation and expensive in realization. Thus, an observer design becomes an attractive approach to estimate the unmeasured states of the HV. Observers appear to be useful in not only system monitoring and regulation but also detecting as well as identifying failures in dynamic systems. There are several well-known observer structures, which are based on different methods such as variable structure approaches. A sliding mode observer yielding insensitivity to unknown parameter variations and noise has been proposed by [18–20]. Second- and third-order sliding mode observers providing finite time states observation, and impact forces estimation, are developed and presented in this work. In most of recent researches, the parameters of the vehicle are supposed to be known and constant. Some of them are given by constructors and manufacturers and others unknown parameters are taken from literature.

On the other hand, the impact forces affect vehicle dynamic performance and behavior properties. These forces are very important to evaluate the rollover risk of heavy vehicle by computing the Load Transfer Ratio (LTR), study and evaluate the damage of the vehicle on the road or bridges and in order to control the weights of the vehicles. These forces can also be used in the control systems in order to assist the driver. It is therefore important to quantify with great precision the magnitude of the impact forces and their variations in real-time. The estimation of the contact forces for heavy duty vehicles is essential in order to obtain the necessary information on the interaction between a vehicle and pavement. This information on contact forces is particularly useful for minimizing road and vehicle damage (tire, suspension, fragile payload) and for enhancing security by preventing the rollover of a vehicle.

The issue of estimating vehicle contact forces has been investigated, to some extent, in the literature. For example, in [21], an Extended Kalman Filter (EKF) was applied to estimate the longitudinal and lateral forces in a bicycle vehicle model. These forces are considered as additional states and their dynamics are modeled by a shaping filter driven by white noise. In [22], a similar estimation algorithm was used for an off-highway mining truck. The main limitation of the approaches adopted in [21,22] is that some assumptions are made on the forces that they are slow varying, which might not be the case in practice. Moreover, these approaches require the exact knowledge of the nature of the forces and the way they vary. In [16], sliding mode observers were applied to estimate contact forces in heavy duty vehicles. In their approaches, the height of the wheel hub was initially estimated. Then, the tire deflection was calculated by measuring the road profile and assuming that the stiffness of a tire varies linearly. Finally, an estimation of the vertical force was obtained by multiplying the tire deflection by its stiffness. The longitudinal and lateral forces were computed using the Pacejka tire model. This method requires an accurate knowledge of the parameters of a vehicle and the pneumatics with the profile of the road. In another work, Imine and Khemoudj [23] have identified the unknown parameters of a heavy vehicle before estimating the vertical forces. An alternative way to estimate the forces is to use strain gauges in the wheel hub which, in fact, provides precise measurement but may prove expensive. Shear strain gauges in the axle bar can be also used to measures forces. This methodology provides reasonable measurement for straight constant speed maneuvers and is ill-suited for turning maneuvers [24]. Moreover, strain gauges in vehicles are complex to install and to calibrate. One can also use optical sensors as described in [25], where the distance between a wheel hub and the ground surface is measured in order to provide the deflection of a tire and the vertical forces acting on it. This measurement configuration is sensitive to errors due to bad placement of sensors and irregularities of the road surface. In [26], a method was developed using an optically dynamic sensor in order to measure the displacement of the tire and through which the tire-forces were estimated. This method used a particular type of optical sensor which was installed on a tire rim. Apart from the difficulty in installing the sensors, the method heavily relies on the accurate knowledge of the tire's parameters. However, the exiting sensors to measure the vertical forces are very expensive and difficult to install [27]. The proposed method is based on sliding mode observer in order to estimate these vertical forces [19,20,28]. For this aim, the contact forces are considered as unknown inputs to the system and use the main embedded sensors in a vehicle to estimate vertical forces. The advantage of this approach is fourfold: (i) the approach avoids using many parameters of a vehicle, which are usually unknown; (ii) it optimizes the sensor's configuration (low-cost sensors); (iii) the proposed approach is better compared to those using strain gauges, especially for vehicles that execute turning maneuvers; (iv) finally, the sensors considered in this work are easy to install and calibrate [28].

5.1.3 Main contributions of the chapter

In the proposed chapter, an observer-controller law using the sliding mode technique is developed in order to achieve good tracking of desired trajectories by ensuring the convergence of the lateral acceleration of the vehicle toward the estimated acceleration limit. This allows the limitation of the load transfer between the right and the left side of the vehicle to its limited value which is set to 0.9. The aim is also to estimate the non-measurable states and the vertical forces of the HV. Design of such observers requires a dynamic model of heavy vehicle which is build up in a first step of this work. This model is coupled with an appropriate wheel road contact model. It has been validated using both PROSPER software [29] and by real measurements carried out with an instrumented tractor [30,31].

This chapter is organized as follows: Section 5.2 deals with the vehicle description and modelling. Section 5.3 concerning the stability control approach description is divided into two parts, first one is devoted to design the sliding mode observers in order to estimate the states and vertical forces and identify the vehicle parameters. The second part of Section 5.3 is devoted to steering control method design in order to avoid rollover and lane departure of the vehicle. Some simulations and experimental results are presented in Section 5.4. Finally, some remarks and perspectives are given in a concluding section.

5.2 Vehicle model description

In order to be able to achieve the aim of this work, namely, stability control of HV, one needs to identify which model of vehicle is appropriate. As indicated before, several types of HVs exist: Tractor, Tractor/semi-trailer, Multi-axles HV, EMS, etc. The best choice HV model is then very important in order to study the stability control and develop the different estimation and control tools.

The vehicle studied in this work and developed by [15] is a non-articulated heavy vehicle with two axles. Therefore, some assumptions were considered: the roll angle is assumed to be small, and the suspension and the tire dynamics are assumed to be linear. The corresponding model has five degrees of freedom and used to represent the roll and lateral dynamics, as shown in Figure 5.6.

The vehicle consists of two bodies. Body 1 which is composed of two axles is represented by the unsprung mass m_u and center of gravity CG_1. Body 2 is the sprung mass m and center of gravity CG_2. The center CG_1 is assumed to be in the road plane above CG_2. In this case, the lateral forces acting on the system can be linearized and computed as follows:

$$\begin{cases} F_{yr} = \mu \; c_r \; \alpha_r \\ F_{yf} = \mu \; c_f \; \alpha_f \end{cases} \tag{5.4}$$

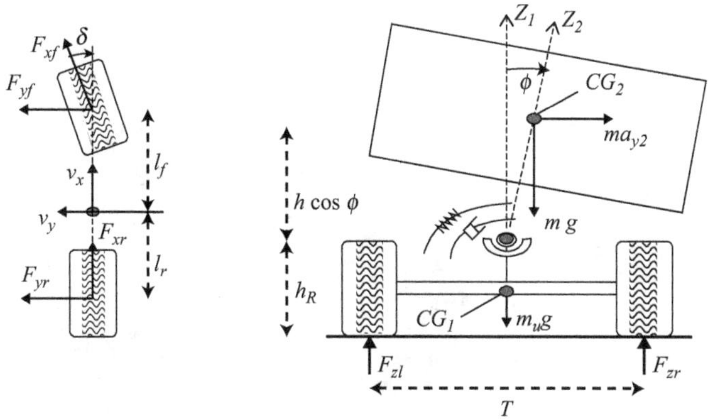

Figure 5.6 Heavy vehicle model

where c_f and c_r are, respectively, the front and the rear cornering stiffness after assuming the equality of the cornering stiffness for the two front wheels and the rear ones. The road adhesion is represented by the variable μ. In this work, the dry road is assumed to be equal to 1. The parameters α_f and α_r are, respectively, the front and rear tire slip angle computed by using the following formula:

$$\begin{cases} \alpha_r = -\left(\beta - l_r \frac{\dot{\psi}}{v}\right) \\ \alpha_f = \delta - \left(\beta + l_f \frac{\dot{\psi}}{v}\right) \end{cases} \tag{5.5}$$

where β is the side slip angle which is assumed equal for the two front wheels and the rear ones, l_f and l_r are, respectively, the distance from CG_1 to front axle and rear axle, v is the vehicle speed, $\dot{\psi}$ is the yaw rate and δ is the front wheel steering angle.

The model is derived using Lagrangian's equations:

$$M(q)\ddot{q} + C(q,\dot{q})\dot{q} + Kq = F_g(q,u) \tag{5.6}$$

where $q = [q_1, q_2, y_l, \varphi, \psi]^T$ represents the coordinate's vector composed of respectively left and right front suspension deflection, lateral displacement, roll angle and yaw angle, $M \in \Re^{5\times5}$ is the inertia matrix, $C \in \Re^{5\times5}$ is the matrix related to the damping effects, $K \in \Re^5$ is the springs stiffness vector, $F_g \in \Re^3$ is a vector of generalized forces and u is the system input.

The suspension is modelled as the combination of spring and damper elements as shown in Figure 5.7.

The vertical acceleration of the chassis is obtained as following:

$$\begin{aligned} \ddot{z} = {} & \left(k_1 q_1 + k_2 q_2 + (k_1 - k_2)\frac{T}{2}\sin(\varphi)\right) \Big/ m \\ & + \left(B_1 \dot{q}_1 + B_2 \dot{q}_2 - (B_1 - B_2)\frac{T}{2}\cos(\varphi)\dot{\varphi}\right) \Big/ m \end{aligned} \tag{5.7}$$

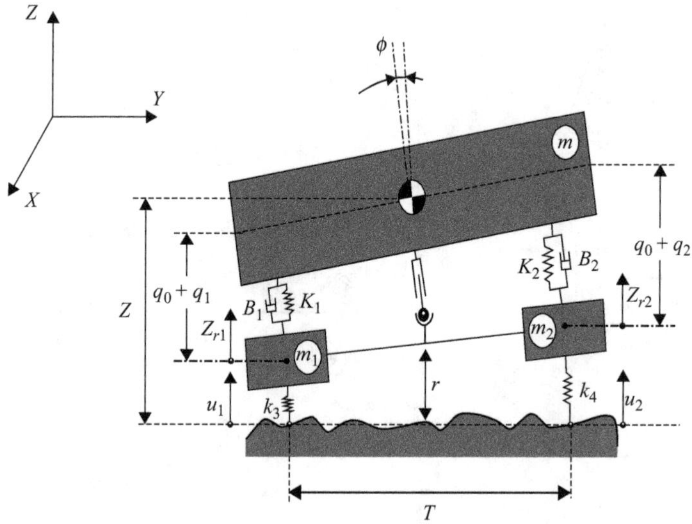

Figure 5.7 Front view of heavy vehicle model

where φ is the tractor roll angle and T is the tractor track width, q_1 and q_2 are, respectively, the left and right front suspension deflection of the tractor, z is the vertical displacement of the tractor sprung mass (center of gravity height). The tractor chassis with the mass m is suspended on its axles through two suspension systems. The suspensions stiffness and damping coefficients are represented, respectively, by $k_i, i = 1, 2$ and by $B_i, i = 1, 2$ in Figure 5.7. The tire is modelized by springs represented by $k_i, i = 3, 4$ in Figure 5.7. The left and right wheels masses are, respectively, m_1 and m_2. At the tire contact, the road profile is represented by the variables u_1 and u_2 which are considered as heavy vehicle inputs. The variables z_{r1} and z_{r2} correspond, respectively, to the vertical displacement of the left and right wheels of the tractor front axle, which can be computed by means of using the following equations:

$$\begin{cases} z_{r1} = z - q_1 - \frac{T}{2}\sin(\varphi) \\ z_{r2} = z - q_2 + \frac{T}{2}\sin(\varphi) \end{cases} \tag{5.8}$$

The lateral and yaw accelerations are computed as follows:

$$\begin{cases} \ddot{y}_l = -\frac{(c_f + c_r)}{mv}\dot{y}_l + \frac{(-c_f l_f + c_r l_r) - mv^2}{mv^2}\dot{\psi} + \frac{c_f}{m}\delta \\ \ddot{\psi} = -\frac{\left(c_f l_f^2 + c_r l_r^2\right)}{I_z v}\dot{\psi} + \frac{l_f c_f}{I_z}\delta + \frac{(-c_f l_f + c_r l_r)}{I_z}\beta \end{cases} \tag{5.9}$$

where Iz is the inertia along the Z-axis. The vertical accelerations of the wheels are given by:

$$\begin{cases} \ddot{z}_{r1} = \left(k_1 q_1 - k_1 \frac{T}{2}\sin(\varphi) + B_1\dot{q}_1\right)/m_1 - \left(B_1\frac{T}{2}\cos(\varphi)\dot{\varphi} + F_{zl}\right)/m_1 \\ \ddot{z}_{r2} = \left(k_2 q_2 + k_2 \frac{T}{2}\sin(\varphi) + B_2\dot{q}_2\right)/m_2 + \left(B_2\frac{T}{2}\cos(\varphi)\dot{\varphi} - F_{zr}\right)/m_2 \end{cases} \tag{5.10}$$

The accelerations of suspensions are given by the following equation system:

$$\begin{cases} \ddot{q}_1 = \left(\frac{1}{m} - \frac{1}{m_1}\right)k_1 q_1 + \frac{k_2 q_2}{m} + (k_1 - k_2)\frac{T}{2}\sin(\varphi) + \left(\frac{1}{m} - \frac{1}{m_1}\right)B_1\dot{q}_1 \\ \quad + \frac{B_2\dot{q}_2}{m} - \left((B_1 - B_2)\frac{T}{2}\cos(\varphi)\dot{\varphi}\right)/m - \frac{F_{zl}}{m_1} \\ \ddot{q}_2 = \left(\frac{1}{m} - \frac{1}{m_2}\right)k_2 q_2 + \frac{k_1 q_1}{m} + (k_1 - k_2)\frac{T}{2}\sin(\varphi) + \left(\frac{1}{m} - \frac{1}{m_2}\right)B_2\dot{q}_2 \\ \quad + \frac{B_1\dot{q}_1}{m} - \left((B_1 - B_2)\frac{T}{2}\cos(\varphi)\dot{\varphi}\right)/m - \frac{F_{zr}}{m_2} \\ \ddot{\varphi} = \left(k_1 q_1 - k_2 q_2 - (k_1 + k_2)\frac{T}{2}\sin(\varphi) + B_1\dot{q}_1 \right. \\ \quad \left. - B_2\dot{q}_2 - (B_1 + B_2)\frac{T}{2}\cos(\varphi)\dot{\varphi}\right)\frac{T}{2}/I_x \end{cases} \tag{5.11}$$

where I_x is the inertia moment in the roll axis, F_{zl} and F_{zr} are, respectively, the vertical forces of the left and the right wheel, which are considered as unknown perturbations to be estimated. These forces can be computed by using the following:

$$\begin{cases} F_{zl} = k_3(z_{r1} - u_1) \\ F_{zr} = k_4(z_{r2} - u_2) \end{cases} \tag{5.12}$$

Model validation

The validation of the HV model is done by using the automotive simulator PROSPER from Oktal [29].

The PROSPER software was itself validated by experiments carried out in collaboration between IFSTTAR and ETAS (Angers Technical Establishment dependent on the General Delegation for Armaments) [32].

In order to achieve this aim, a heavy vehicle of type RVI TRM 10,000 in three configurations was instrumented as shown in Figure 5.8.

A series of tests was performed in transverse stresses in order to correct the settings. The results of these experiments can be found in [32].

Both a straight line maneuver and a double bend maneuver are simulated in order to excite lateral and vertical dynamics of the vehicle. The used road profile is a real acquisition made on the French national road (Route Nationale) RN10.

From Figures 5.9 and 5.10, it can be seen that the model follows correctly the dynamics given by the simulator. For the yaw plane model, the accelerations of the tractor are estimated with errors less than 0.3 $\mathrm{m/s^2}$. For the trailer, the accelerations are estimated with an error less than 1 $\mathrm{m/s^2}$. This last error is higher than for tractor.

Figure 5.8 Experimentation for validation of PROSPER software

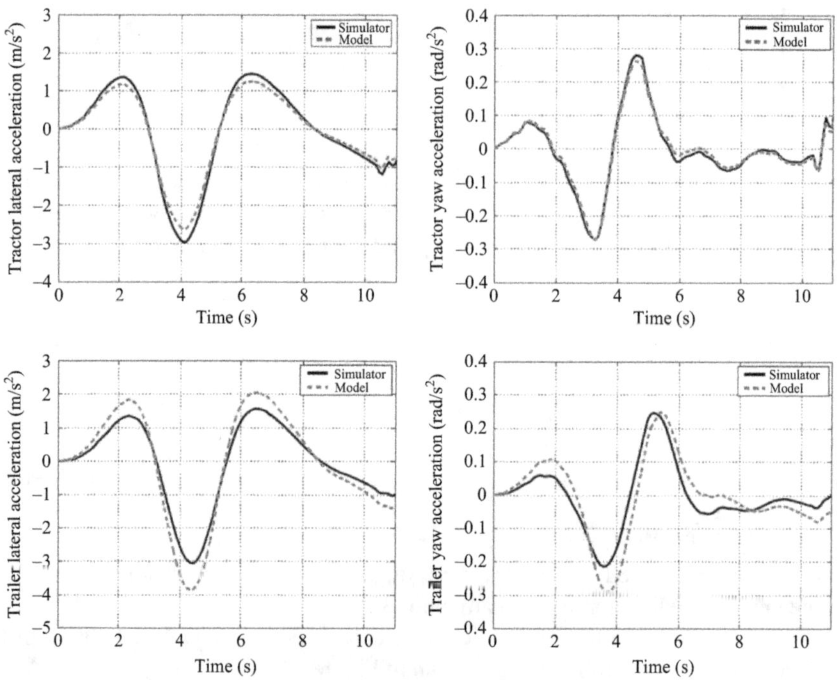

Figure 5.9 Validation of yaw plane model

It is probably due to the fact that a lumped axle-group is considered as a single axle in the model.

Concerning the axle model, first, a straight line maneuver is simulated with an irregular road profile, in order to excite the axle hop. In this case, the model follows the simulator and the peak for vertical force is correctly estimated, as we can see in Figure 5.10(a). The double lane change maneuver in Figure 5.10(b) shows the lateral

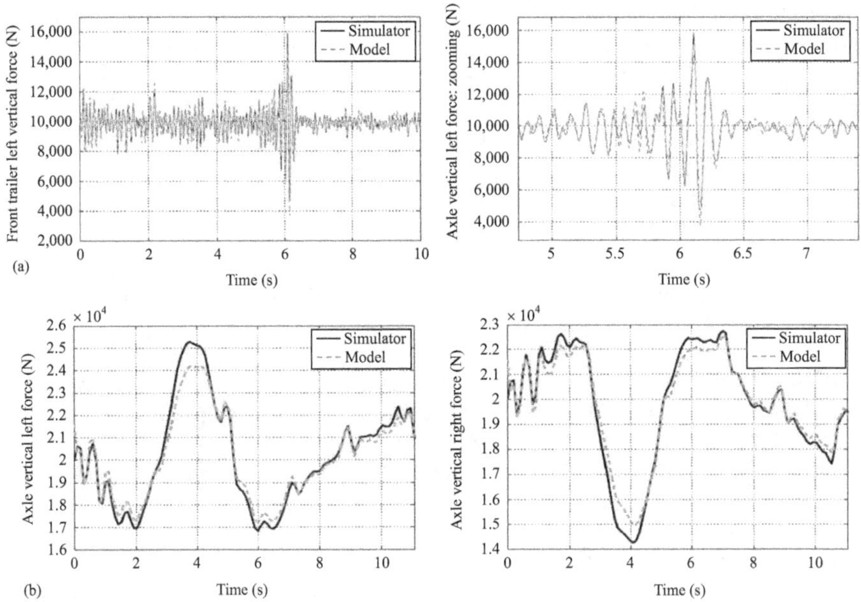

Figure 5.10 Validation of axle roll model: (a) straight line maneuver and (b) double lane change maneuver

load transfer. The model also tracks correctly the simulator with an acceptable error at the maximum load transfer (at 4 s) of 1,000 N less than 5% of the total force 25,000 N. Based on these observations, one can conclude that the model is suitable for developing a state observer to estimate the unknown forces.

5.3 Stability control approach

5.3.1 Sliding mode observer design

The studied model has five degrees of freedom. The speeds and accelerations are considered as unmeasurable variables. Therefore, in order to get all states vector, sliding mode observer is developed. At the same time, the vertical forces which are very important, especially to study the stability of HV, are estimated and presented in this section.

In order to develop the observer, let us rewrite (5.6) in state form as follows:

$$\begin{cases} \dot{x}_1 = x_2 \\ \dot{x}_2 = f(x_1, x_2) + F_g(x_1, u) \\ y = x_1 \end{cases} \tag{5.13}$$

where the variable $x_1 = q$ and $x_2 = \dot{q} \in \Re^5$ represents, respectively, the state vector and its derivative, the variable $y = q \in \Re^5$ is the measured outputs vector of the system, f is a vector of nonlinear analytical function and F_g is an unknown input vector, computed as follows:

$$F_g = \begin{bmatrix} -F_{zl}/m_1 \\ -F_{zr}/m_2 \\ 0 \end{bmatrix} \tag{5.14}$$

Before developing the sliding mode observer, let us consider the following assumptions:

1. The state is bounded ($\|x(t)\| < \infty, \forall t \geq 0$).
2. The inputs of the system are bounded ($\exists$ a constant $\mu \in \Re$ such as $|u_i| < \mu$, $i = 1, 2$).
3. The generalized forces F_g are bounded ($\exists$ a constant $\zeta \in \Re$ such as $F_{gi} < \zeta$, $i = 1, 2$).

5.3.1.1 States observation

The second-order observer developed in [33] has been adapted to the presented model in order to estimate in finite time states and therefore the vertical forces. It has the following form:

$$\begin{cases} \dot{\hat{x}}_1 = \hat{x}_2 + z_1 \\ \dot{\hat{x}}_2 = f(\hat{x}_1, \hat{x}_2) + z_2 \end{cases} \tag{5.15}$$

where $\hat{x}_1$ and $\hat{x}_2$ are, respectively, the estimations of vectors x_1 and x_2. The correction variables z_1 and z_2 represent the output injections of the form:

$$\begin{cases} z_1 = \lambda \Delta(\tilde{x}_1)\mathrm{Sign}(\tilde{x}_1) \\ z_2 = \alpha \mathrm{Sign}(\tilde{x}_1) \end{cases} \tag{5.16}$$

where $\tilde{x}_1 = x_1 - \hat{x}_1 \in \Re^5$ is a vector of the state estimation error. The gains matrices ($\lambda, \alpha \in \Re^{5\times5}$, the matrix $\Delta(\tilde{x}_1)$ and the vector "Sign" are defined as follows:

$$\begin{cases} \lambda = diag\{\lambda_1, \lambda_2, \dots \lambda_5\} \\ \alpha = diag\{\alpha_1, \alpha_2, \dots \alpha_5\} \\ \Delta(\tilde{x}_1) = diag\{|\tilde{x}_{1_1}|^{1/2}, |\tilde{x}_{1_2}|^{1/2}, \dots |\tilde{x}_{1_5}|^{1/2}\} \\ \mathrm{Sign}(\tilde{x}_1) = [\mathrm{sign}(\tilde{x}_{1_1}), \mathrm{sign}(\tilde{x}_{1_2}), \dots \mathrm{sign}(\tilde{x}_{1_5})]^T \end{cases} \tag{5.17}$$

The dynamics estimation errors are calculated as follows:

$$\begin{cases} \dot{\tilde{x}}_1 = \tilde{x}_2 - \lambda \Delta(\tilde{x}_1)\mathrm{Sign}(\tilde{x}_1) \\ \dot{\tilde{x}}_2 = f(x_1, x_2) - f(\hat{x}_1, \hat{x}_2) \\ +F_g(x_1, u) - \alpha \mathrm{Sign}(\tilde{x}_1) \end{cases} \tag{5.18}$$

where $\tilde{x}_2 = x_2 - \hat{x}_2 \in \Re^5$ is the estimation error of the vector x_2. Considering the accelerations of the system as bounded, the elements of the diagonal matrix α can be selected, satisfying the following inequality:

$$\alpha_{ii} > 2|\dot{\hat{x}}_{2_i}|, \quad i = 1, \ldots, 5 \tag{5.19}$$

On the other hand, from [34], the elements of the diagonal matrix λ can be selected as

$$\lambda_{ii} > \sqrt{\frac{2}{\alpha_{ii} - 2|\dot{\hat{x}}_{2_i}|} \frac{(\alpha_{ii} + 2|\dot{\hat{x}}_{2_i}|)(1 + p_i)}{1 - p_i}}, \quad i = 1, \ldots, 5 \tag{5.20}$$

where $p_i \in]0, 1[$ are some constants to be chosen (see proof in [35]). In order to study the observer stability, first, the convergence of $\tilde{x}_1$ in finite time t_0 is proved. Then, some conditions about $\tilde{x}_2$ in order to ensure its convergence to 0 are deduced. Therefore, for $t \geq t_0$, the surface $\tilde{x}_2 = 0$ is attractive, leading $\hat{x}_2$ to converge towards x_2 satisfying the inequalities (5.15) and (5.16). The convergence proof of the second-order observer can be found in [35].

After convergence of the observer (5.15), the variable $\tilde{x}_2$ convergences towards 0 in finite time $t \geq t_0$. In this case, and from (5.18), one obtains:

$$z_2 = \alpha \operatorname{sign}(\tilde{x}_1) = F_{gi}(x_1, u) \tag{5.21}$$

Theoretically, the equivalent output injection is the result of an infinite switching frequency of the discontinuous term. Nevertheless, the realization of the observer produces a high switching frequency which makes the application of a filter necessary. To eliminate the high frequency component, a filter of the following form is used:

$$\tau \dot{\bar{z}}_2(t) = \bar{z}_2(t) + z_2(t) \tag{5.22}$$

where $\tau \in R$ and $s_s \ll \tau \ll 1$, being s_s a sampling step.

The variable is then rewritten as follows:

$$z_2(t) = \bar{z}_2(t) + \xi(t) \tag{5.23}$$

where $\bar{z}_2(t)$ is the filtered version of $z_2(t)$ and $\xi(t) \in R$ is the difference caused by the filtration.

Nevertheless, as is shown in [35,36] that:

$$\lim_{\substack{\tau \to 0 \\ h/\tau \to 0}} \bar{z}_2(\tau, s_s) = z_2(t) \tag{5.24}$$

Thus, it is possible to assume that the equivalent output injection is equal to the output of the filter.

5.3.1.2 Vertical forces estimation

In order estimate the vertical force $F_{gi}(x_1, u)$, the parameters are supposed to be known. Therefore and using (5.21), the vertical force is obtained as follows:

$$F_{gi} = \alpha \text{sign}(\tilde{x}_1) \tag{5.25}$$

One reminds that this vector is composed of the forces F_{zl} or F_{zr} which can be computed using the system equation (5.12). One can then mention the advantages of the proposed method as follows:

1. The measuring of the road profiles u_1 and u_2 is not necessary.
2. The estimation of the vertical displacements of the wheels and its derivative are also not necessary to obtain.

5.3.2 Steering control

In this part, a steering control is developed in order to avoid accidents of heavy vehicle. Two types of accidents have been treated, rollover and lane departure. Sliding mode control (SMC) is proved to be the most interesting tool to achieve this aim. In effect, an SMC and especially a so-called Super-twisting algorithm is robust against uncertainties and perturbations, which can be really present in HVs.

5.3.2.1 Rollover avoidance

Rollover risk evaluation is based on load transfer ratio (LTR) which corresponds to the difference in tire normal forces acting on each side of the vehicle. It can be computed as follows:

$$LTR = \frac{F_{zr} - F_{zl}}{F_{zr} + F_{zl}} = \frac{2m_2}{mT}\left((h_R + h\cos\phi)\frac{a_{y2}}{g} + h\sin\phi\right) \tag{5.26}$$

where $a_{y2} = a_y - h\ddot{\varphi}$ is the lateral acceleration of the sprung mass and $a_y = v(\dot{\psi} + \dot{\beta})$ is the lateral acceleration of the unsprung mass and T is the track width, F_{zl} and F_{zr} are normal forces acting on, respectively, the left and the right side of the vehicle. When *LTR* is equal to 0, the heavy vehicle has a stable roll dynamic. The risk becomes higher when this indicator goes towards ± 1. Both extreme values characterize wheel lift-off. The same model developed before has been used in this section. However, in order to perform the controller, only the lateral part of the model is important. Therefore, the suspension deflections variables are not used here. In this section, an active steering control is developed in order to assist the vehicle in the case of rollover risk. In addition to the steering angle commanded by the driver and noted δ_d, an auxiliary steering angle δ_a is set by an actuator. Therefore, the control input $u = \delta = \delta_a + \delta_d$. In this work, the limit value of LTR is set to 0.9. This chosen value is used arbitrary, less than the limit 1, in order to give sufficient time to the controller/driver to react, before one of wheels lifts-off the road. In this case, the controller has time to avoid the rollover before to obtain high values of lateral accelerations. In the case of small roll angle, one can assume that:

$$h \sin \varphi < \left((h_R + h \cos \varphi) \frac{a_{y2}}{g} \right) \tag{5.27}$$

From (5.27), the acceleration limit can be obtained and approximated by:

$$a_{y2\,\mathrm{lim}} \approx \frac{0.9T\, gm}{2m_2 H}; \quad H = h + H_R \tag{5.28}$$

The aim of the developed steering control is to ensure the convergence of the lateral acceleration a_{y2} of the vehicle to its acceleration limit a_{y2lim}. This will allow the limitation of load transfer between the right and the left side of the vehicle to its limited value 0.9.

The steering control diagram is shown in Figure 5.11:

In order to be able to achieve the aim, let us consider the following sliding mode surface:

$$S = \dot{\tilde{y}}_l + \eta \tilde{y}_l \tag{5.29}$$

where $\dot{\tilde{y}}_l = \dot{y}_l - \dot{y}_{ld}$ and $\tilde{y}_l = y_l - y_{ld}$ are, respectively, the lateral speed error and lateral offset, η is a positive constant, $\dot{y}_{ld}$ and y_{ld} are, respectively, the desired velocity

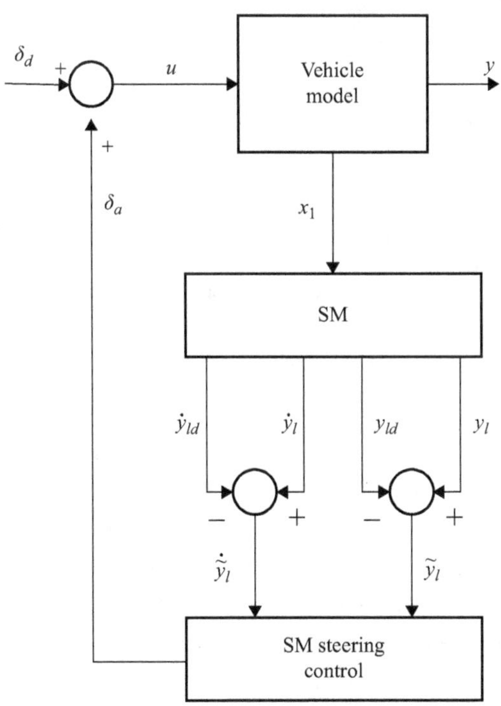

Figure 5.11 Controller diagram

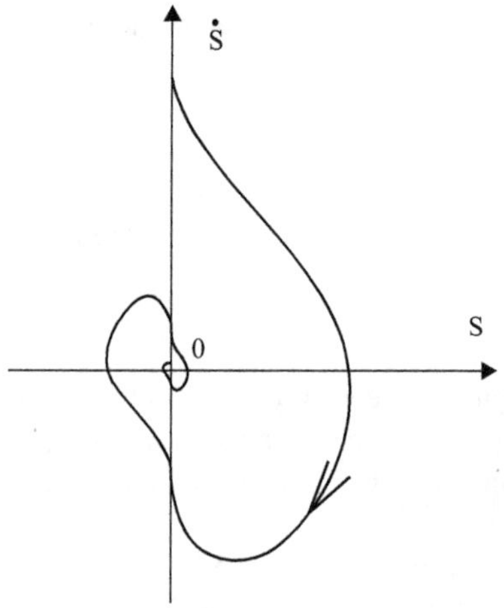

Figure 5.12 Super twisting algorithm trajectory

and the desired lateral displacement obtained, respectively, by first and double integration of the acceleration limit ay_{2lim}, computed earlier using (5.28). Deriving the variable S and using (5.9), it follows:

$$\dot{S} = \ddot{\tilde{y}}_l + \eta\dot{\tilde{y}}_l = \frac{c_f}{m}u + (\eta + a_{55})\dot{\tilde{y}}_l \tag{5.30}$$

The equivalent control u_{eq} is obtained by resolving the equation $\dot{S} = 0$ as follows:

$$u_{eq} = -\frac{m}{c_f}(\eta + a_{55})\dot{\tilde{y}}_l - \delta_d \tag{5.31}$$

The proposed control is based on Super-twisting algorithm. That algorithm has been developed in order to control dynamics of the system while avoiding chattering.

In this case, the trajectories are characterized by twisting around the origin, as shown in Figure 5.12.

The continuous control law u is composed of two terms. The first one is defined by means of its discontinuous time derivative, while the other is a continuous function of sliding variable. Therefore, the proposed control law is defined as follows:

$$\begin{cases} \delta_a = u_{eq} - G_1|S|^{1/2}\text{sign}(S) + u_1 \\ \dot{u}_1 = -G_2\text{sign}(S) \end{cases} \tag{5.32}$$

where G_1 and G_2 are the positive control gains. The convergence proof and analysis of the used super twisting algorithm can be found in [34,36].

5.3.2.2 Lane keeping assistance

Lane keeping assistance is based on control of lateral position, yaw angle and their respective speeds. Therefore, a new variable $y_r = y_l + \psi$ is defined. The estimated value of y_r is defined by $\hat{y}_r$.

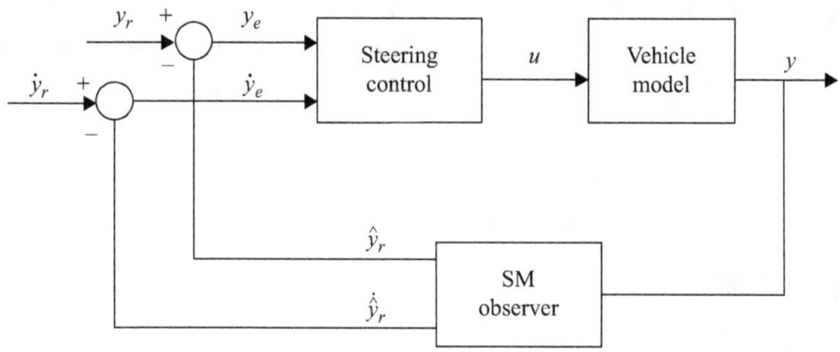

Figure 5.13 Lane keeping controller diagram

This controller makes it possible to maintain the vehicle in the center of the lane by assuming that the lateral offset between the center of the lane and the border of the road is measured.

The proposed steering control diagram is shown in Figure 5.13.

The aim of the developed steering control is to ensure the convergence toward 0 of the variable $y_e = \tilde{y}_l + \tilde{\psi}$ which is the sum of the lateral offset and the relative yaw angle. In order to develop the observer, the dynamic model defined in (5.13) is rewritten in reduced state form as follows:

$$\begin{cases} \dot{x}_1 = x_2 \\ \dot{x}_2 = A_2 x_2 + B_2 u + D_2 \beta \\ y = C_2 x_1 \end{cases} \tag{5.33}$$

where $y = x_1 = [y_l, \varphi, \psi]^T$ is the outputs vector. The matrices $A_2 \in R^{3\times3}$, $B_2 \in R^3$, $C_2 \in R^3$ and $D_2 \in R^3$ are defined as follows:

$$A_2 = \begin{bmatrix} a_{55} & 0 & a_{57} \\ 0 & -B_r/I & hm_2 v/I \\ 0 & 0 & a_{77} \end{bmatrix}; \quad B_2 = \begin{bmatrix} c_f/m \\ 0 \\ l_f c_f/J_z \end{bmatrix}; \quad D_2 = \begin{bmatrix} 0 \\ 0 \\ a_{74} \end{bmatrix};$$

$$C_2 = \begin{bmatrix} 1 & 0 & 0 \\ 0 & 1 & 0 \\ 0 & 0 & 1 \end{bmatrix}$$

On the other hand, the lateral offset $\tilde{y}_l$ and the relative yaw angle $\tilde{\psi}$ can be obtained using the following inequality:

$$\begin{cases} \dot{\tilde{y}}_l = v\beta + l_a \dot{\psi} + v\tilde{\psi} \\ \dot{\tilde{\psi}} = \dot{\psi} - v\phi \end{cases} \tag{5.34}$$

where ϕ is the road curvature and l_d is the look-ahead distance. Note that the road curvature and the look-ahead distance are supposed to be known. In order to be able to ensure that, let us consider the sliding mode surface:

$$S = \dot{y}_e + \lambda y_e \tag{5.35}$$

where λ is a positive constant. Let us define the function $f(x)$ and $g(x)$ such as:

$$\begin{cases} f(x) = A_2 x_2 + D_2 \beta \\ g(x) = B_2 u \end{cases} \tag{5.36}$$

Deriving the variable S and using (5.35), it follows:

$$\begin{aligned} \dot{S} &= \frac{\partial S}{\partial t}(t,x) + \frac{\partial S}{\partial x}(t,x) \cdot (f(x) + g(x,u)) \\ &= \left(\frac{c_f}{m} + \frac{l_f c_f}{J_z}\right) u_a + (\lambda + a_{55})\dot{\tilde{y}}_l + (\lambda + a_{77} + a_{57})\dot{\tilde{\psi}} + a_{74}\beta \end{aligned} \tag{5.37}$$

The equivalent control u_{eq} is obtained by resolving the equation $\dot{S} = 0$ as follows:

$$u_{eq} = \frac{-(\lambda + a_{55})\dot{\tilde{y}}_l - (\lambda + a_{77} + a_{57})\dot{\tilde{\psi}} - a_{74}\beta}{\Gamma} \tag{5.38}$$

where $\Gamma = \left(\frac{c_f}{m} + \frac{l_f c_f}{J_z}\right)$ The control algorithm is defined by the following control law:

$$\begin{cases} u = u_{eq} - k_1 |S|^{1/2} \text{sign}(S) + u_1 \\ \dot{u}_1 = -k_2 \text{sign}(S) \end{cases} \tag{5.39}$$

where $k_1 > 0$ and $k_2 > 0$ are the control gains under some conditions to defined later. The convergence of the controller can easily obtained by verifying the four following assumptions, as described in [37–40].

1. The input $u(t)$ is such that $|u| \leq U_{\max}$, with $U_{\max}$ is a real positive constant and any solution of the system (5.6) is defined for all t.
2. There exists $u_1 \in [0 \quad U_{\max}]$ such that for any continuous function $u(t)$ with $|u(t)| > u_1$, there exists time t_1 such that, $S(t)u(t) > 0$ for all $t > t_1$.
   Then, the control input $u = -U_{\max}\text{sign}(S(t_0))$, where t_0 is the initial value, permits to attend the sliding surface $S = 0$ in finite time.
3. There are positive constants $S_0, G_1, G_2, u_0 < 1$, such that if $|S(t,x)| < S_0$, then $0 < G_1 \leq \frac{\partial \dot{S}}{\partial u}(t,x,u) \leq G_2 \; \forall \, x, u$.
4. There exists a constant H such that if $|S(t,x)| < S_0$, then:

$$\left| \frac{\partial \dot{S}}{\partial t}(t,x,u) + \frac{\partial \dot{S}}{\partial x}(t,x,u) \cdot (f(x) + g(x,u)) \right| \leq H$$

The corresponding sufficient conditions for the finite time convergence to the sliding manifold are:

$$\begin{cases} K_2 > \frac{H}{G_1} \\ K_1 > \frac{2}{G_1}\sqrt{\frac{G_2 H(K_2+H)}{G_1(K_2-H)}} \end{cases} \tag{5.40}$$

It is easy to show here that all the assumptions given before are satisfied. Therefore, the convergence of the controller is ensured and the sliding surface S convergences toward 0 and $\dot{S} = 0$ in finite time t_0 ensuring then in the same time the convergence toward 0 of lateral offset and the relative yaw angle.

5.4 Validation results

The tractor of Figure 5.14 has been instrumented on behalf of VIF project [41]. The vehicle was equipped with different sensors to measure the dynamic states of the vehicle (gyrometers, accelerometers, LVDT, Laser, etc.) as shown in Figure 5.15. The installation and positions of these sensors in the tractor are illustrated in Figure 5.14.

As shown in Figure 5.16, different sensors have been installed in order to validate and calibrate the whole system:

- four accelerometers installed on the chassis in order to measure the vertical accelerations of wheels,
- four sensors LVDT in order to measure the suspensions deflections (q_1 and q_2 have been used in the observer),
- three axial gyroscopes installed on the chassis, to measure the angular rates (roll, pitch and yaw rates).

Figure 5.14 Instrumented vehicle

The roll angle is deduced from integration of the measured roll rate or by computation using the following formula:

In this work, this last method was preferred and used in order to avoid errors which can result from roll rate integration.

- two lasers in order to measure the vertical displacements of the chassis.

Many tests and scenarios have been performed with the instrumented vehicle driving at various speeds. In this work, the results obtained from zigzag and ramp tests are presented in order to show the robustness of the estimation and identification. While, the presented results for steering control to avoid rollover are obtained from

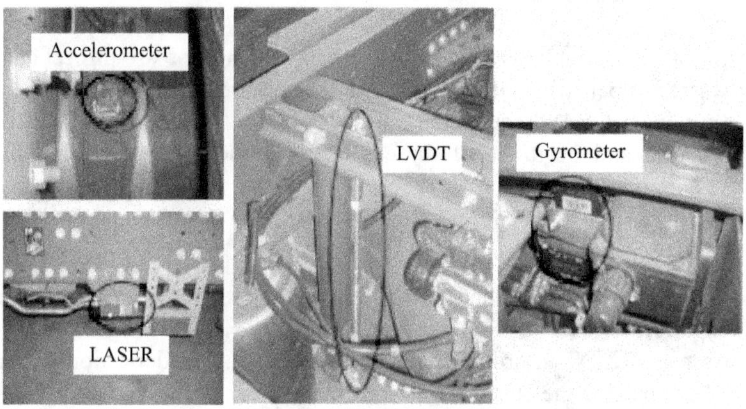

Figure 5.15 Sensors in the vehicle

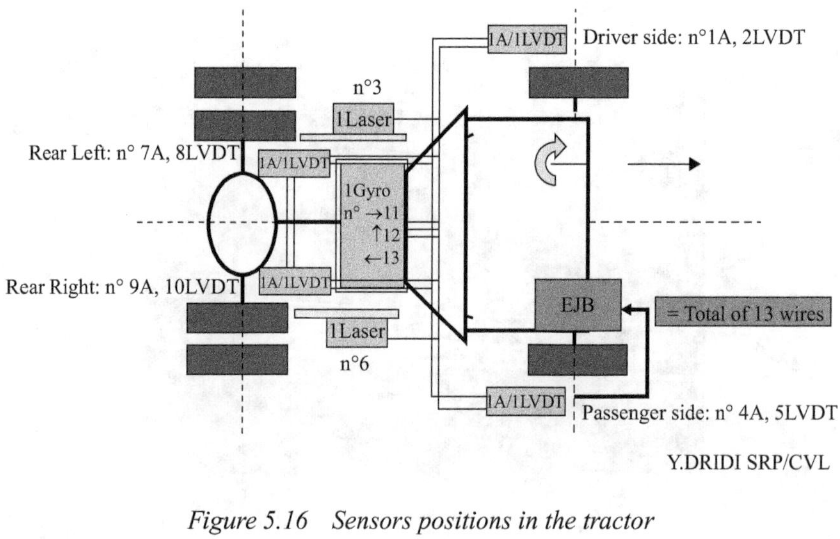

Figure 5.16 Sensors positions in the tractor

simulation. The nominal dynamic parameters and the static vertical forces are measured before the tests. The static values of front left and front right vertical forces are, respectively, 24,200 N and 25,250 N. The static values of rear left and rear right vertical forces are, respectively, 9,450 N and 12,050 N. The nominal values of the unsprung masses m_1 and m_2 are, respectively, 100 and 95 kg and the nominal values of suspensions stiffness k_1 and k_2 are, respectively, 194,680 N/m and 188,540 N/m. The results presented in this work are coming from two different sources: simulation and experimentation. The estimation results are obtained using the real signals coming from sensors, while control results are coming from simulation using the software tool PROSPER. This software developed by the company OKTAL (www.oktal.fr) can analyze the dynamic behavior of heavy trucks and articulated vehicles. It is based on the most accurate, detailed and efficient methods. Prosper is universally the preferred tool for analyzing vehicle dynamics, developing passive and active safety systems (controller, etc.) performance characteristics evaluation, etc.

5.4.1 Rollover avoidance results

5.4.1.1 Chicane test

Chicane test is presented in this section. This test is very important in order to show the reaction of controller when the driver changes at short time, often and brutally the direction of the vehicle. The simulation results are shown in Figure 5.17.

One remarks that the rollover risk appears at 14 s and lasts approximately 1 s. The value of the steering angle at this time is about −0.06 rad. This value decreases

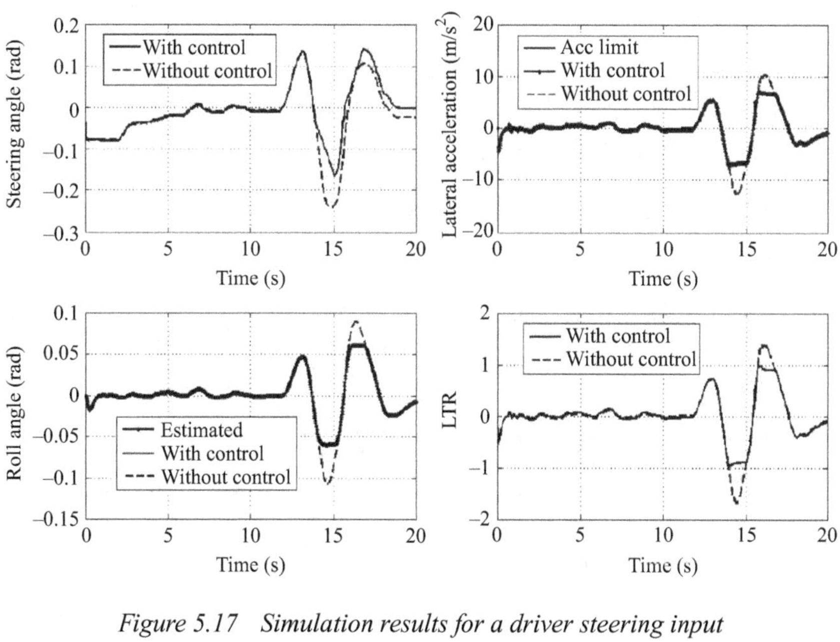

Figure 5.17 Simulation results for a driver steering input

until −0.25 rad when the controller is *OFF* (dashed line). Otherwise, and when the controller is *ON* (solid line), this value is decreased to −0.15 rad. At the critical time 14 s, and in order to stabilize the value of the LTR at its limit 0.9, the computed lateral acceleration limit is around $-8\ \mathrm{m/s^2}$. When the controller is activated (solid line), the lateral acceleration is still equal to $-8\ \mathrm{m/s^2}$. Otherwise and if the controller is still *OFF*, the lateral acceleration decreases until $-12\ \mathrm{m/s^2}$ (dashed line). During the risk time interval [14, 15] s, the controller is activated and the load transfer ratio (LTR) is stabilized to −0.9 (solid line). Without control, the rollover risk decreases and the LTR tends towards −1.8 (dashed line). The same situation occurs at the time interval [16, 17] s. In this case, the LTR is also stabilized to the limited value 0.9 when the controller is active. Otherwise, the LTR tends towards 1.5. The use of sliding mode observer allowed a well and quick estimation of roll angle as presented in Figure 5.17. Without control (dashed line), the absolute value of this variable increases from 0.06 to 0.11 rad during the first time interval [14, 15] s and from 0.06 to 0.09 rad in the time interval [16, 17] s.

5.4.1.2 Ramp test

The results of ramp test are presented in this section. The steering angle increased until a maximum value of 0.5 rad, during 5 s before to stabilize, as shown in Figure 5.18.

During the time interval [0, 1.5] s, there is no rollover risk, as one can remark in the left side of Figure 5.18. Although during this time, the estimated steering angle coming from control block is the same than that coming from the model without control. The LTR in this case is less than 1 as shown in the right side of Figure 5.18.

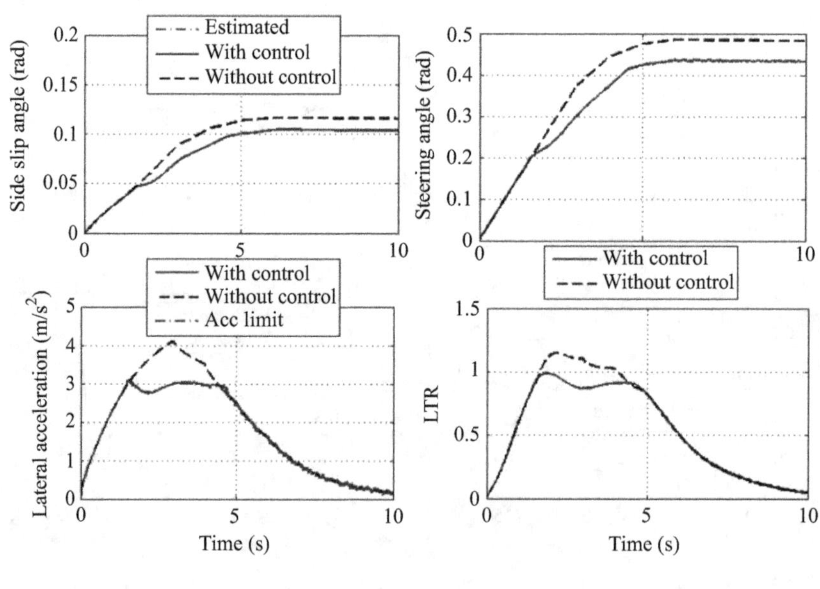

Figure 5.18 Simulation results for a driver steering input

After 1.5 s, the risk appears and the LTR reaches the limit value 1, which corresponds to the situation where one of the wheels of the same axle lifts-off. In order to avoid this risk situation, the controller is then activated in order to avoid the rollover of the vehicle. With the active controller, the steering angle is reduced and its value becomes less than the original one coming from the model. In this case, the lateral acceleration limit is estimated and shown in Figure 5.18 (solid line). Without control (dashed line), the lateral acceleration increases until 4 m/s^2. Otherwise, when the control is activated (solid line), this acceleration does not exceed 3 m/s^2. Therefore, the value of LTR is reduced and it becomes less than 1. On the other hand, the sliding mode observer allows to estimate in finite time and quickly the different variables of the system. In Figure 5.18, one notices also the well estimation of side slip angle when it's compared to the variable coming from the model.

5.4.2 Lane keeping results

5.4.2.1 First test

The lane keeping control based on super twisting algorithm is tested in this section. The road curvature that the vehicle should follow is shown in Figure 5.19.

The estimated lateral offset and the relative yaw angle are presented in Figure 5.20(a).

One notices that the absolute value of lateral offset does not exceed 0.1 m and the maximum value of the relative yaw angle is less than 0.1 rad. The unmeasured states of the vehicle are well estimated in finite time using third-order sliding mode observer. The convergence of the yaw rate and the roll angle are quick and the estimation is of quality as shown in Figure 5.20(b). The robustness of such observer is proved since there is no chattering.

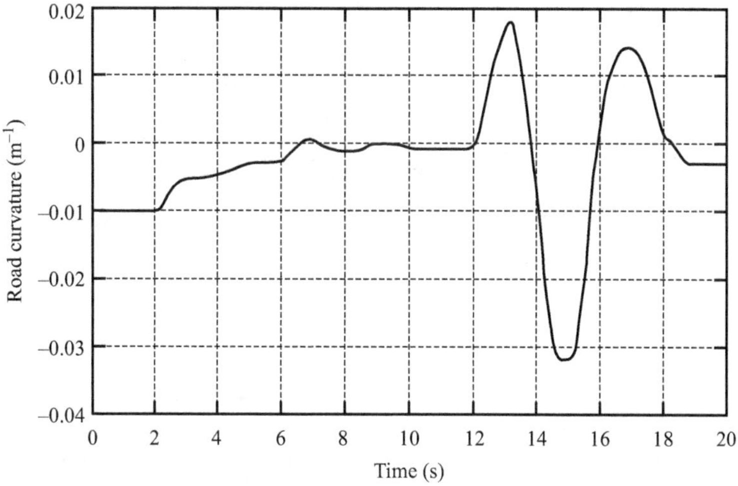

Figure 5.19 Road curvature

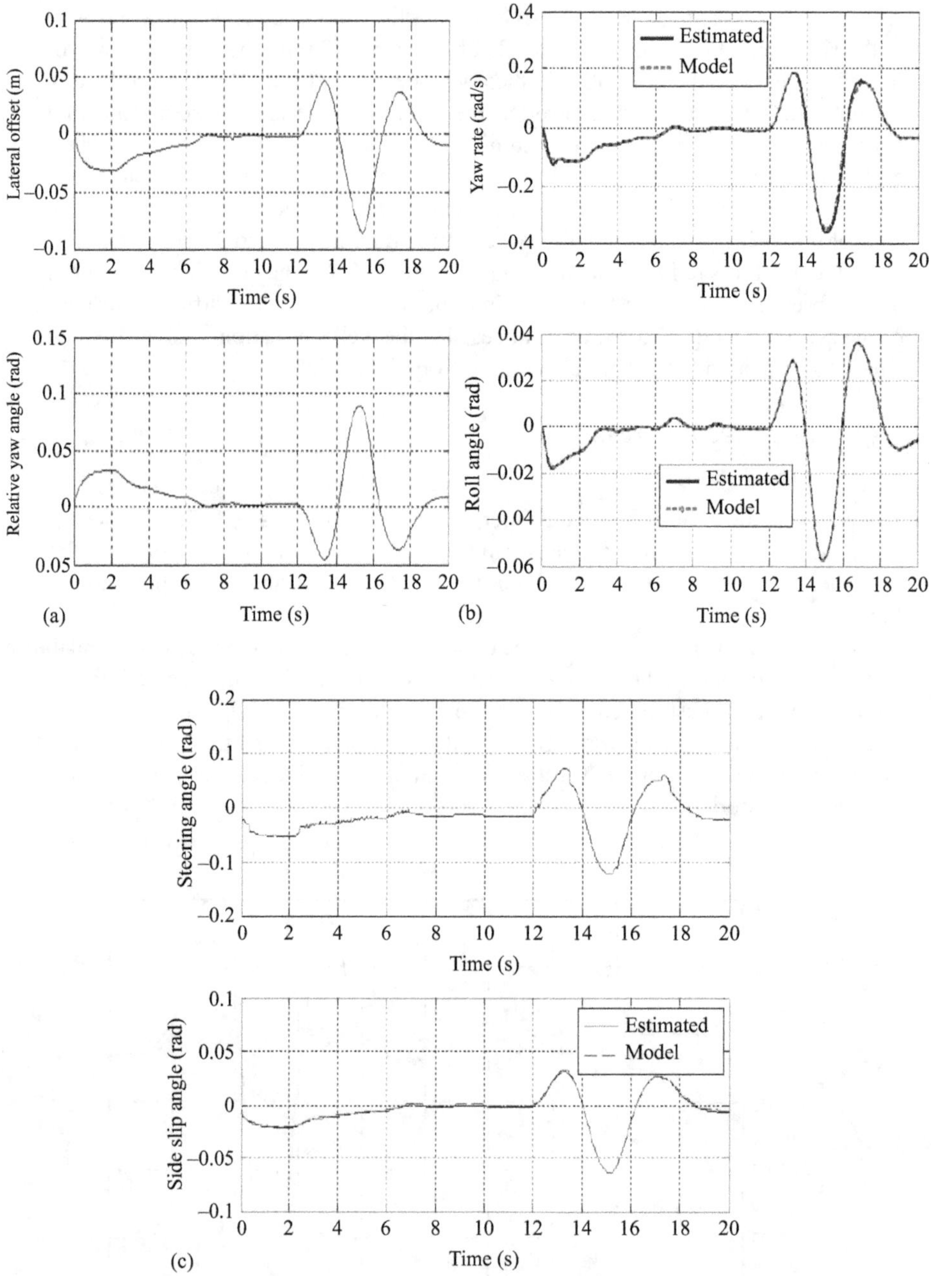

Figure 5.20 Simulation results for a lane keeping maneuver: first test. (a) Lateral offset and relative yaw angle: first test; (b) Yaw rate and roll angle: first test; (c) Steering and side slip angle: first test

The steering angle coming from steering control block and the side slip angle are shown in Figure 5.20(c). One remarks that the calculated steering angle has the same pace than the road curvature, with a maximum value of 0.1 rad. The estimation of side slip angle compared to the one coming from the model is of quality with an error close to 0.

5.4.2.2 Second test

The aim of the second test presented in this section, is to confirm the validation of the proposed approach. Another type of road curvature has been applied as shown in Figure 5.21.

As one can remark, the driver changes the direction abruptly and straightens the vehicle after around 2.5 s. Even in such complex situation, the positive effect of proposed control should be proven.

Figures shown in Figure 5.22 present the obtained results of this test.

According to Figure 5.22(a), the vehicle drives on straight line before to take a turn at 2 s. One notices that the lateral offset is very low, less than 0.03 m. Same remark can be given about the relative yaw angle with a maximum absolute value of 0.03 rad. One can also notice that the convergence of the observer is quick and the dynamic states are well estimated, with errors close to *zero*. This result is confirmed by comparison between the estimated and the measured roll angle and yaw rate presented in Figure 5.22(b). The maximum value of estimated steering angle shown in Figure 5.22(c) is very low and it is equal to 0.037 rad. Same remark can be given to side slip angle estimation, with a maximum value of about 0.022 rad.

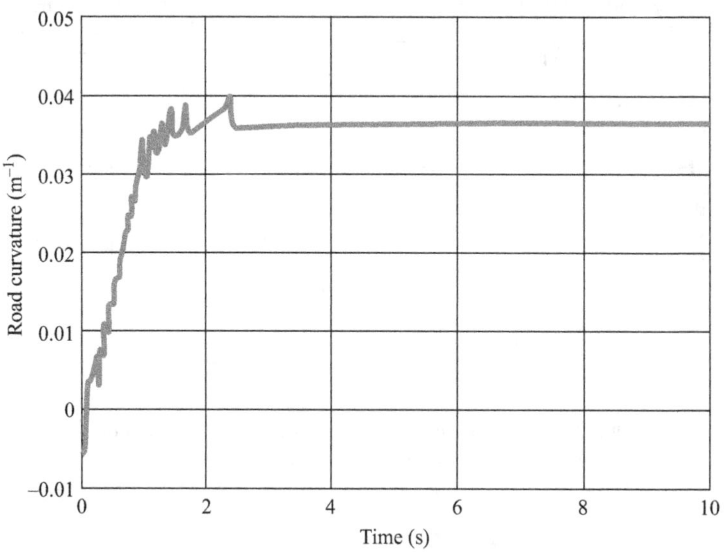

Figure 5.21 Road curvature of the second test

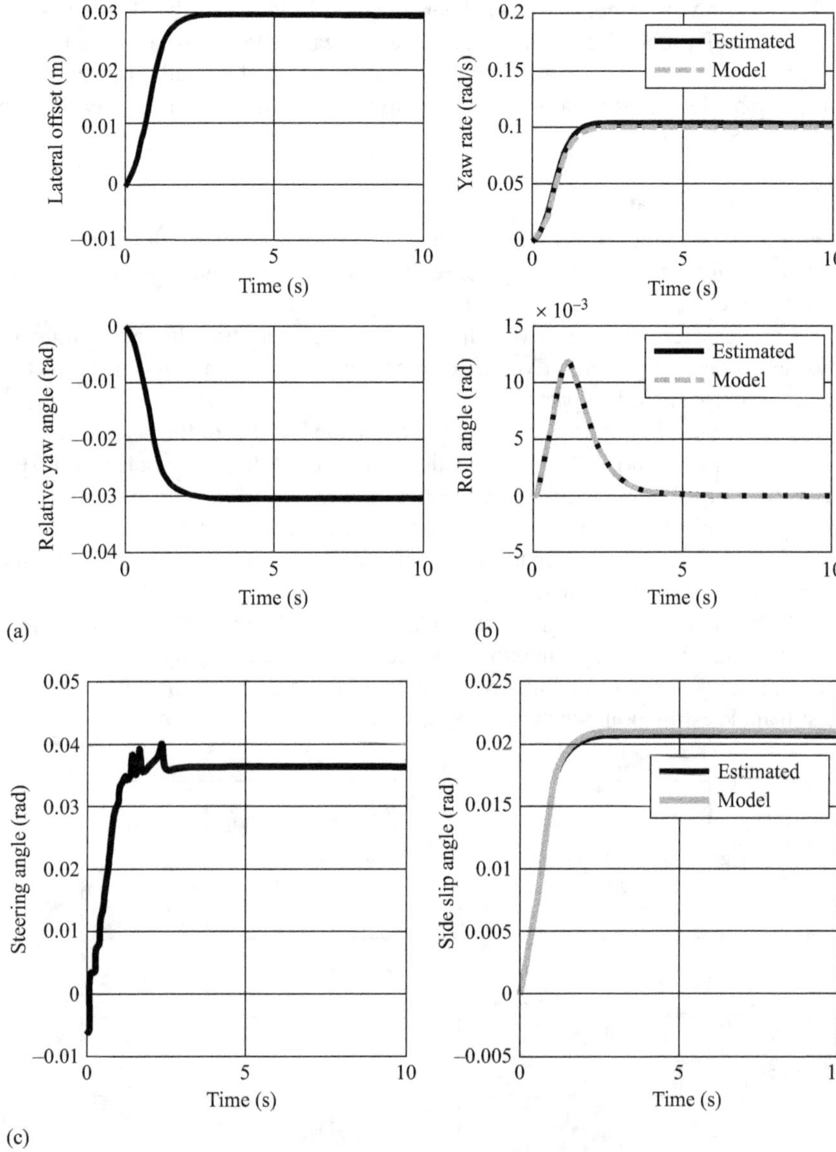

Figure 5.22 Simulation results for a lane keeping maneuver: second test. (a) Lateral offset and relative yaw angle: second test; (b) Yaw rate and roll angle: second test; (c)): Steering and side slip angle: second test

From these last results, one can then conclude that the combination between estimation and control developed system is well worked, since the vehicle is maintained in the center of the road, avoiding then the lane departure of the HV.

5.5 Conclusion

In this chapter, a steering control system was developed in order to avoid the rollover and lane departure of heavy vehicle. It is based on control of lateral acceleration of the vehicle. The lateral acceleration limit is computed using the LTR, given in Ackerman expression. The LTR limit is set to be equal to 0.9. Previously, an estimator based on sliding mode was implemented. It is made possible to estimate the unmeasured dynamics of the vehicle. Lateral acceleration and roll angle estimation are presented in this work. Comparing to existing method, the proposed approach is based on robust controller-estimator.

A real chicane scenario is tested and presented. The results showed that this system is effective and made it possible to control the vehicle and to avoid its rollover. The lateral acceleration limit is stabilized. Therefore, the steering angle of the vehicle is modified in order to force this latter to stay on a safety trajectory. The LTR is maintained around a maximum value of 0.9. The experimental tests done on an instrumented truck showed the quality of this approach since the convergence of the observer is quick and it occurred in finite time.

A second ramp test is presented in this paper. Compared to the first one, the identification process has been carried out with errors quite close to zero and the controller is activated quickly in order to avoid the accident.

The second part of this chapter is devoted to develop a steering control in order to avoid the lane departure of heavy vehicle. It is based on the control of lateral position. In this case also, the estimation the dynamic states of the vehicle such as lateral acceleration, side slip angle, roll angle and yaw rate is necessary. Simulation results show that the states estimation using sliding mode observer is of quality.

Combining this observer with the steering controller based on sliding mode control, permits to maintain the heavy vehicle at the center of the road and therefore avoid the lane departure, at less than 2 s.

The next work will be also devoted to the application of the developed approach to the tractor-semi trailer model. In this case, both the tractor and the trailer should be controlled in order to avoid lane departure. In this case, some other outputs should be considered, namely the outputs coming from sensors to be installed in the trailer in order to measure its dynamics, such as using the dynamo wheel in order to measure the impact forces, which will be the reference for a better validation of the estimation.

Acknowledgement

This work was developed by the French IFSTTAR laboratory (ex LCPC: Laboratoire Central des Ponts et chaussées) in collaboration with French industrial partners,

Renault Trucks, Michelin and Sodit in the framework of French project VIF (Véhicule Lourd Interactif du Futur). This work was supported by the French Ministry of Industry and the Lyon Urban Trucks & Bus competitiveness cluster; the authors gratefully acknowledge their contributions. Hocine Imine and Leonid Fridman gratefully acknowledge the financial support from by Programa deApoyo a Proyectos de Investigacion e Innovacion Tecnologica (UNAM) 113216, DGAPA PASPA Program and CLOVER H2020-EU:734832 project of European Union.

Antonella Ferrara gratefully acknowledges the financial support by ITEAM Project, under the Marie Skodowska-Curie grant agreement No. 675999.

References

[1] T.V. Fossum and G.N. Lewis, "A mathematical model for trailer-truck jackknifing", *SIAM Review*, vol. 23, no. 1, pp. 95–99, 1981.

[2] M. Bussières and B. Falah, "Outil permettant d'analyser la stabilité des véhicules lourds", Rapport Technique du Service de la Normalisation Technique, Ministère des Transports du Québec, 2001.

[3] D.P. John, P. Baas, D. Hutchinson and D. Lalasih, "Including performance measure in dimensions and mass regulations", *Seventh International Symposium on Heavy Vehicle Weights & Dimensions*, Delft, the Netherlands, Europe, June 16–20, 2002.

[4] M. Haldane and J. Bunker, "Assessing the impacts of multi-combination vehicles on traffic operations and safety", A Literature Review, Queensland Government, Department of Main roads.

[5] W.H. Douglas, B.P. Ingrid, J.T. Darren and D.G. William, "Highway/heavy vehicle interaction", A Synthesis of Safety Practice, Commercial Truck and Bus Safety, July 2003.

[6] B. Johansson and M. Gafvert, "Untripped SUV rollover detection and prevention", *43rd IEEE Conference on Decision and Control*, pp. 5461–5466, Atlantis, Paradise Island, Bahamas, December 14–17, 2004.

[7] H. Imine, S. Srairi, D. Gil and J. Receveur, "Heavy vehicle, vehicle management center and infrastructure interactivity to manage the access to infrastructure", *ICHV'08, International Conference on Heavy Vehicle*, May 19–22, 2008, Paris, France.

[8] H. Imine, A. Benallegue, T. Madani and S. Srairi, "Rollover risk prediction of an instrumented heavy vehicle using high order sliding mode observer", *ICRA'09, IEEE International Conference on Robotics and Automation*, pp. 523–527, May 12–17, 2009, Kobe, Japan.

[9] H. Imine and V. Dolcemascolo, "Rollover risk prediction of heavy vehicle in interaction with infrastructure", *International Journal of Heavy Vehicle Systems*, vol. 14, no. 3, pp. 294–307, 2007.

[10] S. Mammar, "Tractor-semitrailer lateral control robust reduced order two-degree-of-freedom", *Proceedings of the American Control Conference*, San Diego, California, June, 1999.

[11] J. Wang and M. Tomizuka, "Robust H{221e} lateral control of heavy-duty vehicles in automated highway system," *IEEE American Control Conference*, San Diego, California, USA, pp. 3671–3675, June 2–4, 1999.
[12] M. Tomizuka, M. Tai, J.-Y. Wang and P. Hingwe, "Automated lane guidance of commercial vehicles", *International Conference on Control and Applications*, Koala Coast-Island of Hawaii, USA, pp. 1359–1364, August 22–21, 1999.
[13] W.H. Douglas, B.P. Ingrid, J.T. Darren and D.G. William, "Highway heavy vehicle interaction", *Transportation Research Board, A Synthesis of Safety Practice, Commercial Truck and Bus Safety*, vol. 1, Washington, DC, USA, July 2003.
[14] J. Ackermann and D. Odenthal, "Damping of vehicle roll dynamics by speed-scheduled active steering", *Proceedings of the European Control Conference*, Karlsruhe, August 31–September 3, 1999.
[15] D. Odenthal, T. Bunte and J. Ackermann, "Nonlinear steering and braking control for vehicle rollover avoidance", *Proceedings of the European Control Conference*, Karlsruhe, August 31–September 3, 1999.
[16] H. Imine and V. Dolcemascolo, "Sliding mode observers to heavy vehicle vertical forces estimation", *IJHVS, International Journal of Heavy Vehicle Systems*, vol. 15, no. 1, pp. 53–64, 2008.
[17] A.J.P. Miège, "Active roll control of an experimental articulated vehicle", PhD Thesis, Department of Engineering, University of Cambridge, UK, 2003.
[18] C. Kim and P.I. Ro, "A sliding mode controller for vehicle active suspension systems with non-linearities", *Proceedings of the Institution of Mechanical Engineers, Part D: Journal of Automobile Engineering*, vol. 212, pp. 79–92, 1998.
[19] O. Khemoudj, H. Imine, M. Djemai and L. Fridman, "Variable gain sliding mode observer for heavy duty vehicle contact forces", *IEEE Variable structure systems conference, VSS*, June 26–28, 2010, Mexico.
[20] O. Khemoudj, H. Imine, and M. Djemai, "Robust observation of tractor–trailer vertical forces using model inverse and numerical differentiation", *International Journal of Materials and Manufacturing*, vol. 3, no. 1, pp. 278–289, August, 2010.
[21] L.R. Ray, "Nonlinear state and tire force estimation for advanced vehicle control", *IEEE Transactions on Control Systems Technology*, vol. 3, pp. 117–124, 1995.
[22] P. M. Siegrist, "A methodology for monitoring tyre-forces on off-highway mining trucks", PhD Thesis, University of Queensland, 2003.
[23] H. Imine and O. Khemoudj, "Dynamic parameters identification and estimation of the vertical forces of heavy vehicle", *SAE'11 World Congress on Automotive*, Detroit, USA, April 13–15, 2011.
[24] L. Davis and J. Bunke, "Suspension testing of 3 heavy vehicles-methodology and preliminary frequency analysis", report, Queensland University of Technology and Queensland Department of Mainroads, 2008.
[25] C. Blanksby, R. George, B. Peters, A. Ritzinger and L. Bruzsa, "Measuring dynamic wheel loads on tri and quad axle groups", *Proceedings of International Conference on Heavy Vehicles*, pp. 223–236, Paris, France, 2008.

[26] A. Tuononen, "On-board estimation of dynamic tyre forces from optically measured tyre carcass deflections", *International Journal of Heavy Vehicle Systems*, vol. 16, no. 3, pp. 362–378, 2009.

[27] Fatigue Strength Measurement from Kistler – Peak Performance in Endurance Mode, Kistler, Suisse. https://www.kistler.com/fr/en/applications/automotive-research- test/vehicle-dynamics-durability/durability-testing/

[28] H. Imine, M. Djemai, O. Khemoudj, and A. Germanchev, "Sliding mode observers for tyre vertical forces estimation: experimental results using ALF", *HVTT12, International Symposium on Heavy Vehicle Transport Technology*, Stockholm, Sweden, September 16–19, 2012.

[29] Automotive simulator PROPSER, from OKTAL. http://www.oktal.fr/en/automotive/range-of-simulators/range-of-simulators.

[30] H. Imine, "Plan d'expérience, instrumentation, traitement des signaux et vérification du système au laboratoire", rapport Projet VIF2, Juin 2007.

[31] H. Imine, S. Srairi and T. Madani, "Experimental validation of rollover risk prediction of heavy vehicles", *TRA'08, Transport Research Arena*, April 21–25, 2008, Ljubljana, Slovenia.

[32] Y. Delanne, V. Schmitt and V. Dolcemascolo, "Heavy truck rollover simulation", *18th International Conference on the Enhanced Safety of Vehicles*, Nagoya, Japan, pp. 1233–1237, 2003.

[33] F. Floret-Pontet and F. Lamnabhi-Lagarrigue, "Parameter identification methodology using sliding mode observers", *International Journal of Control*, vol. 74, no. 18, pp. 1743–1753, 2001.

[34] L. Fridman, A. Levant and J. Davila, "High-order sliding-mode observation and identification for linear systems with unknown inputs", *45th Conference on Decision in Control*, San Diego, 2006.

[35] I. Boiko and L. Fridman, "Analysis of chattering in continuous sliding-mode controllers", *IEEE Transactions on Automatic Control*, vol. 50, no. 9, pp. 1442–1446, September, 2005.

[36] H. Imine, L. Fridman, H. Shraim and M. Djemai, "Sliding mode based analysis and identification of vehicle dynamics", *Lecture Notes in Control and Information Sciences*, vol. 414, Springer-Verlag, London, 2011.

[37] A. Levant, "Robust exact differentiation via sliding mode technique", *Automatica*, vol. 34, no. 3, 379–384, 1998.

[38] L. Fridman, "The problem of chattering: an averaging approach", in K.D. Young and U. Ozguner, editors, *Variable Structure, Sliding Mode and Nonlinear Control*, Lecture Notes in Control and Information Science, vol. 247, pp. 363–386, Springer-Verlag, London, 1999.

[39] V.I. Utkin, *Sliding Modes in Control and Optimization*, Springer, Berlin, 1992.

[40] M.L. Hautus, "Strong detectability and observers", *Linear Algebra and Its Applications*, vol. 50, pp. 353–368, 1983.

[41] H. Imine, S. Srairi and T. Madani, "Experimental validation of rollover risk prediction of heavy vehicles", *TRA'08, Transport Research Arena*, Ljubljana, Slovenia, April 21–25, 2008.

Chapter 6
Sliding mode approach in semi-active suspension control

Dzmitry Savitski[1], Dmitrij Schleinin[1], Valentin Ivanov[1], and Klaus Augsburg[1]

Ride of the vehicle envelops the heave, pitch and rolling motion in forced vibration caused by road unevenness and roughness. Ride comfort is mainly concerned with the ability of chassis to cope with the various forms of the vertical dynamic excitation, which are unpleasant to the driver and passengers. Such excitations can be suppressed by the proper adjustment of passive suspension or control of semi-active and active suspension systems. This chapter introduces recent developments in the area of suspension control methods, overviews basics of vehicle vertical dynamics modelling, provides survey on existing control methods and proposes new hierarchical control architecture to reduce vertical, pitch and roll accelerations during driving over uneven road. This control strategy is based on application of integral sliding mode approach and optimal distribution of virtual demand between four semi-active shock absorbers. Obtained simulation results confirm positive effects in ride comfort compared to the vehicle with passive suspension and continuous Skyhook control.

6.1 Introduction

Ride of the vehicle envelops the heave, pitch and rolling motion in forced vibration caused by road unevenness and roughness. Ride comfort [1] is mainly concerned with the ability of chassis to cope with the various forms of the vertical dynamic excitation, which are unpleasant to the driver and passengers. Ride comfort [2] is related to the range of 0–25 Hz, while higher frequencies of 25 Hz–20 kHz are considered as noise. The ride comfort can be divided into three categories [3], namely, Primary-Ride (vehicle body motion such as pitch, roll and heave in range from 0 to 5 Hz), Secondary-Ride (5–25 Hz), Harshness and Noise (20–500 Hz).

A suspension system is responsible for the ride quality of the vehicle. The suspension consists of components providing spring linkages between sprung and unsprung

[1]Automotive Engineering Group, Department of Mechanical Engineering, Technische Universität Ilmenau, Ehrenbergstr. 15, 98693 Ilmenau, Germany

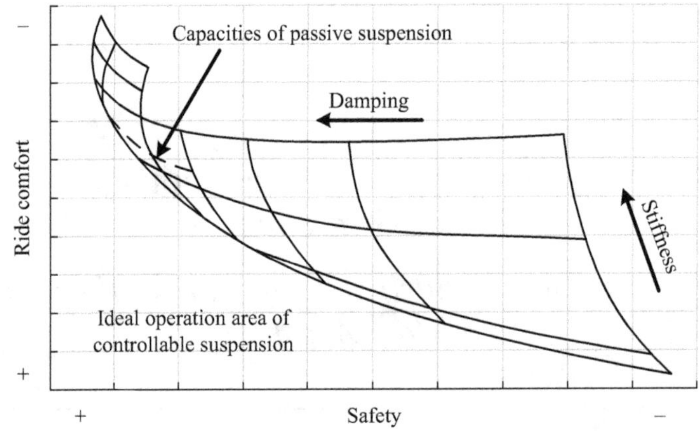

Figure 6.1 Influence of suspension parameters on comfort and safety

masses of the vehicle. Hence, the suspension reduces dynamic loads influencing the sprung mass. Generally, the automotive suspension includes the spring, linkage and damping parts. The spring part transfers vertical forces from the road to the vehicle sprung mass in order to reduce the dynamic loading and to improve the ride quality in general. The damping part (usually a shock absorber) quenches oscillations of the vehicle body and the wheels by converting the energy of oscillations into the heat energy and its subsequent dissipation in environment. The linkage part receives longitudinal and lateral forces and torques on the wheel. Kinematics of the guide part defines the character of wheel motion relative to the vehicle undercarriage. The suspension can also be advanced with roll bars responsible for improving the vehicle roll stability through reduction of the vehicle roll angle. Generally, the suspension either has an axle or independent design. By the axle suspension, the motion of one wheel on the axle depends on the motion of another wheel. In this case the forces and torques are transferred from wheels to the vehicle body through bow springs or rods. By the independent suspension, wheels on the same axle have independent motion. The independent suspension can be of a wishbone, telescoping and McPherson-type. Detailed information about the suspension design can be found in [4–6].

Within the framework of this chapter, the focus will be given to electronically controlled suspensions. Damping characteristics of such suspensions can be automatically adjusted to current road conditions and dynamics of the vehicle manoeuvre. The goal of the suspension control is not only overall improvement of the ride quality but also finding a reasonable compromise between the ride and handling performance of the vehicle for actual driving conditions. As can be seen from Figure 6.1, increased damping is better for handling, but reduced damping allows improving the ride quality.

The suspension control is realized through adjustment of shock absorber damping characteristics that can be done in a semi-active or active mode. In the semi-active suspension, power from the wheel motion caused by the road profile is absorbed by the controlled shock absorber. Hence, a pure dissipation process takes place here. Contrastingly, the active suspension has added power from external

Table 6.1 Characteristics of passive, semi-active and active suspension systems [7]

Type	**Passive**	**Adaptive**	**Semi-active**	**Slow active**	**Full active**
Bandwidth	–	<1 Hz	<5 Hz	<5 Hz	<25 Hz
Force–velocity diagram	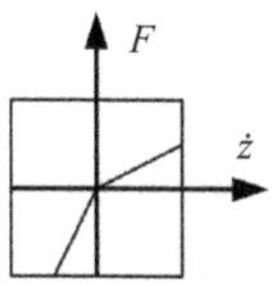	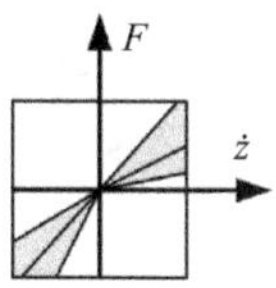	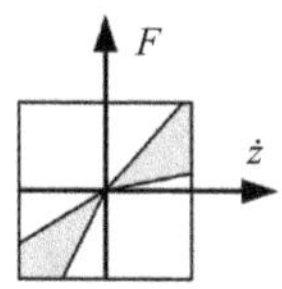	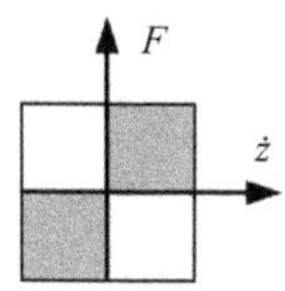	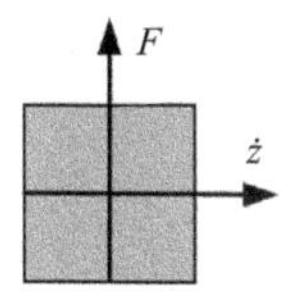
Scheme	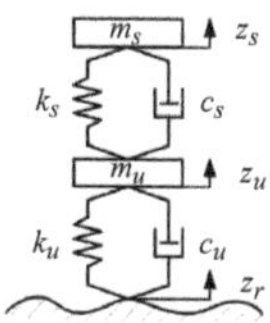	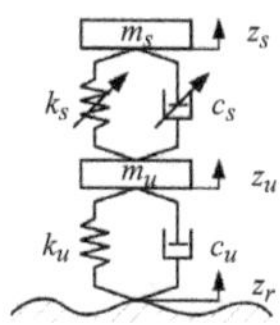	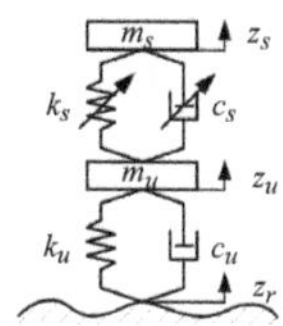	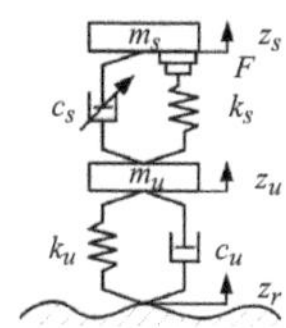	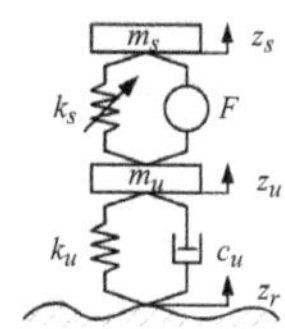
Energy demand	–	Low	Low	Medium	High

sources. The shock absorbers as actuators of semi-active or active suspension usually can be of hydraulic, pneumatic, electro-rheological and magneto-rheological type. Some recent studies are proposing also electric motors as possible damping actuators, but such technology is not sufficiently matured now for the implementation on mass-production vehicles. General comparison of several suspension types is summarized in Table 6.1. Next sections provide a short overview of recent solutions for semi-active and active suspension.

6.2 Semi-active suspension system

The overall functional target of a semi-active suspension is to achieve maximum possible ride quality for actual road conditions through alteration of the damping without deterioration of the vehicle handling. For further consideration, general terms ‘low damping’ and ‘high damping’ will be used to describe boundary characteristics of shock absorber.

The required damping level is closely connected with the wheel excitation frequency. In a simple case, semi-active suspension can be classified as high- and low-frequency bandwidth systems [8]. Within the framework of this classification, low frequencies of 3–4 Hz can be covered by most of the semi-active suspension systems. Systems with high-frequency bandwidth are dedicated for operating also at frequencies of 10–12 Hz. However, the low or high damping level cannot be simply attributed to a particular frequency band. As was illustrated in [10], four characteristic regions can be allocated for the frequency band by the tyre vertical force oscillations, Figure 6.2. It can be seen that the low and high damping calls for different tyre behaviour in each region. Following the recommendation from [10], the high damping

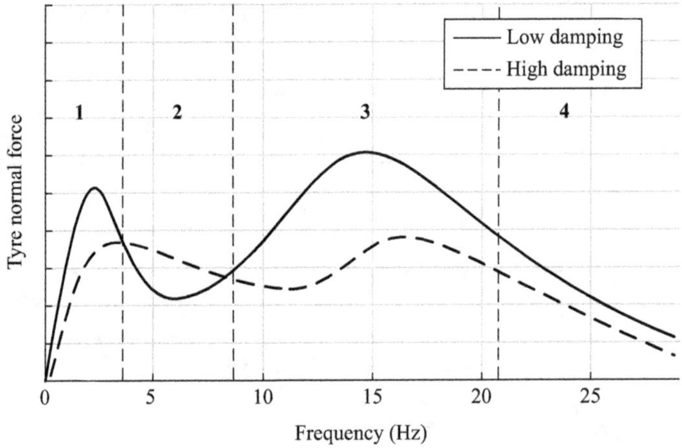

Figure 6.2 Tyre vertical force variation in different frequency domains

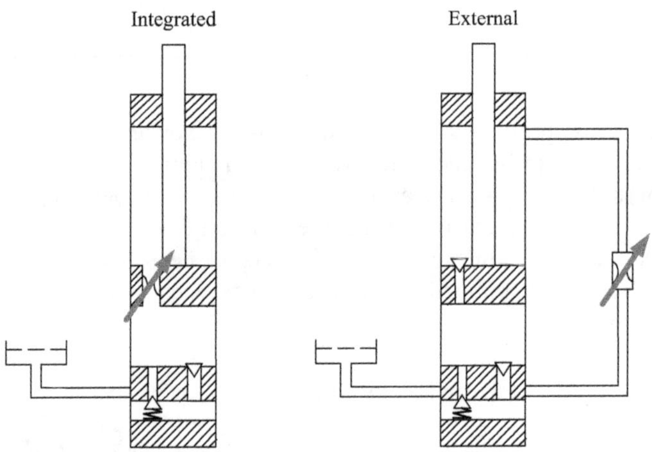

Figure 6.3 Schematic representation of CDC shock absorber (reproduced from [9])

should be selected for the regions 1 (1–2 Hz, corresponds to the sprung mass mode) and 3 (8–20 Hz, where unsprung mass resonance can be expected). In the intermediate ride dynamics (2–8 Hz) or harshness (>20 Hz) modes the low damping has to be selected.

Constructively the damping level can be controlled through the change of:

- the aperture related to the fluid flow;
- the fluid viscosity in the shock absorber or the strut.

The aperture-related control in a shock absorber can be usually realized through (i) valve with the variable orifice or (ii) bypass valve with the solenoid, Figure 6.3.

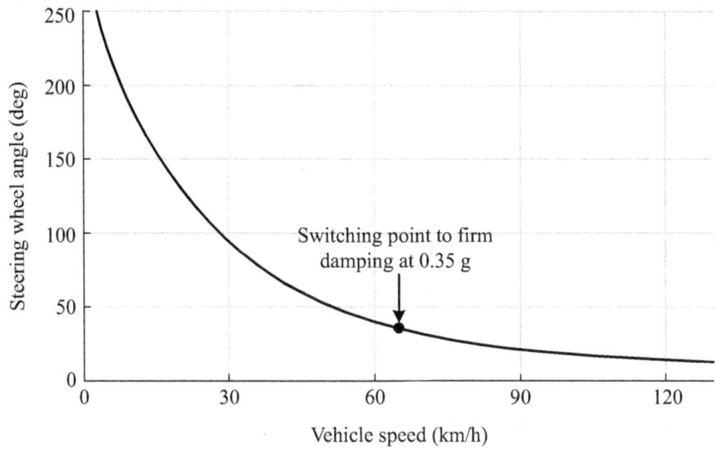

Figure 6.4 Example of switching thresholds of semi-active shock absorber (reedited from [10])

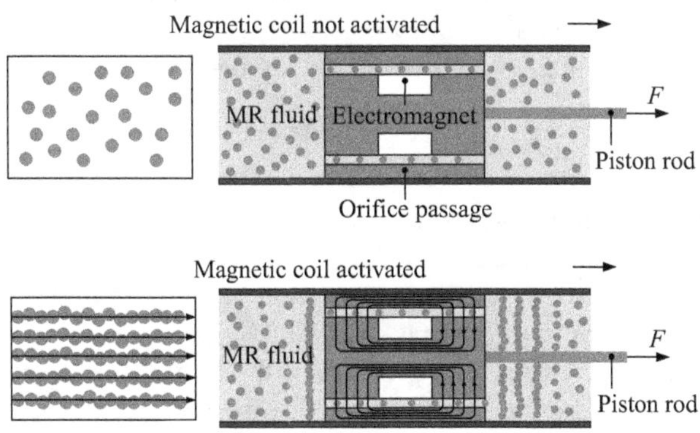

Figure 6.5 Principle of magneto-rheological damping

Figure 6.4 shows an example of operational characteristic of the semi-active shock absorber with the aperture-related control. Such design of the semi-active suspension received a wide distribution for mass-production vehicles. One of the most popular relevant variants is the Continuous Damping Control (CDC) developed by ZF Sachs company. Its main element is the magnetically controlled proportional valve, which changes the damping in accordance with the actual driving situation using the Skyhook algorithm [11]. The vehicle dynamics is estimated in the CDC-controller on the basis of information provided by the acceleration sensors of wheels and vehicle body.

Semi-active suspensions using the shock absorbers of variable fluid viscosity were introduced for the serial vehicles from early 2000s. The electro-rheological or the magneto-rheological fluid is used in such shock absorbers. Figure 6.5 shows the

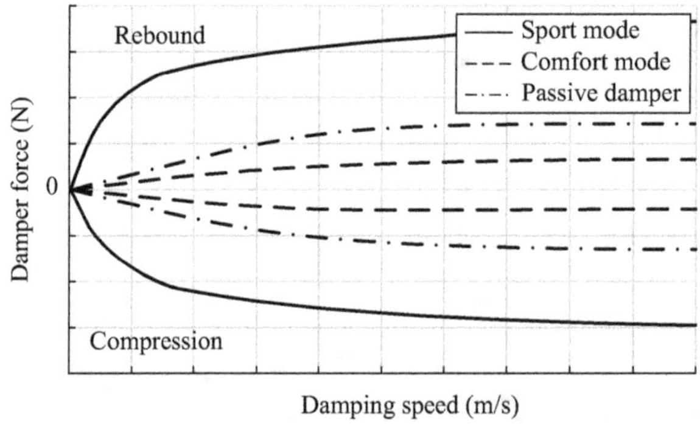

Figure 6.6 Example of operational characteristic of a magneto-rheological shock absorber

principle of magneto-rheological damping. Figure 6.6 gives the example of typical operational characteristic of a magneto-rheological shock absorber. The semi-active suspension based on the variable fluid viscosity principle still have certain technological and cost-related limitations and, therefore, are mainly considered for sports cars and the luxury vehicles.

6.3 Active suspension system

The step from semi-active systems with variation of damping and stiffness characteristics to active control functions was done with introduction of electromechanical and electrohydraulic actuators. The problem of the fully active suspension is relatively high energy consumption, especially in the case of electromechanical approach. Nevertheless, some suspension actuators can also recuperate energy from road oscillations contributing to better energy efficiency of the system. Due to these advantages, active suspension systems are very demanded by carmakers [12].

The active valves in the *electrohydraulic shock absorber* allows controlling the damping ratio or even generating the force. The ACOCAR active suspension developed by Tenneco utilizes electrohydraulic actuation principle and was patented in 1997 [13]. This development allows controlling wheel and body motion up to 12 Hz and higher. Besides that the ACOCAR design is more energy-efficient and compact compared to the previously known solutions. These advantages were confirmed through experimental tests on test rig and vehicle [14]. Another variant was proposed by the Levant Power patented the electrohydraulic system GenShock in 2014 [15]. Besides the main functions related to ride dynamics and comfort, this system provides also possibility to recuperate energy [16]. One of the drawback of the electrohydraulic design is the low system bandwidth as compared to electromechanical solutions, which will be described further.

Table 6.2 Comparison of electromechanical and electrohydraulic active suspension system [7]

Distance	**Electromechanical**		**Hydraulic system**
	Current (A)	**Forecasted additional consumption (l/100 km)**	**Additional consumption (l/100 km)**
Rough test track	28	0.5	1.0
Rough country road	8	0.2	0.6
Highway 160 km/h	2.5	0.15	0.4

The main advantage of active *electromechanical actuators* is a higher system bandwidth compared to other actuator variants. Nevertheless, such benefits in control performance lead at the same time to higher energy consumption. Nowadays, several configurations of such systems were developed but did not find application on serial vehicles.

For example, the electromagnetic suspension developed by TU Eindhoven and SKF has showed promising results on the test rig [17]. These actuators were designed for the BMW 5 Series vehicle, but at the moment their functionality was not fully validated on the road tests. Another variant of active suspension developed by ZF and Volkswagen is aimed first of all at minimizing energy consumption simultaneously providing agile system dynamics [7,18]. This system already tested on the vehicle. The corresponding benchmarking [7] confirms that energy consumption of electromechanical actuator in comparison with electrohydraulic system can be reduced. Table 6.2 shows positive results by this criteria expressed in l/100 km.

Moreover, an electromechanical active suspension is also able to regenerate the energy from vertical vehicle oscillations.

Thus, the choice between electromechanical and electrohydraulic approaches leads to the trade-off between the energy consumption and performance.

6.4 Methods on suspension control

Invention of controllable suspension systems produced a demand in development of appropriate control strategies. They need to consider not only predefined control objectives regarding ride quality but also type of the selected suspension actuator. In this section a basic knowledge about vehicle suspension models and the most-known control strategies will be given.

6.4.1 Vehicle modelling

Nowadays various vehicle models are used in simulation and for formulation of the suspension control laws. These models can be very simple, like quarter car model, as well as more complex, e.g., with consideration of multi-body dynamics of chassis elements. A trade-off between precision and complexity should be kept in balance depending on the control task and objectives. In terms of understanding vehicle dynamics, it is important to keep in mind basics of vertical vehicle motion starting from very simple mathematical models. Good examples of such models and analysis can be found in [19,20]. These sources are used for basic analysis introduced below.

6.4.2 Single degree of freedom model

Single degree of freedom (DoF) system is the simplest representation of the vehicle vertical motion, Figure 6.7. Here the vehicle body is interacting with the road surface through the tyre considered as the stiffness and damping element.

Motion of the sprung mass in this vibration system is governed by the following equation:

$$m_s\ddot{z}_s + c_t(\dot{z}_s - \dot{z}_r) + k_t(z_s - z_r) = 0, \tag{6.1}$$

where m_s is the sprung mass, c_t and k_t are, respectively, damping and stiffness of the tyre.

Neglecting initial displacement z_r, this equation has a general solution:

$$z_s = Ze^{st}, \tag{6.2}$$

where s is the Laplace operator and Z is the amplitude.

Substituting (6.2) in (6.1) leads to

$$(m_s s^2 + c_t s + k_t)Ze^{st} = 0. \tag{6.3}$$

Solution of this characteristic equation produces two roots:

$$s_{1,2} = -\frac{c_t}{2m_s} \pm \sqrt{\left(\frac{c_t}{2m_s}\right)^2 - \frac{k_t}{m_s}}. \tag{6.4}$$

Considering initial conditions for deriving constants C_1 and C_2, a general solution of the system is:

$$z_s(t) = C_1e^{s_1t} + C_2e^{s_2t}. \tag{6.5}$$

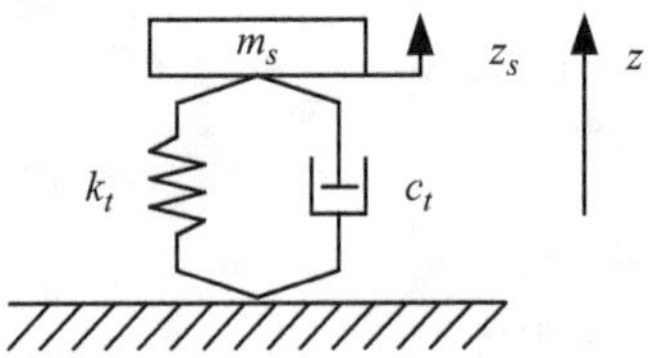

Figure 6.7 Single degree of freedom model

Natural frequency of the system is defined as

$$\omega_n = \sqrt{\frac{k_t}{m_s}}. \tag{6.6}$$

Considering natural frequency expressed in the previous equation, the critical damping can be defined as

$$c_c = 2m_s\omega_n. \tag{6.7}$$

Therefore, for analysing border damping conditions, a damping ratio can be represented as

$$\xi = \frac{c_t}{c_c}. \tag{6.8}$$

Depending on the value of damping ratio ξ, three special cases of system behaviour are determined:

- underdamped if $\xi < 1$;
- critically damped if $\xi = 1$;
- overdamped if $\xi > 1$.

Such classification allows one to understand the importance of proper stiffness and damping adjustment in a vibration system. Neglecting initial displacement z_r and setting initial displacement $z_s = 0.1$, behaviour of the system with $m_s = 1$ kg is represented in Figure 6.8 in time domain. As can be seen, the underdamped system with $\xi = 0$ never comes to the equilibrium position oscillating periodically over it. In reality its value cannot be equal to zero as there is always energy dissipation in physical systems. The overdamped system with infinite value of $\xi = \inf$ remains at the initial displacement. High values of damping $\xi > 5$ lead to very slow dynamics of

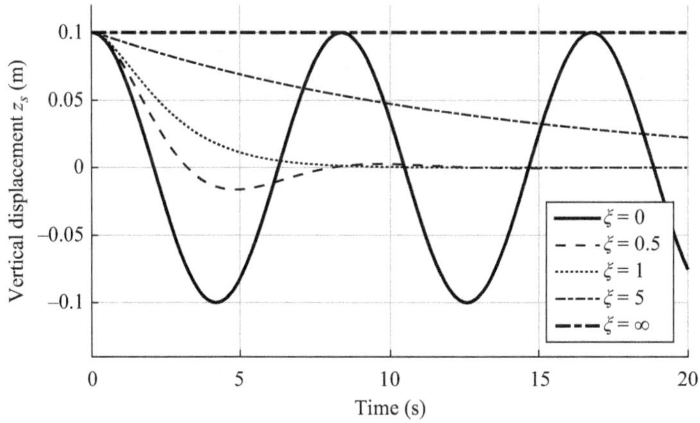

Figure 6.8 System behaviour in time domain with different damping ratio

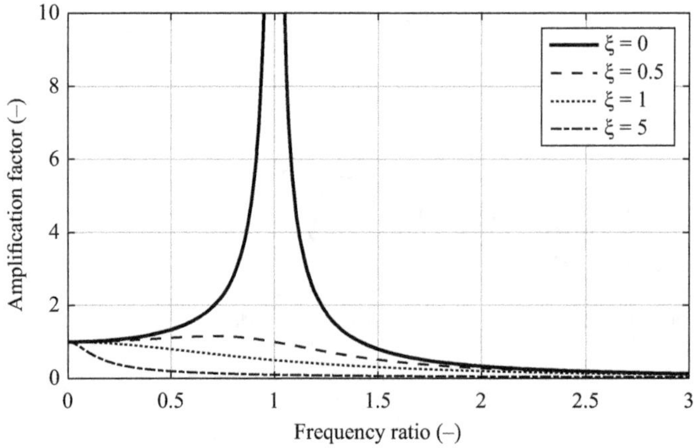

Figure 6.9 Amplification factor over frequency ratio

the system whereas more realistic exponential tending to equilibrium is characterized by cases of $0 < \xi < 1$.

Besides analysis in time domain, it is always important to assess system behaviour as a response to the periodical excitation with different frequencies. In fact it represents the frequency response function of the system.

The system (6.1) can be written as:

$$m_s\ddot{z}_s + c_t\dot{z}_s + k_t z_s = f, \tag{6.9}$$

where $f = k_s z_r + c_s\dot{z}_r$ is the excitation force applied to the system. In our case this is the force occurring due to the road roughness.

Therefore, considering non-zero excitation force, a forced vibration problem can be formulated. To analyse the response of the system due to excitation, the amplification factor may be analysed and calculated as:

$$\frac{X}{F/k} = \frac{1}{\sqrt{(1-\beta^2)^2 + (2\psi\beta)^2}}, \tag{6.10}$$

where β is the frequency ratio expressed as the relation between actuation frequency and natural frequency of the system.

From (6.10) the system response is shown as the amplification factor in Figure 6.9. As can be seen, response of the system with $\xi = 0$ at the natural frequency comes to an infinity producing resonance. Such huge increase of oscillations amplitude has a negative influence on mechanical systems and human and has to be reduced. Increase of damping ratio solves this problem by decreasing an amplitude at the resonance frequency.

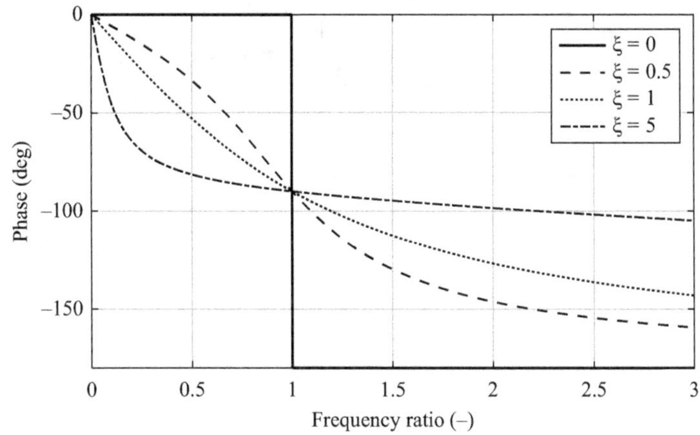

Figure 6.10 Phase angle of the system response

To assess the delay of the system response to the excitation force, a phase is derived as:

$$\theta = \operatorname{atan}\left(\frac{-\omega c_t}{k_t - \omega^2 m_s}\right). \tag{6.11}$$

Plotting the face against the frequency ratio, it can be observed in Figure 6.10 that for $\xi = 0$ the phase changes its value from 0 to $-180°$ at the resonance frequency. In theory this case is characterized by instantaneous phase shift whereas it becomes gradual with increase of the damping ratio.

6.4.3 Quarter car model

Another simple formulation of the vertical vehicle dynamics is the two DoF *quarter car model*, Figure 6.11. It includes dynamics of sprung and unsprung masses joined by spring and damper. Unsprung mass interacts with the road as the stiffness and damping elements.

A quarter car model has following mathematical representation:

$$\begin{cases} m_s\ddot{z}_s + c_s(\dot{z}_s - \dot{z}_u) + k_s(z_s - z_u) = 0 \\ m_u\ddot{z}_u + c_s(\dot{z}_u - \dot{z}_s) + k_s(z_u - z_s) + c_t(\dot{z}_u - \dot{z}_r) + k_t(z_u - z_r) = 0 \end{cases}, \tag{6.12}$$

where m_s is the sprung mass, m_u is the unsprung mass, k_s and c_s is the stiffness and damping of suspension system, k_t and c_t is the stiffness and damping of the tyre.

Considering solution $z = Ze^{i\omega t}$ and neglecting damping and force terms in (6.12), the following determinant is derived:

$$\begin{vmatrix} k_s - m_s\omega^2 & -k_s \\ -k_s & k_s + k_t - m_u\omega^2 \end{vmatrix} = 0. \tag{6.13}$$

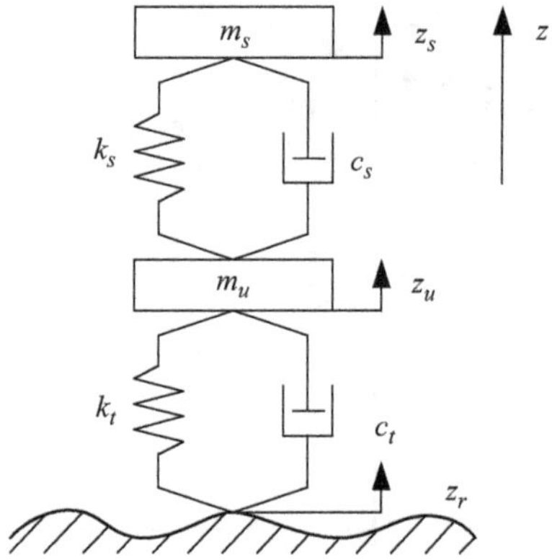

Figure 6.11 Quarter car model

Expanding this determinant leads to the characteristic equation:

$$\omega^4 m_s m_u - \omega(m_s k_s + m_s k_t + m_u k_s) + k_s k_t = 0. \tag{6.14}$$

Its solution allows finding natural frequencies:

$$\omega_{s,u} = \sqrt{\frac{m_s k_s + m_s k_t + m_u k_s \pm \sqrt{(m_s k_s + m_s k_t + m_u k_s)^2 - 4 m_s m_u k_s k_t}}{2 k_s k_t}}. \tag{6.15}$$

Using parameters of typical sport utility vehicle (SUV) and (6.15), natural frequency for sprung $f_s = 1.8$ Hz and unsprung $f_u = 12$ Hz mass can be calculated.

This system may be analysed in frequency domain representing three important criteria. First of them is the *transmissibility ratio*, which is the relation of unsprung mass travel to the excitation from the ground. It shows ability of the system to isolate vibrations coming from road disturbance. In Figure 6.12 two resonance peaks can be observed at frequencies of 1.8 and 12 Hz, which correspond to motion of sprung and unsprung mass, respectively. Oscillations can be reduced with higher damping amplitude of sprung mass whereas a lower damping ratio is required to reduce wheel hop in higher frequencies.

The suspension travel is limited in the vehicle by design and packaging specifications. To avoid both damage of the suspension elements and deterioration of ride comfort, suspension should operate in "safe" range of suspension travel. With increase of damping the suspension travel reduces, as can be seen in Figure 6.13. Compared to the transmissibility ratio, such effect can be observed through the entire frequency range.

Suspension system is responsible not only for the ride comfort but also for handling and safety properties of the vehicle. This may be assessed by dynamic tyre deflection, Figure 6.14. It can be seen that a higher damping corresponds to better road holding ability.

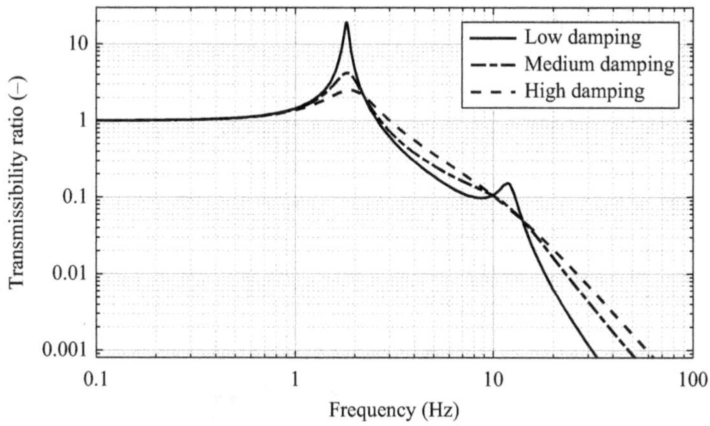

Figure 6.12 Transmissibility ratio

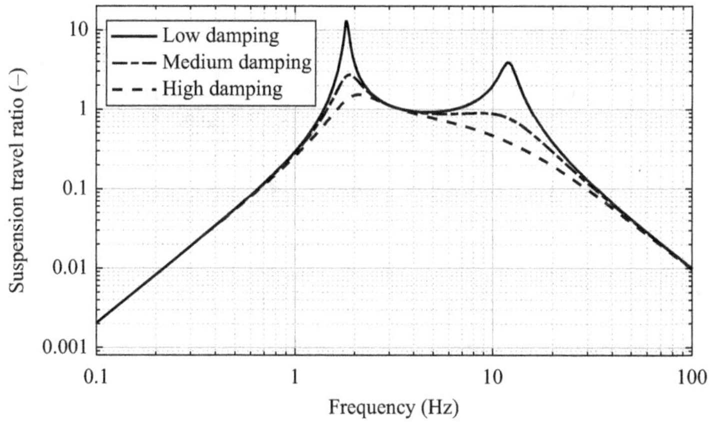

Figure 6.13 Suspension travel ratio

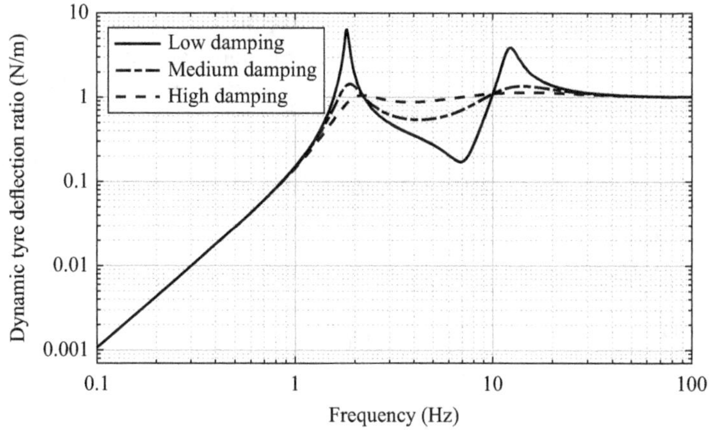

Figure 6.14 Dynamic tyre deflection ratio

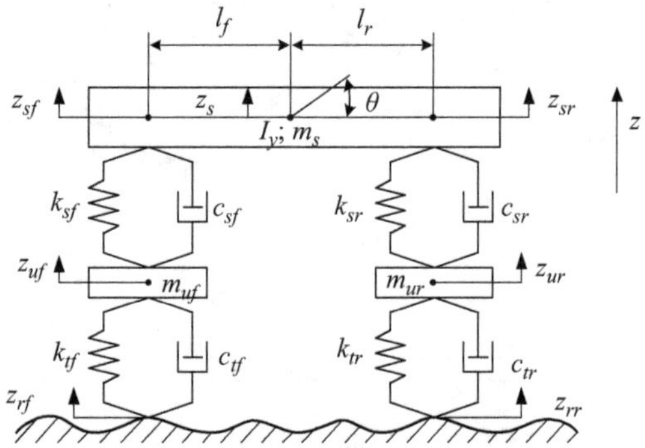

Figure 6.15 Half car model

6.4.4 Half car model

Pitching is at least not less annoying for the driver than heave motion [19]. Hence, it is important to consider this motion component for the analysis of ride comfort. One of the simplest models, which includes pitch motion of vehicle body, is the *half car* model, Figure 6.15.

This model is governed by the following set of equations:

$$\begin{cases} m_s\ddot{z}_s = -c_{sf}(\dot{z}_{sf} - \dot{z}_{uf}) - k_{sf}(z_{sf} - z_{uf}) - c_{sr}(\dot{z}_{sr} - \dot{z}_{ur}) - k_{sr}(z_{sr} - z_{ur}) \\ I_y\ddot{\theta} = l_f c_{sf}(\dot{z}_{sf} - \dot{z}_{uf}) + l_f k_{sf}(z_{sf} - z_{uf}) - l_r c_{sr}(\dot{z}_{sr} - \dot{z}_{ur}) - l_r k_{sr}(z_{sr} - z_{ur}) \\ m_{uf}\ddot{z}_{uf} = c_{sf}(\dot{z}_{sf} - \dot{z}_{uf}) + k_{sf}(z_{sf} - z_{uf}) - c_{uf}(\dot{z}_{uf} - \dot{z}_{rf}) - k_{uf}(z_{uf} - z_{rf}) \\ m_{ur}\ddot{z}_{ur} = c_{sr}(\dot{z}_{sr} - \dot{z}_{ur}) + k_{sr}(z_{sr} - z_{ur}) - c_{ur}(\dot{z}_{ur} - \dot{z}_{rr}) - k_{ur}(z_{ur} - z_{rr}) \end{cases}, \quad (6.16)$$

$$\begin{cases} z_{sf} = z_s - l_f\theta \\ z_{sr} = z_s + l_r\theta \end{cases}, \quad (6.17)$$

where θ is the pitch angle, I_{yy} is the inertia of the sprung mass, l_f and l_r is the distance between vehicle centre of gravity and front and rear wheels respectively.

Considering a general solution $z = Ze^{i\omega t}$ and equations of heave and pitch motion from (6.16), a determinant is derived:

$$\begin{vmatrix} \frac{1}{m_s}(k_f + k_r) - \omega^2 & \frac{1}{m_s}(k_r l_r - k_f l_f) \\ \frac{1}{I_y}(k_r l_r - k_f l_f) & \frac{1}{I_y}(k_r l_r^2 + k_f l_f^2) - \omega^2 \end{vmatrix} = 0. \quad (6.18)$$

Expanding this determinant leads to the characteristic equation:

$$\omega^4 - \left(\frac{1}{m_s}(k_f + k_r) + \frac{1}{I_y}(k_r l_r^2 + k_f l_f^2)\right)\omega^4 + \left(\frac{1}{m_s}(k_f + k_r)\frac{1}{I_y}(k_r l_r^2 + k_f l_f^2) - \frac{1}{m_s I_y}((k_r l_r - k_f l_f))^2\right) = 0. \quad (6.19)$$

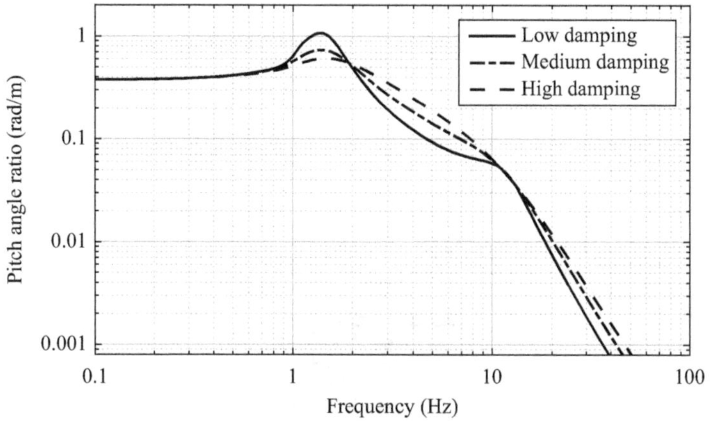

Figure 6.16 Pitch angle ratio

Natural frequencies can be found by solving (6.19):

$$\omega_{\theta,z}^2 = \frac{1}{2}\left(\frac{1}{m_s}(k_f + k_r) + \frac{1}{I_y}(k_r l_r^2 + k_f l_f^2)\right)$$
$$\pm\sqrt{\frac{1}{4}\left(\frac{1}{m_s}(k_f + k_r) - \frac{1}{I_y}(k_r l_r^2 + k_f l_f^2)\right)^2 + \frac{1}{m_s I_y}(k_r l_r - k_f l_f)^2} \tag{6.20}$$

Representing the response of the system in terms of pitch angle, it can be observed that it has similar behaviour as vertical vehicle body displacement, Figure 6.16. Increase of damping reduces magnitude of pitch oscillations.

6.4.5 Control methods overview

Application of semi-active and active suspension can bring potentially an effect around 30% in ride comfort and around 25% in safety compared to the passive suspension [21]. These improvements depend not only on the configuration of suspension components itself but also on the applied control strategy. In this section the most widespread control strategies for controlling semi-active and active shock absorbers are overviewed.

The first group of the control approaches is aimed at improvement of the *ride comfort* reducing the vertical acceleration of the vehicle body. For this purpose, a Skyhook approach was firstly introduced in [11]. This principle is based on the imaginary damper between the 'sky' and the vehicle body to reduce vertical body oscillations induced by road roughness, Figure 6.17. It is realized through transmitting the force from the imaginary damper to the physical semi-active or active damper.

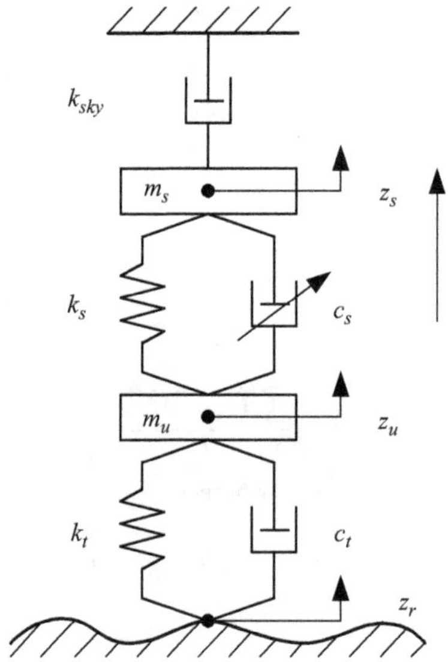

Figure 6.17 Schematic representation of Skyhook control principle

To fulfil aforementioned tasks, a two states Skyhook control strategy was introduced and applied to switchable semi-active damper [11]. This control is based on the following rule:

$$c_d = \begin{cases} c_{\max}, & \text{if } \dot{z}_s \dot{z}_{st} > 0 \\ c_{\min}, & \text{otherwise} \end{cases}, \tag{6.21}$$

where z_{st} is the suspension travel.

Controlling the semi-active damper, where no additional energy can be applied to the system, the Skyhook control adjusts the damping ratio based on body acceleration and suspension travel. Consequently, when the damper works in the rebound direction and therefore the velocity is negative, the resultant force pushes the vehicle body upwards. Reversed, the relative velocity is negative, when the damper works in the compression direction. Hence, the generated force pulls the vehicle body down.

The control law shown in (6.21) is applicable for the switchable dampers as soon this strategy utilizes two damper states. Its extension known as *continuous Skyhook* allows to consider intermediate damper characteristics and is applicable also to CDC dampers:

$$c_d = \begin{cases} sat\left(\frac{\alpha_{sh} c_{max} \dot{z}_{st} + (1-\alpha_{sh}) c_{max} \dot{z}_s}{\dot{z}_{st}}\right), & \text{if } \dot{z}_s \dot{z}_{st} > 0 \\ c_{min}, & \text{otherwise} \end{cases}, \tag{6.22}$$

where damping c_d is saturated as $c_d \in [c_{min}; c_{max}]$, α_{sh} is the tuning parameter. It can be varied between values of 0 and 1 where 0 turns the control to two states logic.

Similar to this classical approach, a *balance control* was proposed and discussed in [22,23]. It utilizes measured information about suspension travel and expressed as:

$$c_d = \begin{cases} c_{\min}\dot{x}_{st} + 2\alpha_b k_s |x_{st}| \operatorname{sign} \dot{x}_{st}, & \text{if } x_{st}\dot{x}_{st} \leq 0 \\ c_{\min}\dot{x}_{st}, & \text{otherwise} \end{cases} . \tag{6.23}$$

This control logic may be tuned by adjustment of balance parameter α_b. As a drawback of this method, the use of the spring stiffness k_s requires knowledge of this parameter in the control.

Another variant of the semi-active suspension control was proposed in [24] and known as *Acceleration Driven Damper Control (ADD)*, where the damping switching rule is governed by:

$$c_d = \begin{cases} c_{\max}, & \text{if } \ddot{z}_s \dot{z}_{st} > 0 \\ c_{\min}, & \text{otherwise} \end{cases} , \tag{6.24}$$

where $c_{\max}$ and $c_{\min}$ are the maximal and minimal damping of the shock absorber.

This control law is very similar to the classical two states Skyhook control but the acceleration $\ddot{z}_s$ is used instead of vertical velocity of vehicle body $\dot{z}_s$. Compared to Skyhook, this strategy considerably improves ride comfort in higher frequencies while has worse results in lower frequencies.

To overcome these drawbacks of ADD strategy, the *mixed Skyhook-ADD* control was proposed in [25] and represents following switching logic:

$$c_d = \begin{cases} c_{\max}, & \text{if } (\ddot{z}_s^2 - \alpha \dot{z}_s^2 \leq 0 \wedge \dot{z}_s \dot{z}_{st} \geq 0) \vee (\ddot{z}_s^2 - \alpha_{sh-add} \dot{z}_s^2\, 0 \wedge \dot{z}_s \dot{z}_{st} \geq 0) \\ c_{\min}, & \text{if } (\ddot{z}_s^2 - \alpha \dot{z}_s^2 \leq 0 \wedge \dot{z}_s \dot{z}_{st} < 0) \vee (\ddot{z}_s^2 - \alpha_{sh-add} \dot{z}_s^2\, 0 \wedge \dot{z}_s \dot{z}_{st} < 0) \end{cases} . \tag{6.25}$$

In this control law, α_{sh-add} is the crossover frequency at which frequency responses with Skyhook and ADD intersect. Including such switching allows using benefits both of Skyhook as well as of ADD in a defined frequency range.

Proposed *Power Driven Damper Control* addresses a problem of controlling suspension as a port Hamiltonian system. As soon as this approach is based on modelling the system with dissipative components, this approach fits well to the suspension control tasks. Derived control strategy calculates damping of shock absorber according to the following set of rules:

$$c_d = \begin{cases} c_{\max}, & \text{if } k\dot{z}_{st} z_{st} + c_{\max} \dot{z}_{st}^2 < 0 \\ c_{\min}, & \text{if } k\dot{z}_{st} z_{st} + c_{\max} \dot{z}_{st}^2 \geq 0 \\ -kx_{st}/\dot{x}_{st}, & \text{otherwise} \end{cases} , \tag{6.26}$$

where $c_{\max}$ and $c_{\min}$ are the maximal and minimal damping of the shock absorber. Advantageous in this strategy is elimination of chattering, which is specific to ADD and two states Skyhook strategies. It allows reducing vertical jerk of the vehicle body that improves ride comfort.

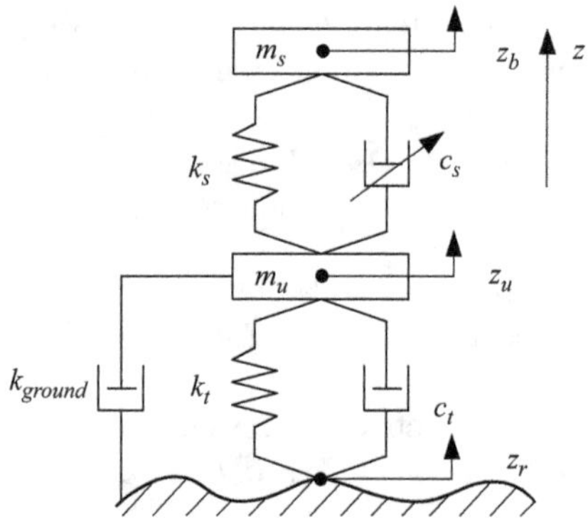

Figure 6.18 Schematic representation of Groundhook control principle

All these strategies are aimed at improved ride comfort by selecting damping of semi-active shock absorbers correspondingly. Another group of control strategies were introduced to minimize tyre deflection for a better road holding capability of the vehicle. This group of control methods is known as *Groundhook* control and was introduced in [26]. As shown in Figure 6.18, the imaginary damper is placed between the unsprung mass and the road to minimize tyre deflection. By analogy with Skyhook, this force is further transmitted to the semi-active suspension system to enhance road holding and vehicle stability.

A classical approach of the *Groundhook* control is well suitable to the two-state semi-active shock absorbers, where a control rule is governed by:

$$c_d = \begin{cases} c_{\max}, & \text{if } -\dot{z}_{st}\dot{z}_u < 0 \\ c_{\min}, & \text{otherwise} \end{cases}. \tag{6.27}$$

Similar to the Skyhook control, the *continuous Groundhook control* is governed by

$$c_d = \begin{cases} sat\left(\frac{\alpha_{gh} c_{\max} z_{st} + (1-\alpha_{gh}) c_{\max} \dot{z}_u}{\dot{z}_{st}}\right), & \text{if } -\dot{z}_u\dot{z}_{st} > 0 \\ c_{\min}, & \text{otherwise} \end{cases} \tag{6.28}$$

where α_{gh} is the tuning parameter.

Skyhook and Groundhook control strategies deal with one particular control objective that is applicable for a theoretical consideration but encounters problems in practice. In other words, improving comfort by abandoning road holding is counterproductive. Therefore, the control for semi-active and active suspension should at least achieve the same performance (as passive suspension) for one control objective by improving the other.

To overcome the shortcomings out of introduced control strategies, a fusion of both is presented in the literature, Figure 6.19. The often called *Hybrid control* gives a possibility to use Skyhook or Groundhook independently and in combination.

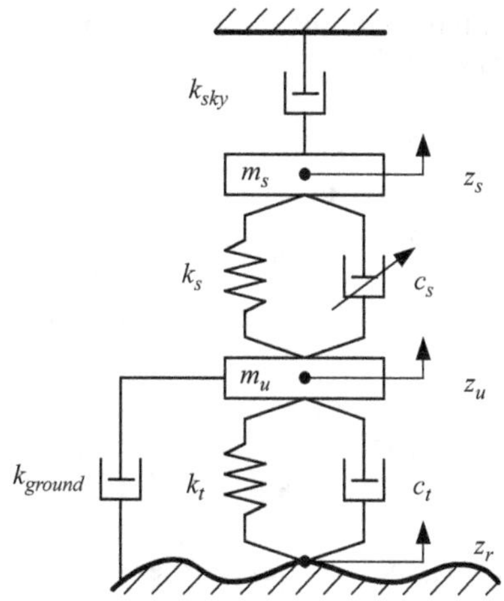

Figure 6.19 Schematic representation of Hybrid control principle

The mathematical expression of hybrid control is a linear combination of Skyhook and Groundhook control and is formulated as follows [27]:

$$c_d = G(\alpha_{hyb}\sigma_{sky} + (1 - \alpha_{hyb})\sigma_{gnd}), \tag{6.29}$$

with Skyhook and Groundhook components defined as

$$\sigma_{sky} = \begin{cases} \dot{x}_s, & \text{if } \dot{x}_s\dot{x}_{st} > 0 \\ 0, & \text{otherwise} \end{cases}. \tag{6.30}$$

$$\sigma_{gnd} = \begin{cases} \dot{x}_u, & \text{if } -\dot{x}_s\dot{x}_{st} > 0 \\ 0, & \text{otherwise} \end{cases}. \tag{6.31}$$

The contribution of Skyhook force component F_{sky} and Groundhook force component F_{ground} can be adjusted by the relative ratio α_{hyb}. If α_{hyb} takes a value of 1, then the hybrid control changes to the pure Skyhook control and vice versa by $\alpha_{hyb} = 0$ to the Groundhook control.

Most of control strategies mentioned before involve tuning parameters. They can be tuned manually considering each operation case or by minimizing the following cost function, for example, as proposed in [28]:

$$J = \int_0^T \{[w_1(z_u - z_r)]^2 + [w_2(z_s - z_u)]^2 + (w_3\ddot{z}_s)^2\}dt. \tag{6.32}$$

where w_i are the weighting coefficients of the cost function.

Besides represented traditional approaches some other control techniques were also used in the area of semi-active and active suspension control. First of all, they are aimed at the improvement of the ride comfort and road holding providing robust and fault-tolerant operation. Many efforts have been done in this area in recent decade and various control techniques have been applied.

For example, the mixed H_2/H_{inf} is used to provide an optimal suspension control. This algorithm may be adjusted to change a trade-off between ride comfort and road holding [29]. Such techniques as sliding mode control are usually used as the high-level controller and provides a robust control of the vehicle suspension [30]. They may be combined also with neural networks or fuzzy logic methods to enhance adaptability [31,32]. Nevertheless, most of these approaches are using suboptimal allocation of the control demand ignoring suspension limits. In this regard, model predictive control techniques allow finding and optimal trade-off between ride comfort and road holding considering also limitations of the suspension actuators [33,34].

Taking into account known issues in the area of semi-active and active suspension control, a multi-layered controller based on the integral sliding mode control and optimal control effort distribution will be formulated in Section 6.5.

6.5 Semi-active suspension control using sliding mode approach

6.5.1 Problem statement

Proposed control approach is focused on suppression of vertical oscillations of the vehicle induced by road irregularities. The aim of the developed control is to enhance ride comfort of the vehicle. Even considering that the ride comfort is a subjective criterion and based on the driver's and passenger's opinion, several assessment techniques were proposed in recent decades. They have a good correlation with subjective criteria and may be used for validation in simulation. It includes analysis of the results both in time and frequency domains supplied by evaluation of discomfort and ride index.

The agile and robust control should be developed to provide control demand to the suspension system. To attain these ends, the integral sliding mode control is considered as a promising candidate for the formulation of suspension control functions. As soon as deflection of the suspension and possible control force are limited, this demand should be properly redistributed between suspension actuators. Consideration of the vehicle with four semi-active dampers leads to the control allocation problem, which can be solved by online optimization. Besides already mentioned advantages, such approach can provide fault-tolerant operation of the suspension system.

Knowing requirements to the control, vehicle model used for the controller formulation, model of semi-active dampers and detailed specifications of the controller will be introduced in next subsection. Results will include numerical assessment of control quality and comparison of developed control strategy with passive suspension as well as with the linear Skyhook control.

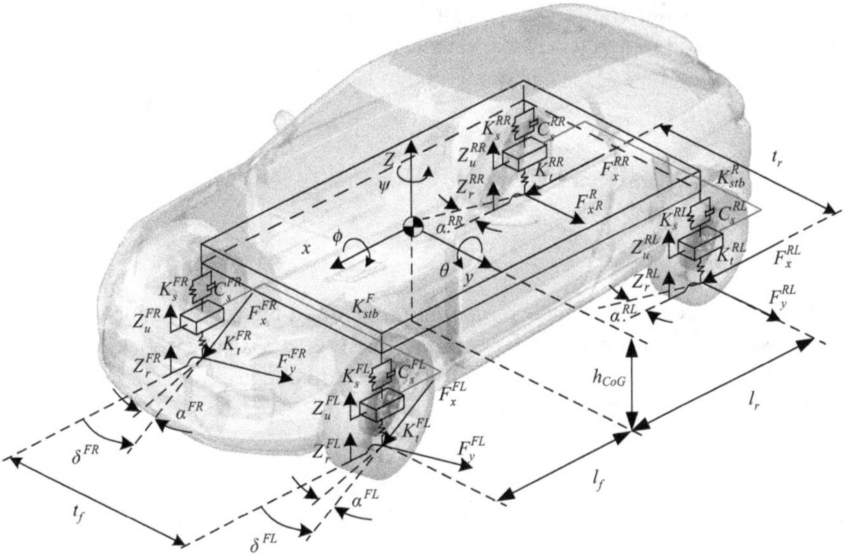

Figure 6.20 Seven DoF vehicle dynamics

6.5.2 Vehicle dynamics model

In Section 6.4.1 several types of vehicle model have been presented. To describe properly vertical vehicle dynamics in terms of heave, pitch and roll motion, seven DoF model can be utilized as a good compromise between the model fidelity and complexity. This model is depicted in Figure 6.20, where arrows on axis show the positive direction of translational and rotational motion.

In such representation three DoFs relate to heave, pitch and roll motion of the vehicle's sprung mass, and each unsprung mass performs vertical motion representing remained four DoFs. Mathematically this model is described by the following equations:

$$
\begin{cases}
m_s \frac{\partial^2 z_b}{\partial^2 t} = \sum F_s^{ij} + \sum F_c^{ij} \\
I_{xx} \frac{\partial^2 \phi}{\partial^2 t} = -\sum\limits_{i=F;R} t_{iL}(F_s^{iL} + F_c^{iL}) + \sum\limits_{i=F;R} t_{iR}(F_s^{iR} + F_c^{iR}) \\
I_{yy} \frac{\partial^2 \theta}{\partial^2 t} = -l_F \sum\limits_{j=L;R} F_s^{Fj} + F_c^{Fj} + l_R \sum\limits_{j=L;R} F_s^{Rj} + F_c^{Rj} \\
m_u^{ij} \frac{\partial^2 z_u}{\partial^2 t} = F_t(\dot{z}_{td}, z_{td}) - (F_s^{ij} + F_c^{ij})
\end{cases}, \quad (6.33)
$$

where i and j denote a wheel position, F_s^{ij} are vertical suspension forces, m_s and m_u are the sprung and unsprung masses, respectively, F_c^{ij} are control forces from active suspension systems, t_{ij} is the position of the centre of gravity from left and right wheels, l_i is the position of the centre of gravity from front and rear wheels, I_{xx} and

I_{yy} are moments of inertia for roll and pitch, respectively, z_b, θ_b and ϕ_b is the vertical position, pitch and roll angles of the vehicle body.

Suspension forces F_s^{ij} under small motion assumption are:

$$F_s^{ij} = -c_d^{ij}\dot{z}_{st}^{ij} - k_s^{ij}z_{st}^{ij}, \tag{6.34}$$

where z_{st} is the suspension travel equal to the difference between sprung and unsprung mass displacement:

$$z_{st} = z_s - z_u. \tag{6.35}$$

Control forces F_c^{ij} and their limitations depend on the type of controllable suspension system. A mathematical description of the semi-active shock absorber is explained in Section 6.5.3 and approach for controlling is given in Section 6.5.7.

The system (6.33) can be represented in the state space as:

$$\dot{x} = Ax + Bu + h(x,t), \tag{6.36}$$

where $h(x,t)$ represents a non-linear perturbation.

In such formulation the state vector consists of 14 state variables:

$$x = \begin{bmatrix} \dot{z}_b & \dot{\phi}_b & \dot{\theta}_b & z_b & \phi_b & \theta_b & z_u^{ij} & \dot{z}_u^{ij} \end{bmatrix}^T. \tag{6.37}$$

Applying a following transformation, mathematical relation between suspension forces and equivalent control inputs can be obtained:

$$\begin{bmatrix} F_c^z \\ M_c^y \\ M_c^x \end{bmatrix} = \begin{bmatrix} 1 & 1 & 1 & 1 \\ -l_f & -l_f & l_r & l_r \\ t_{fl} & -t_{fr} & t_{rl} & -t_{rr} \end{bmatrix} \begin{bmatrix} F_c^{FL} \\ F_c^{FR} \\ F_c^{RL} \\ F_c^{RR} \end{bmatrix} = HF_c. \tag{6.38}$$

Now a more evident form of the input vector can be proposed. Therefore, input vector u consists of three equivalent control forces F_z, F_θ and F_ϕ related to heave, pitch and roll motion, respectively, and four control forces related to the motion of each unsprung mass:

$$u = \begin{bmatrix} H \\ diag(1,1,1,1) \end{bmatrix} \begin{bmatrix} F_c^{FL} \\ F_c^{FR} \\ F_c^{RL} \\ F_c^{RR} \end{bmatrix} = \begin{bmatrix} F_c^z & M_c^y & M_c^x & F_c^{ij} \end{bmatrix}^T. \tag{6.39}$$

In the next section modelling and validation of semi-active damper will be explained.

6.5.3 Actuator model

A semi-active suspension control is performed by continuously controlled damper. This damper type switches not only between two states but can use also some intermediate damper characteristics. Damper characteristics were experimentally derived

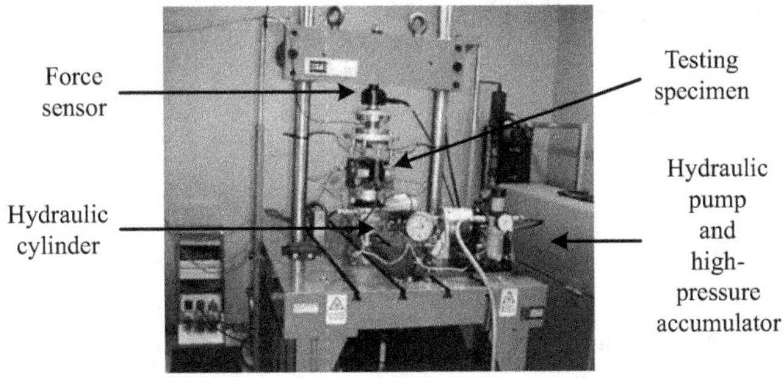

Figure 6.21 Hydropulser test bench

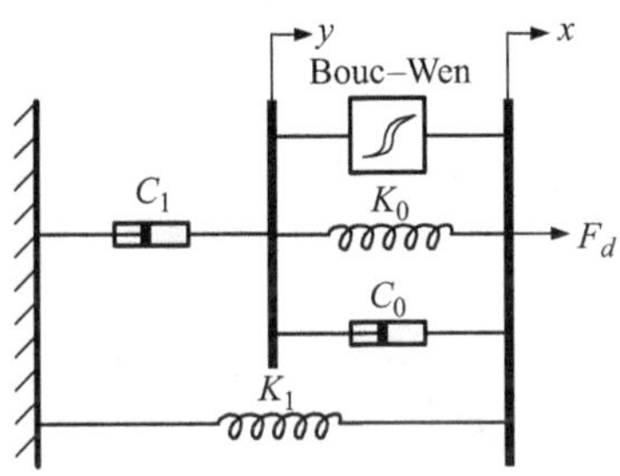

Figure 6.22 Bouc–Wen model of a shock absorber

at the hydro-pulser test rig at Technische Universitaet Ilmenau (Figure 6.21) to provide a realistic model for the simulation as well as to design low-level controller. Considering a variation of damper characteristics due to different operational modes, the validation was performed by applying actuation frequencies from 0.1 to 25 Hz with amplitude of 38; 55 and 70 mm. Besides that, these modes were investigated with current of 0.65 and 2 A, where the former corresponds to the nominal current for active suspension mode and the latter corresponds to the maximally possible current.

Description of hysteresis phenomenon in shock absorbers can be done by the use of nonlinear physical models, computation intelligence tools, Bingham, Dahl, Bouc–Wen and other models. For the class of mechanical systems as the shock absorbers it is admitted that Bouc–Wen model adequately describes system behaviour [35].

Among its various modifications the one introduced in [36] can also more accurately describe the roll-off effect at small velocities compared to other approaches (Figure 6.22). Besides that a special attention during modelling should be paid to the hysteresis asymmetry and inclusion of the control current in the representation. Following scheme in Figure 6.22, the damper force is governed by the equation:

$$F_d = K_0(x - y) + C_0(\dot{x} - \dot{y}) + K_1(x - x_0) + \alpha z, \tag{6.40}$$

where internal displacement is calculated as

$$\dot{y} = \frac{1}{C_0 + C_1}(\alpha z_{bw} + C_0\dot{x} + K_0(x - y)). \tag{6.41}$$

An auxiliary variable z is ruled by

$$\dot{z}_{bw} = -\gamma|\dot{x} - \dot{y}||z|^{n-1}z - \beta(\dot{x} - \dot{y})|z|^n + A(\dot{x} - \dot{y}). \tag{6.42}$$

In these equations parameters A, C_a, C_b, K_0, C_1, K_1, α_a, α_b, β, γ are describing a basic hysteresis shape. Nevertheless, due to the presence of significant asymmetry in the damper characteristics, it is proposed to vary parameters α and C_0 accordingly:

$$\alpha = \begin{cases} \alpha_a + I_c\alpha_a^I, & \text{if } z > 0 \\ \alpha_b + I_c\alpha_b^I, & \text{otherwise} \end{cases}, \tag{6.43}$$

$$C_0 = \begin{cases} C_a + I_cC_a^I, & \text{if } \dot{x} > 0 \\ C_b + I_cC_b^I, & \text{otherwise} \end{cases}, \tag{6.44}$$

adding four new parameters and adjusting them according to the applied current. Here, I_c is the applied current, α_a^I, α_b^I, C_a^I and C_b^I are additional parameters for fitting damper characteristics.

Finally, a set of parameters θ in this model should be identified applying any of passing optimization techniques:

$$\theta = \begin{bmatrix} A & C_a & C_b & K_0 & C_1 & K_1 & \alpha_a & \alpha_b & \beta & \gamma \end{bmatrix}. \tag{6.45}$$

Using a genetic optimization algorithm, these parameters were fitted to get a validated model of the shock absorber and depicted in Figures 6.23(a) and 6.23(b).

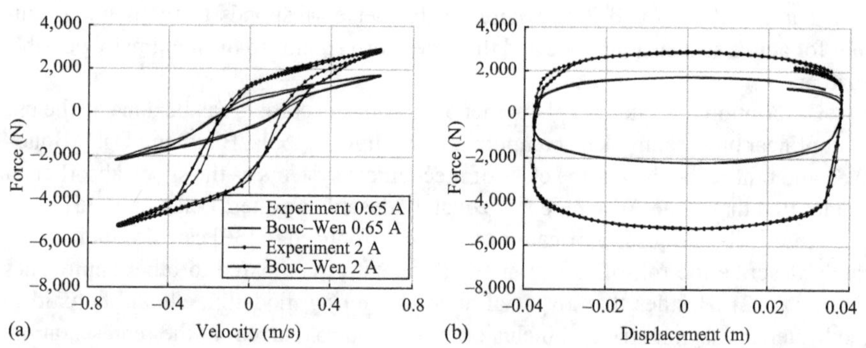

Figure 6.23 Bouc–Wen model validation: (a) damper force–velocity diagram and (b) damper force–displacement diagram

Dynamics of the shock absorber is described as the following expression:

$$\dot{I}_c = \tau(I_c - I_{ref}). \tag{6.46}$$

In the active mode, a nominal current I_n is always applied. If the current is lower than this value, the shock absorber is switched to the fail-safe mode and operates as the passive damper.

6.5.4 *Overall control specifications*

Proposed controller is aimed at reduction of control error between actual and desired vertical vehicle motion. Its main part is represented by an integral sliding mode approach (ISM) and supplemented with the controller responsible for allocation of generated control effort between semi-active shock absorbers. An overall structure of the controller is given in Figure 6.24 and explained in this section.

A six degree of freedom inertial measurement unit (IMU) feeds information to the controller about the vertical acceleration $\ddot{z}_b$, pitch $\dot{\theta}$ and roll $\dot{\phi}$ rates. To obtain the vertical velocity $\dot{z}_b$, acceleration signal is integrated and considered to be a zero-mean. Additionally, an average value is subtracted over integration time.

A *high-level controller* utilizes information about the body motion and based on the integral sliding mode approach. This controller generates overall or so-known virtual control input ν, which consists of the generalized vertical force F_c^z, pitch M_c^y and roll M_c^x moments.

A linear variable differential transformer (LDTV) is considered to be installed at each vehicle corner to obtain information about the suspension travel z_{st}. From this signal the suspension velocity $\dot{z}_{st}$ is derived applying a low-pass filter to eliminate measurement noise. These two signals are used as inputs to the Bouc–Wen model in the middle-layer control part for lower u_{min} and upper u_{max} boundaries of potential suspension force.

Due to the presence of constraints on force amplitude, a *middle-layer controller* redistributes virtual control input for heave, pitch and roll motion of the vehicle between four suspension actuators. In fact, it solves control allocation problem through the interior-point online optimization.

Lower-level controller is responsible for sending appropriate control signal to the suspension actuators based on the demanded control force. This control input is obtained by the use of the inverse Bouc–Wen model and represented by the model-based feed-forward control.

Such architecture of the controller with proper tuning guarantees robust fault-tolerant operation of suspension system with providing desired ride comfort. Principle of each control parts is explained below in more details.

6.5.5 *High-level controller*

A purpose of the high-level controller is to obtain generalized or virtual control demands for providing required ride comfort characteristics. As was mentioned in Section 6.4.5, several control techniques may be applied for this purpose. To cover an objective of robust control, the ISM control can be successfully applied. Compared

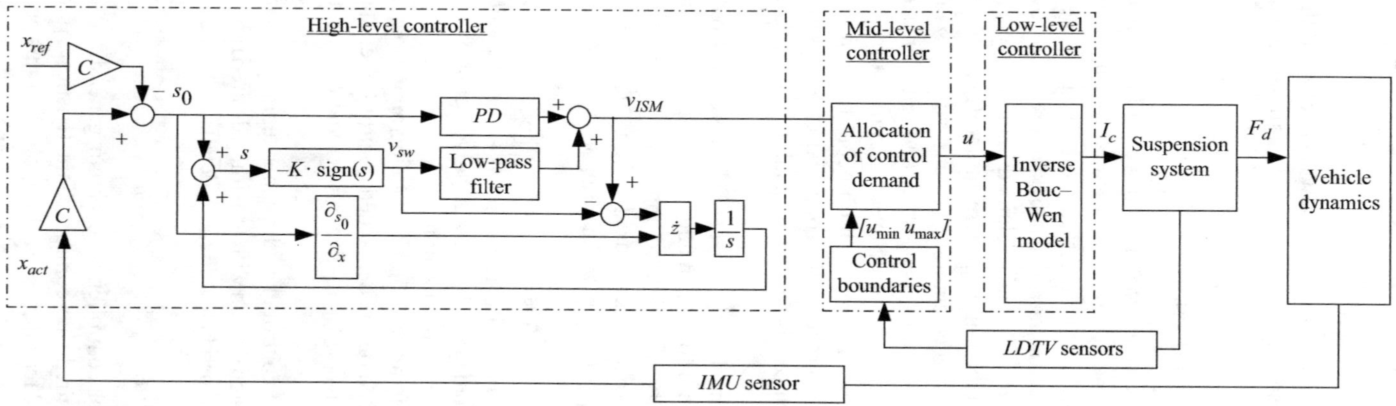

Figure 6.24 Overall controller specifications

to conventional sliding mode approach, the order of the motion equation in integral sliding mode is the same as in the original system [37]. An advantage of this method lies in a smooth control action without deterioration of the main advantages of sliding mode control provided by the application of first-order filter to the discontinuous control contribution.

According to general formulation of ISM control law, the ideal closed-loop system is represented as

$$\dot{x} = A(x)x + B(x)\nu_0, \tag{6.47}$$

A control is designed as follows:

$$\nu = \nu_0 + \nu_1, \tag{6.48}$$

where ν_0 is the ideal discontinuous control law and ν_1 is continuous part to reject perturbations terms $h(x, t)$.

A PD-control represents a feedback part ν_0:

$$\nu_0 = K_p(e_x + t_d e_x). \tag{6.49}$$

To develop a control strategy aimed at improvement of the ride comfort and omitting complex estimation procedure, the output matrix C is defined as

$$C = \begin{bmatrix} I_{3x3} & 0_{11x3} \end{bmatrix}. \tag{6.50}$$

Considering such C the control error e_x can be derived as follows:

$$e_x = C(x_{act} - x_{ref}) = \begin{bmatrix} \dot{z}_{act} - \dot{z}_{ref} & \dot{\theta}_{act} - \dot{\theta}_{ref} & \dot{\phi}_{act} - \dot{\phi}_{ref} \end{bmatrix}^T \tag{6.51}$$

where x_{act} and x_{ref} is the vector of actual and reference states, respectively.

The discontinuous control part is defined as

$$\nu_1 = -K(x)\text{sign}(s). \tag{6.52}$$

where gain K is represented by a time-invariant matrix:

$$K = diag(K_z, K_\theta, K_\phi). \tag{6.53}$$

Switching function can be designed as:

$$s = s_0 + z, \tag{6.54}$$

where s_0 is equal to the control error and defined as:

$$s_0 = \begin{bmatrix} s_0^z & s_0^\theta & s_0^\phi \end{bmatrix}^T = \begin{bmatrix} \dot{z}_{act} - \dot{z}_{ref} & \dot{\theta}_{act} - \dot{\theta}_{ref} & \dot{\phi}_{act} - \dot{\phi}_{ref} \end{bmatrix}^T. \tag{6.55}$$

The integral term z is represented as:

$$\dot{z} = -\frac{\partial s_0}{\partial x} C[Ax(t) + Bu(t) - Bu_1(t)], \text{ where } z(0) = -s_0(x(0)). \tag{6.56}$$

For this particular case, an integral term can be formulated as

$$\dot{z} = -\frac{\partial s_0}{\partial x}\begin{bmatrix} \frac{\sum F_s^{ij}}{m_s} + \frac{1}{m_s}(F_{ISM}^z - F_{sw}^z) \\ -\frac{1}{I_{yy}} l_F \sum_{j=L;R} F_s + \frac{1}{I_{yy}} l_R \sum_{j=L;R} F_s + \frac{1}{I_{yy}}(M_{ISM}^y - M_{sw}^y) \\ -\frac{1}{I_{xx}} \sum_{i=F;R} t_{iL} F_s^{iL} + \frac{1}{I_{xx}} \sum_{i=F;R} t_{iR} F_s^{iR} + \frac{1}{I_{xx}}(M_{ISM}^x - M_{sw}^x) \end{bmatrix}, \tag{6.57}$$

where initial conditions $z(0) = -s_0(x(0))$.

To proof system stability, Lyapunov candidate is calculated as:

$$V = \frac{1}{2} s(t)^T s(t). \tag{6.58}$$

Time derivative of the Lyapunov function should be negative to guarantee asymptotically stable system, i.e.:

$$\begin{aligned} \dot{V} &= \frac{1}{2} s(t)^T \dot{s}(t) = s(t)^T [C\dot{x}(t) - C(Ax(t) + Bu_0(t))] = s(t)^T [CB(v_1(t) + h(t))] \\ &= [s(t)_z s(t)_\theta s(t)_\phi] \begin{bmatrix} -\frac{1}{m_s}(K_z \operatorname{sign}(s_z) + h_z(t)) \\ -\frac{1}{I_{yy}}(K_\theta \operatorname{sign}(s_\theta) + h_\theta(t)) \\ -\frac{1}{I_{xx}}(K_\phi \operatorname{sign}(s_\phi) + h_\phi(t)) \end{bmatrix} < 0. \end{aligned} \tag{6.59}$$

Therefore, to guarantee stability and robustness without estimation of vertical suspension forces, following inequalities should be satisfied:

$$K_z > |h_z(t)|, \qquad K_\theta > |h_\theta(t)|, \qquad K_\phi > |h_\phi(t)|. \tag{6.60}$$

6.5.6 Allocation of the control demand

A task of redistribution of the equivalent control demand between suspension actuators is solved in the middle-layer controller. This can be done in a straightforward way by neglecting constraints applied to the actuators. Nevertheless, due to operational boundaries of the semi-active suspension, its force is limited in amplitude:

$$-u_{\min} \leq u \leq u_{\max}, \tag{6.61}$$

where $u_{\min}$ and $u_{\max}$ are control boundaries.

Such formulation follows to control allocation problem where number of controlled degrees of freedom (heave, pitch and roll) is lower than the number of involved actuators. Such task is very common in modern vehicle chassis control systems and reader may refer to the paper [38]. In general form this problem can be written as:

$$\nu(t) = Bu(t), \tag{6.62}$$

where B is the effectiveness matrix.

Effectiveness matrix for the case of controlling heave, pitch and roll motion is taken from (6.38) and calculated as:

$$B = \begin{bmatrix} 1 & 1 & 1 & 1 \\ -l_f & -l_f & l_r & l_r \\ t_{fl} & -t_{fr} & t_{rl} & -t_{rr} \end{bmatrix}. \tag{6.63}$$

A quadratic programming problem can be formulated as:

$$u = \min(\gamma ||W_u u||^2 + ||W_\nu(Bu - \nu)||^2), \tag{6.64}$$

where γ characterizes minimization of control effort applied by actuators, W_ν is the weight matrix for putting priorities between heave, pitch and roll control, W_u sets priority between four suspension actuators. In this particular study W_u is taken as diagonal identity matrix $I(1,4)$.

Potentially, the use of optimal demand allocation is also very useful in terms of redistributing weights for different manoeuvres by shifting control priorities. Besides that some additional safety objectives may be implemented, e.g., limited suspension travel.

6.5.7 Low-level controller

To derive an inverse model of the Bouc–Wen model, its simpler representation depicted in Figure 6.25 is utilized.

Governing equations of this model are:

$$F_d = K_0(x - x_0) + C_0\dot{x} + \alpha z, \tag{6.65}$$

$$\dot{z}_{bw} = -\gamma|\dot{x}||z|^{n-1}z - \beta\dot{x}|z|^n + A\dot{x}. \tag{6.66}$$

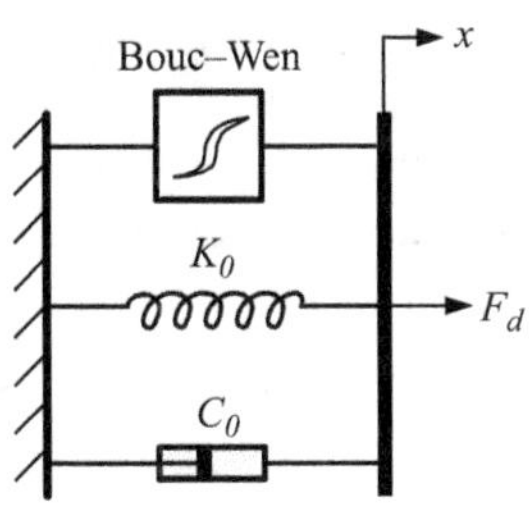

Figure 6.25 Bouc–Wen model of a shock absorber

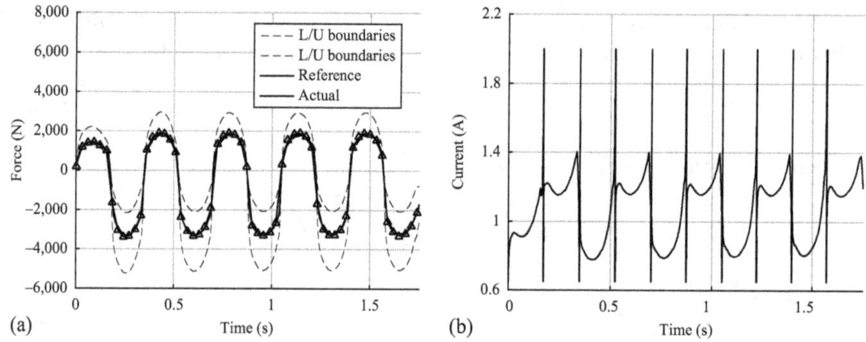

Figure 6.26 Validation of inverse Bouc–Wen model: (a) damper force and (b) control current

Table 6.3 Vehicle specifications

Parameter	Value	Units
m_s	1,963.3	kg
I_{xx}	560.7	kg m^2
I_{yy}	2,525	kg m^2
I_{zz}	2,761	kg m^2
Tyres	235/55 R19	

An inverse Bouc–Wen model describes a current required to achieve desired control force considering limitations of the semi-active suspension:

$$I_d = \frac{F_d - K_0 x + C_0 \dot{x} - \alpha z}{C_I \dot{x} + \alpha_I z}(I_{\max} - I_n) + I_n \tag{6.67}$$

This formulation induces also asymmetric form of hysteresis by (6.43, 6.44) and parametrized accordingly. To omit too low or excessive current, its signal is bounded by 0.65 and 2 A.

To validate a proposed control approach, a simulation was performed by using validated hysteresis model and proposed model-based feed-forward control. Results represent precise tracking of the desired suspension force by semi-active shock absorber, Figures 6.26(a) and 6.26(b). It shows that a root mean square error is equal to 11 N during 1.75 s of operation with proposed control approach.

6.5.8 Results and discussions

For investigation a sport utility vehicle (SUV) was selected. Having relatively high centre of gravity, such type of car is exposed to roll and pitch motion on rough road surfaces. Parameters of the vehicle prototype are represented in Table 6.3.

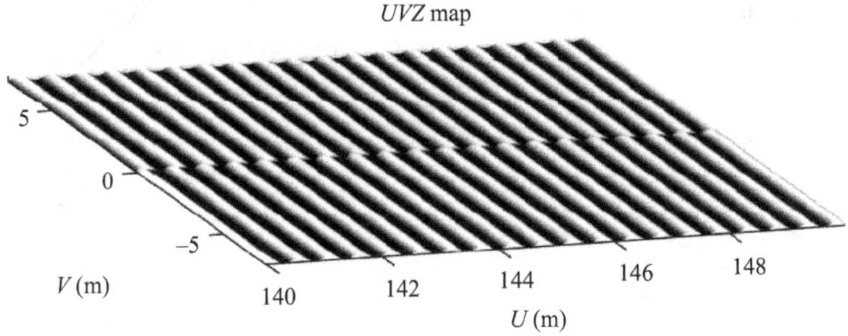

Figure 6.27 Road profile for sine sweep test

This parametrization is based on the real SUV prototype investigated in European project EVE [39]. All simulations were performed in commercial software with validated kinematics and compliance and suspension system.

Developed control strategy was compared not only to the passive suspension, but also linear (or continuous) Skyhook was taken into account. As is known from Section 6.4.5, this control strategy is based on principles of conventional Skyhook approach and extended for application with continuously controlled dampers.

Frequency response function. Typical road tests were selected for the controller validation. To obtain frequency response function (FRF) and to evaluate ride index, in the first test an out-of-phase chirp road profile with varying magnitude was used. Taking into account frequencies of interest it has frequency contents from 0 to 18 Hz. Such road profile induces heave and pitch as well as roll motion due to difference in disturbance under the left and right wheels. A part of such road profile is shown in Figure 6.27. During this testing procedure each trial was performed with constant vehicle velocity of 90 km/h.

Developed control strategy and linear Skyhook approach were tuned to obtain first of all positive effects in reduction of magnitude of the vertical vehicle body acceleration. Figure 6.28 represents obtained results and clearly shows that up to 9 Hz magnitude of vertical acceleration was significantly reduced by application of developed control. Much smaller reduction of the magnitude was obtained with linear Skyhook, which was effective in the frequency range from 0 to 1 Hz and above 8 Hz.

Response in terms of pitch acceleration shows evident benefits of developed control over suboptimal linear Skyhook algorithm, Figure 6.29. They have comparable behaviour up to 1 Hz, but reduction of magnitude by developed control is more significant above this frequency.

Both control strategies have less positive effects in terms of roll acceleration due to physical limitations of semi-active suspension, Figure 6.30. Moreover, linear Skyhook control a priori based on the vertical velocity of sprung and unsprung mass induces higher roll acceleration magnitudes in frequencies lower than 1 Hz.

Aforementioned results provide a qualitative assessment, evaluation of the developed control and characterization of suspension system. To provide a quantitative

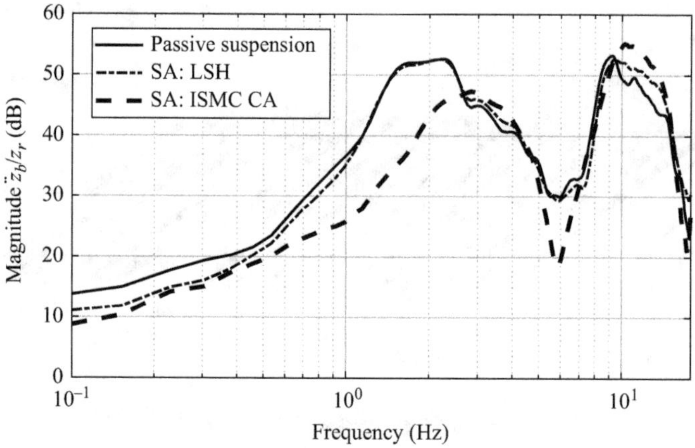

Figure 6.28 Frequency response function – vertical acceleration of the vehicle body

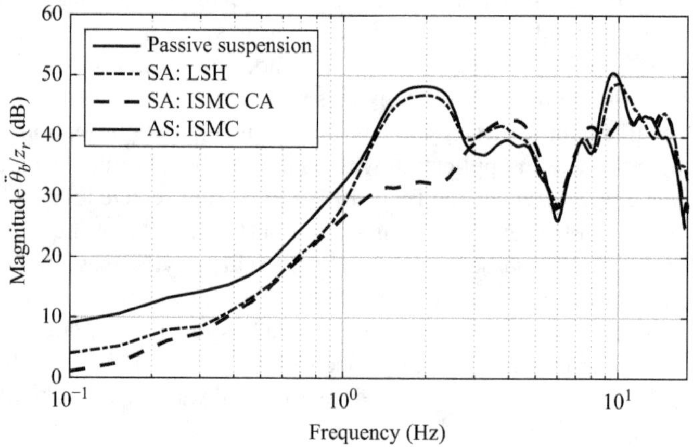

Figure 6.29 Frequency response function – pitch rate

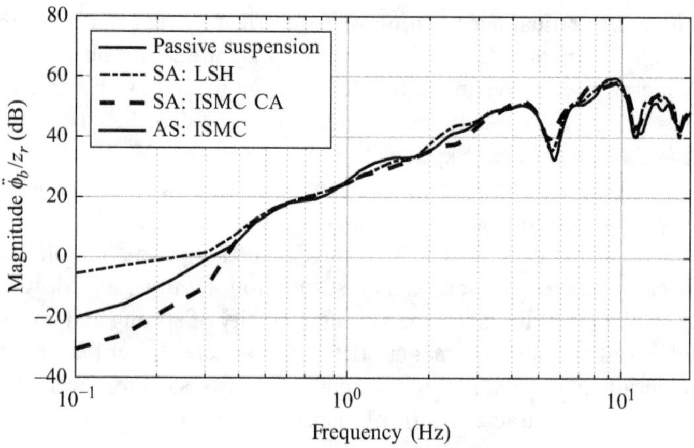

Figure 6.30 Frequency response function – roll rate

Table 6.4 Ride index

Control art	$\ddot{z}_w$ (dB/Hz)	$\ddot{\theta}_w$ (dB/Hz)	$\ddot{\phi}_w$ (dB/Hz)
Passive	13.6	5.0	48.8
Linear Skyhook	11.2	3.5	43.7
ISMC CA	8.9	2.1	42.3

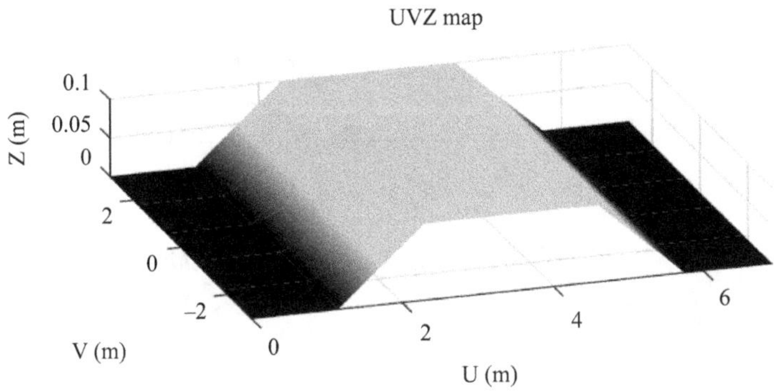

Figure 6.31 Profile of road bump

assessment it is proposed to apply ride index [40] to three considered states $\ddot{z}_b$, $\ddot{\theta}_b$ and $\ddot{\phi}_b$. In its typical formulation ride index is the root-mean square of weighted power spectral density PSD_x of the state $\ddot{x}$:

$$x_w = \sqrt{\frac{\sum_{f=0}^{f_m} (w_f PSD_x)^2}{n}}, \tag{6.68}$$

where frequency f is considered from 0 to $f_m = 18$ Hz and consists of n elements, PSD_x is the power spectral density of x state, vector of weights w_f expresses control priorities.

The developed control is aimed at the reduction of amplitudes in frequencies mainly lower than 8 Hz related to vehicle body motion. Therefore, vector of weights is selected in accordance to control objectives and recommendations in ISO 2631-1. Obtained results show evident benefits of developed control over passive suspension and linear Skyhook control. Especially, these benefits are clearly seen in control of vertical as well as pitch acceleration, Table 6.4. As was mentioned before, less effects in relation to passive suspension were obtained for vehicle roll motion due to physical limitations of the suspension system.

To evaluate performance of the suspension control in terms of plant response, a road bump in height of 10 cm was used. Its profile is shown in Figure 6.31. This test was performed at a constant vehicle velocity of 30 km/h.

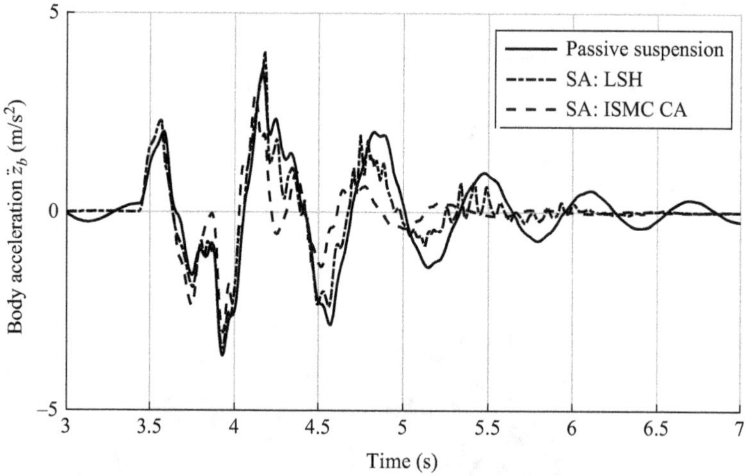

Figure 6.32 Bump test – vertical acceleration of the vehicle body

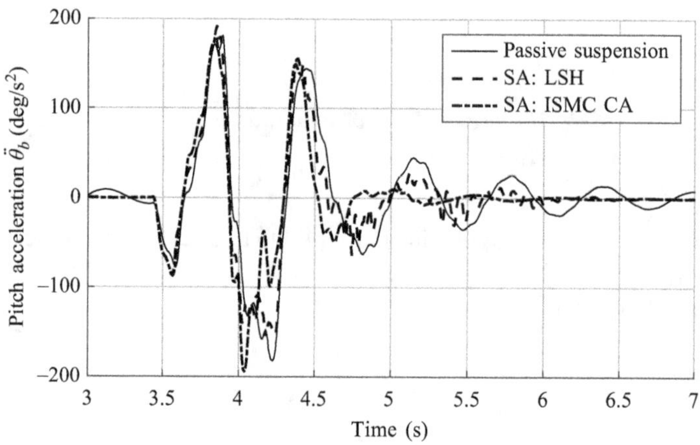

Figure 6.33 Bump test – pitch acceleration

As can be seen in Figure 6.32, acceleration of vehicle body under such disturbance at the initial phase is closely similar to three investigated configurations of the suspension control. At the same time, the developed control and linear Skyhook further effectively suppress vertical oscillations reducing deviation of the acceleration over desired zero value.

Similar tendencies in the pitch acceleration are obtained and depicted in Figure 6.33. In contrary to very effective control of vertical deceleration, developed control produces at 4 s the highest peak value of pitch acceleration. Nevertheless, in scope of complete test, it provides the smallest deviation from desired value of the pitch acceleration.

Table 6.5 Road bump test

Control art	a_z^{peak} (m/s^2)	a_z^{RMS} (m/s^2)	$\ddot{\theta}_b^{peak}$ (deg/s^2)	$\ddot{\theta}_b^{RMS}$ (deg/s^2)
Passive	3.6	1.0	183.5	53.6
Linear Skyhook	4.2	0.9	194.2	50.6
ISMC CA	3.0	0.7	194.6	46.8

The analysis above can be supplemented by typical quantitative criteria. Peak value in this case is the maximal nominal amplitude of oscillations. To evaluate variation of states from the desired values, an RMS deviation was calculated. This numerical data is concluded in Table 6.5.

As can be seen, developed control has evident advantages over traditional approach in continuous control of the semi-active suspension. This shows effective suppression of vertical vehicle oscillations in terms of vehicle's heave, pitch and roll motion. Besides that, it ensures a fault-tolerant and robust operation by optimal redistribution of control effort between four suspension actuators. This control is applicable also for the use in active suspension systems.

6.6 Conclusions

In this chapter problematic of the semi-active and active suspension control has been overviewed. It provides basic knowledge to the reader about hardware components, typical characteristics and known control approaches. The proposed control strategy is based on the seven DoF vehicle model. Validation of developed control is performed in commercial vehicle software with validated vehicle and suspension models. Proposed control is compared to the passive suspension as well to linear Skyhook control. Results show evident advances in the integral sliding mode control with optimal allocation of control demand. It is confirmed by assessment of frequency response function, ride index and deviation from the desired values. By including optimal allocation of control demand, the represented control strategy provides fault-tolerant operation. It may be applied not only to the semi-active suspension system, but also to the active suspension control.

Acknowledgements

The project leading to this application has received funding from the European Union's Horizon 2020 research and innovation programme under the Marie Skodowska-Curie grant agreement No. 645736. Authors are sincerely grateful to Jaguar Land Rover and ZF TRW for their kind support. Authors would like to express thanks to Dr. Barys Shyrokau (TU Delft) for very useful discussions on topics of vehicle dynamics and control allocation.

References

[1] E. M. Heißing, Bernd and Ersoy, Metin, *Chassis Handbook: Fundamentals, Driving Dynamics, Components, Mechatronics, Perspectives*. Wiesbaden, Germany: Springer Science & Business Media, 2010.

[2] T. Gillespie, *Fundamentals of Vehicle Dynamics*. Warrendale, USA: SAE International, 1992.

[3] H.-J. Büchler, "Comfort simulation: Goals – issues – solutions," in *at VI-grade User's Conference*, Marburg, Germany, 2007.

[4] J. Reimpell, H. Stoll, and J. Betzler, *The Automotive Chassis: Engineering Principles*. Warrendale, USA: SAE International, 2001.

[5] J. Dixon, *Tires, Suspension and Handling*. Warrendale, USA: SAE International, 1996.

[6] D. Bastow, G. Howard, and J. Whitehead, *Car Suspension and Handling*. Warrendale, USA: SAE International, 2004.

[7] M. Muenster, U. Mair, J. Gilsdorf, *et al.*, "Electromechanical active body control," *ATZautotechnology*, vol. 9, no. 3, pp. 24–29, 2009.

[8] B. Fijalkowski, *Automotive Mechatronics: Operational and Practical Issues (Vol. 1)*. Dordrecht, Netherlands: Springer Science & Business Media, 2010.

[9] B. Heißing, M. Ersoy, and S. Gies, "Fahrwerkhandbuch: Grundlagen, fahrdynamik, komponenten, systeme, mechatronik, perspektiven, dritte auflage," Wiesbaden, Germany: *Vieweg+Teubner Verlag*, 2011.

[10] D. Crolla, *Automotive Engineering*. Oxford, UK: Butterworth-Heinemann, 2009.

[11] D. Karnopp, M. Crosby, and R. Harwood, "Vibration control using semi-active force generators," *Journal of Engineering for Industry*, vol. 96, no. 2, pp. 619–626, 1974.

[12] Research and Markets, "The global chassis sector report: An analysis of the braking, steering and suspension markets," 2015.

[13] K. Reybrouck, "Active suspension system," 4 November 1997.

[14] K. Reybrouck, B. Vandersmissen, and K. Six, "ACOCAR: ultimate comfort and safety through the energy-efficient active damping system of Tenneco," in *Proceedings of the 21st Aachen Colloquium Automobile and Engine Technology*, Aachen, Germany, 2012.

[15] M. Giovanardi, C. Tucker, R. Wendell, *et al.*, "Active suspension with on-demand energy flow," 18 September 2014.

[16] J. Ekchian, W. Graves, Z. Anderson, *et al.*, "A high-bandwidth active suspension for motion sickness mitigation in autonomous vehicles," *SAE Technical Paper 2016-01-1555*, 2016.

[17] B. Gysen, T. van der Sande, J. Paulides, and E. Lomonova, "Efficiency of a regenerative direct-drive electromagnetic active suspension," *IEEE Transactions on Vehicular Technology*, vol. 60, no. 4, pp. 1384–1393, 2011.

[18] A. Thomae, J. Gilsdorf, M. Muenster, *et al.*, "Electromechanical active body control," in *FISITA World Automotive Congress*, Munich, Germany, 2008.

[19] J. Wong, *Theory of Ground Vehicles*. New York, USA: John Wiley & Sons, 2001.

[20] N. Maia and J. Silva, *Theoretical and Experimental Modal Analysis*. Hertfordshire, UK: Research Studies Press, 1997.

[21] D. Fischer and R. Isermann, "Mechatronic semi-active and active vehicle suspensions," *Control Engineering Practice*, vol. 12, no. 11, pp. 1353–1367, 2004.

[22] C. Stammers and T. Sireteanu, "Vibration control of machines by using semi-active dry friction damping," *Journal of Sound and Vibration*, vol. 209, pp. 671–684, 1997.

[23] S. Rakheja and S. Sankar, "Vibration and shock isolation performance of a semi-active 'on-off' damper," *Journal of Vibration, Acoustics, Stress, and Reliability in Design*, vol. 107, pp. 671–684, 1985.

[24] S. Savaresi, E. Silani, and S. Bittanti, "Acceleration-driven-damper (add): An optimal control algorithm for comfort-oriented semiactive suspensions," *Journal of Dynamic Systems, Measurement, and Control*, vol. 127, no. 2, pp. 218–229, 2005.

[25] S. Savaresi and C. Spelta, "Mixed sky-hook and add: Approaching the filtering limits of a semi-active suspension," *Journal of Dynamic Systems, Measurement, and Control*, vol. 129, no. 4, pp. 382–392, 2007.

[26] M. Valášek and M. Novák, "Groundhook for semi-active damping of truck's suspension," in *Proc. of CTU Workshop*, vol. 96, pp. 467–468, Brno, Czech Republic, 1996.

[27] M. Valášek, M. Novák, Z. Šika, and O. Vaculin, "Extended ground-hook-new concept of semi-active control of truck's suspension," *Vehicle System Dynamics*, vol. 27, no. 5–6, pp. 289–303, 1997.

[28] E. Guglielmino, T. Sireteanu, C. Stammers, G. Ghita, and M. Giuclea, *Semi-active Suspension Control: Improved Vehicle Ride and Road Friendliness*. London, UK: Springer Science & Business Media, 2008.

[29] J. Lu and M. DePoyster, "Multiobjective optimal suspension control to achieve integrated ride and handling performance," *IEEE Transactions on Control Systems Technology*, vol. 10, pp. 807–821, 2002.

[30] S. Toyama and F. Ikeda, "Integral sliding mode control for active suspension systems of half-vehicle model," *Mechanical Engineering Journal*, vol. 2, pp. 1034–1046, 2015.

[31] K. Spentzas and S. Kanarachos, "Design of a non-linear hybrid car suspension system using neural networks," *Mathematics and Computers in Simulation*, vol. 60, pp. 369–378, 2002.

[32] K. Spentzas and S. Kanarachos, "A neural network approach to the design of a vehicle's non-linear hybrid suspension system," *Proceedings of the Institution of Mechanical Engineers, Part B: Journal of Engineering Manufacture*, vol. 216, pp. 833–838, 2002.

[33] N. Giorgetti, A. Bemporad, H. Tseng, and D. Hrovat, "Hybrid model predictive control application towards optimal semi-active suspension," *International Journal of Control*, vol. 79, pp. 521–533, 2006.

[34] M. Canale, M. Milanese, and C. Novara, "Semi-active suspension control using fast model-predictive techniques," *IEEE Transactions on Control Systems Technology*, vol. 14, pp. 1034–1046, 2006.

[35] A. Dominguez, R. Sedaghati, and I. Stiharu, "Modelling the hysteresis phenomenon of magnetorheological dampers," *Smart Materials and Structures*, vol. 13, pp. 1351–1361, 2004.

[36] B. Spencer, S. Dyke, M. Sain, and J. Carlson, "Phenomenological model for a magnetorheological damper," *Journal of Engineering Mechanics of the American Society of Civil Engineering*, vol. 123, p. 230–252, 1997.

[37] V. Utkin and J. Shi, "Integral sliding mode in systems operating under uncertainty conditions," in *Proceedings of the 35th IEEE Conference on Decision and Control, 1996*, vol. 4, pp. 4591–4596, Kobe, Japan: IEEE, 1996.

[38] B. Shyrokau, D. Wang, D. Savitski, K. Hoepping, and V. Ivanov, "Vehicle motion control with subsystem prioritization," *International Journal of Mechatronics*, vol. 30, pp. 297–315, 2015.

[39] D. Savitski, V. Ivanov, K. Augsburg, M. Dhaens, C. Sandu, and S. Els, "State-of-the-art and future developments in integrated chassis control for ground vehicles," in *Proceedings of the 13th European Conference of the ISTVS*, pp. 268–272, Rome, Italy: ISTVS, 2015.

[40] B. Kim and H. Yoo, "Ride comfort uncertainty analysis and reliability design of a passenger vehicle undergoing random road excitation," *Proceedings of the Institution of Mechanical Engineers, Part D: Journal of Automobile Engineering*, vol. 123, pp. 433–442, 2012.

Chapter 7
Observer-based parameter identification for vehicle dynamics assessment

Robert Tafner[1], *Markus Reichhartinger*[2], *and Martin Horn*[2]

Assessment of the vehicle handling especially with respect to its lateral dynamics is an important aspect of the overall vehicle design and development process. However, an increasing rate of vehicle update cycles, a massively growing number of variants and a required high-quality comfort and/or driving reward render the evaluation process a challenging task. Consequently, virtual methods support the overall development process and increase time and cost efficiency significantly. The so-called model-based (objective) handling methodology aims to extract certain vehicle and/or driver model parameters from measurement data. These can then be used to simulate standard handling maneuvers, rather than performing them on a test track. State-of-the-art parameter identification mechanisms are commonly performed offline and require extensive instrumentation of the test vehicle.

This chapter presents a novel approach exploiting observer-based parameter identification techniques. It introduces the advantages of online capability, time-efficient experiment execution and reduction of sensing devices due to estimation of specific system states. The joint estimation of states and parameters is formulated as an unknown input recovery problem. Using sliding mode mechanisms allows formulation of state observers that are invariant with respect to certain classes of perturbations. Furthermore, the attractiveness is increased considering the property of finite time convergence. Herein, higher-order sliding mode concepts are used for the task of parameter identification providing robustness, finite convergence time and stability even for non-persistently excited systems.

Evaluation of the concepts is performed in a twofold way: (a) An industrial vehicle dynamics simulation tool provides data for the observation concepts. The resulting parameter estimates are integrated into the offline simulation of standard handling maneuvers, e.g., step input steering. Comparing these results with the reference data allows to draw conclusions on the expected accuracy of the method. (b) The selected

[1]AVL DiTEST GmbH, Graz, Austria
[2]Institute of Automation and Control, Graz University of Technology, Graz, Austria

concepts are evaluated on experimental facilities, i.e., standard vehicles and an electric power steering test bench.

7.1 Introduction

Mobility can be seen as a basic requirement of human beings especially by considering the whirligig of time. Referring to the most flexible and individual means of transportation, i.e., the self-propelled vehicle, there exists an unbowed demand for innovative and exciting products by customers. On the one hand, more and more stringent emission legislations are introduced that reduce the environmental burden, but also increase the complexity of the developed internal combustion engines. And, on the other hand, increasing demands on vehicle safety require the manufacturers to tackle the up-coming problems. Overall, the modern vehicle development process needs to adapt to ever-changing customer requirements resulting from global trends such as *urbanization*, inevitable increase in the world's *natural resources demand*, *environmental awareness*, *ageing of the society* and an increase in *wealth* [1]. Especially the latter results in a request for *exclusive* and *individual* products without any compromises allowed. The automotive industry answers to that by e.g., offering the customer a tremendous number of vehicles variants. As long waiting times for innovative products are not accepted the vehicle update cycles are cut down constantly. Otherwise, market shares might be lost to competitors. These conflicting requirements of decreasing product update cycles and increasing number of vehicle variants led to a more extensive use of the virtual development paradigm. In fact, numerous challenges arising during the product development process are supported by computer-aided engineering tools. As a consequence, the quality level of developments during the conceptual phase improves, whereas the necessity to build hardware prototypes decreases. Clearly, this also reduces the overall development costs and times.

An integral and inevitable part of the vehicle development task represents the *handling evaluation*. It focuses not only on safety-related aspects, but also takes into account the fun factor, enthusiasm and fulfilling of driver expectations that potential customers experience during a vehicle ride. Besides its design and image, this is a decisive factor for the buying decision of a specific vehicle [2]. In order to understand the general notion of *vehicle handling* better it shall be clarified what a *good* handling refers to. Actually, if the vehicle reacts on specific driver inputs, such as steering actuations, accelerator pedal tip-ins, braking, etc. in an expected, predictable way and the vehicle follows exactly the trajectory the driver dictates, it is said to behave *good* [3].

Methodology-wise there exist different approaches for a systematic analysis of the vehicle handling evaluation. Aside the classical approach that relies on driving experts and exploits their feedback as rating basis, more systematic methods, known as objective and model-based objective approaches, were introduced a couple of decades ago. These methods will be discussed in the first part of this chapter.

The subsequent work focuses on potential improvements of the parameter identification task that is part of the model-based handling evaluation process. More specifically, it discusses novel approaches to estimate state and parameters in a joint fashion. For further details the interested reader is referred to [4].

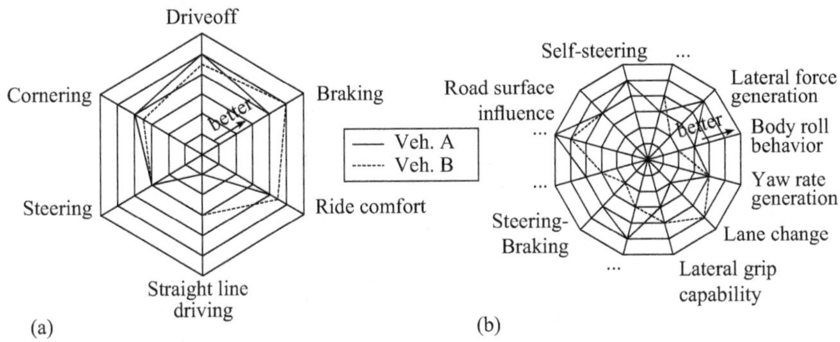

Figure 7.1 Graphical evaluation of handling disciplines: (a) overall vehicle behavior and (b) cornering behavior

This chapter is organized as follows: the remainder of Section 7.1 reviews the general approaches how to evaluate the lateral handling characteristics and introduces the physical aspects as well as mathematical modeling of the vehicle's lateral and roll dynamics. Section 7.2 provides an overview of research work that has been conducted within the field of vehicle handling assessment and joint state and parameter estimation in the automotive area. In Section 7.3, the contributions of the authors are presented and analyzed. All proposed mechanisms are evaluated in simulations as well as real-world applications in order to demonstrate their capability to support the overall evaluation process. Finally, the chapter is concluded by Section 7.4.

Subjective handling evaluation

The traditional approach, referred to as *subjective* paradigm, aims to evaluate a vehicle's handling relying on the expert knowledge of highly skilled test drivers. Those intend to extract the required information from vehicle responses as a reaction to certain maneuver performed with test vehicles. Generally, there is no consensus whether the freedom of maneuver selection is at the discretion of the driver or formulated by the process designer [5]. Moreover, also the selection of drivers (trained vs. untrained) is still under discussions, e.g., [6,7]. Beside these challenges it is the design of the so-called *questionnaire* that plays a crucial part within the process. Clearly, the formulation of these questions determines the technical accuracy of the provided feedback. In other words, asking the drivers only restricted[1] questions harmonizes the answers' variation, but also reduces the information fed back from the drivers. On the contrary, open questions increase the difficulty to harmonize answers from numerous drivers immensely.

Without discussion of too many details related to the subjective handling evaluation process a possible outcome is depicted in Figure 7.1. It shows a general behavior assessment as well as a specific one related to the vehicle's cornering characteristics.

[1]A *restricted* question reduces the space of possible answers. In contrast, an *open* question maintains the complete freedom which details are provided within an answer.

Table 7.1 Recommended measurements and transducer accuracies

Sensor principle	Measured quantity	Measured range	Required accuracy	Source
Measured steering wheel	Steering wheel angle	$\pm 180°$	$\pm 2°$	[9]
	Steering wheel torque	± 30 N·m	± 0.3 N·m	[9]
Accel. sensor	Longitudinal acceleration	± 15 m·s^{-2}	± 0.15 m·s^{-2}	[10]
	Lateral acceleration	± 15 m·s^{-2}	± 0.15 m·s^{-2}	[9]
Gyroscopic platform	Roll angle	$\pm 15°$	$\pm 0.15°$	[9]
	Pitch angle	$\pm 15°$	$\pm 0.15°$	[10]
	Yaw velocity	$\pm 50°$/s	$\pm 0.5°$/s	[9]
Optical speed sensor	Longitudinal velocity	0–50 m·s^{-1}	± 0.5 m·s^{-1}	[9]
	Lateral velocity	± 10 m·s^{-1}	± 0.1 m·s^{-1}	[9]
	Sideslip angle	$\pm 15°$	$\pm 0.15°$	[9]

Despite the obvious advantage of fully integrating the human being into the evaluation process this is simultaneously its main drawback. In fact, the driver's *mood*, its *personal preferences* as well as *memory of perception* and *power of discrimination* render the process error prone in a sense that it suffers from weak repeatability.

To overcome these drawbacks a more systematic approach, supported extensively by standardized driving maneuvers as well as measurement equipment, is discussed within the next paragraph.

Objective handling evaluation

The objective evaluation aims to identify the handling characteristics in a more systematic way in order to reduce the influence of human variability and to increase repeatability. Switching from closed-loop operation, where the driver dictates the trajectories of the vehicle, to an open-loop paradigm, standardized[2] driving maneuvers are exploited for extraction of the so-called *objective metrics*. These metrics describe important signal characteristics related to the vehicle responses. An example might be the overshoot percentage or the time from 50% (w.r.t. to steady-state values) of the system excitation to 90% of the steady-state vehicle responses when being excited by a step steering input.

Clearly, in order to extract objective metrics certain vehicle responses need to be measured. Traditionally, test vehicles are heavily instrumented, as summarized by Table 7.1. Therein, the suggested accuracies are extracted from specific standards and literature. Closed-loop maneuvers, such as the ISO lane change maneuver [11], are not considered in here, as the driver is again part of the experiment and the vehicle responses depend highly on the human controller.

[2]Well-known maneuvers are the steady-state circular drive [8] or step steering and sinusoidal inputs [9]. The latter might be performed with constant and/or varying frequency. Standards such as [8,9] not only describe exactly how to perform these maneuvers, but also offer suggestions on the vehicle responses to be measured and the extraction of interesting signal characteristics.

From a more general perspective, the objective handling evaluation not only aims to annihilate the disadvantages of the subjective evaluation paradigm, but eventually strives to replace it completely. Generation of synthetic evaluations uses identified correlations between extracted objective criteria and subjective ratings obtained for a considered vehicle. However, finding these links in a systematic way to be further used for generation of synthetic (subjective) ratings also for new vehicles, is still an active field of research, see e.g., [12]. In fact, identifying these correlations is not focus of the presented work and therefore not further discussed.

Model-based objective handling evaluation

Even though the objective handling evaluation partially solves the repeatability problem of the subjective rating process, the standardized driving maneuvers are mostly conducted by human drivers. Theoretically, installation of steering robots would reduce any influences of human beings on the performed experiments, but these robots are cost-intensive and installation is time-consuming. The *model-based* approach strives for high repeatability and low influence of the human driver on the maneuver execution. The method aims to identify model parameters related to vehicle and/or driver dynamics. These parameters are valuable for the following two reasons:

- Aside the objective metrics (extracted from vehicle responses as reaction to standardized driving maneuvers) the model parameters increase the size of the database. Correlation with subjective ratings results eventually in synthetic evaluations.
- The model parameters (in conjunction with the underlying model structure) condense the vehicle characteristics in some sense. Clearly, in order to obtain vehicle responses (to standardized driving maneuvers) simulations exploit these model parameters and allow fully repeatable execution of the handling evaluation task.

The herein discussed modeling of the lateral vehicle dynamics[3] relies on physical effects, but can also be based on a phenomenological description, e.g., [15,16]. These physically motivated models will be presented in Section 7.1.1.

A review of the work that has been conducted in the areas of objective vehicle handling evaluation and more specifically model-based paradigm will be discussed in Section 7.2. Without anticipating the detailed results of the review the overall motivation of the presented work is to identify model parameters related to the lateral and roll dynamics of the vehicle and its chassis respectively. These parameters have a great impact on the model-based handling evaluation. The proposed methods are required to operate time-effectively in both parameter identification and handling maneuver conduction. Consequently, the parameter identification methods will be based on observation techniques providing for online estimation capability as well as low requirements on installed measurement equipment as some of the vehicle responses can be estimated rather than measured. Furthermore, the online estimation of the parameters allows to provide the driver with some feedback on the matching

[3]This work focuses on vehicle dynamics exclusively. Modeling of the driver dynamics and identification of its parameters can be found in, e.g., [13,14].

between identified characteristics and the real vehicle behavior. Evaluations of the capability to predict vehicle responses within open-loop handling maneuvers based on the identified model parameters will be demonstrated in simulations (Section 7.3.3) and experiments (Section 7.3.4).

From the foregoing discussion on various approaches how to evaluate the vehicle handling characteristics it is clear that the model-based paradigm represents the most efficient (w.r.t. time and costs) method. Moreover, it also provides the highest potential in terms of repeatability. Clearly, the request for time efficiency motivates an online capability of the identification techniques. Therefore, the authors' contribution refers to state observation mechanisms that are robust with respect to unknown input signals. In fact, higher-order sliding mode techniques robustly estimate the system state variables and also allow finite time recovery of unknown parameters. Moreover, it is the use of low-cost measurement equipment the state observers operate on, that renders the mechanisms highly attractive for the proposed field of operation.

7.1.1 Modeling of selected vehicle dynamics

7.1.1.1 Lateral vehicle motion

Excitation of the lateral vehicle dynamics can be conducted by actuation of the steering wheel (for simplicity assuming a flat, non-inclined ground). Turning the wheels out of their heading direction causes the build-up of lateral tire forces. Consequently, the vehicle rotates around its vertical axis also known as a yaw motion. One of the simplest mathematical approaches to describe these dynamics is the classical single-track model. There exist a number of assumptions for its validity, see e.g., [17], which will not be listed here explicitly. In summary, the vehicle's center of gravity (CoG) is assumed at ground level, i.e., the degrees of freedom are reduced to a planar motion and a rotation around the vertical axis. Moreover, the individual tire forces are lumped to axle-related characteristics and the vehicle is considered as a point mass concentrated at the CoG. Figure 7.2 depicts the kinematics, kinetics and geometry of the assumed single-track model.

Under the assumption of a rigid vehicle body the translational/rotational motion can be described by

$$m\frac{dv_x}{dt} = mv_y\omega_z + F_{x,v,f} + F_{x,v,r} \tag{7.1a}$$

$$m\frac{dv_y}{dt} = -mv_x\omega_z + F_{y,v,f} + F_{y,v,r} \tag{7.1b}$$

$$J_z\frac{d\omega_z}{dt} = l_fF_{y,v,f} - l_rF_{y,v,r} \tag{7.1c}$$

Therein, the longitudinal and lateral velocities are denoted by v_x and v_y, respectively. Moreover, ω_z represents the yaw rate, m the total vehicle mass, J_z the moment of inertia w.r.t. the vertical axis, l_f, l_r the distances between CoG and front/rear axles.

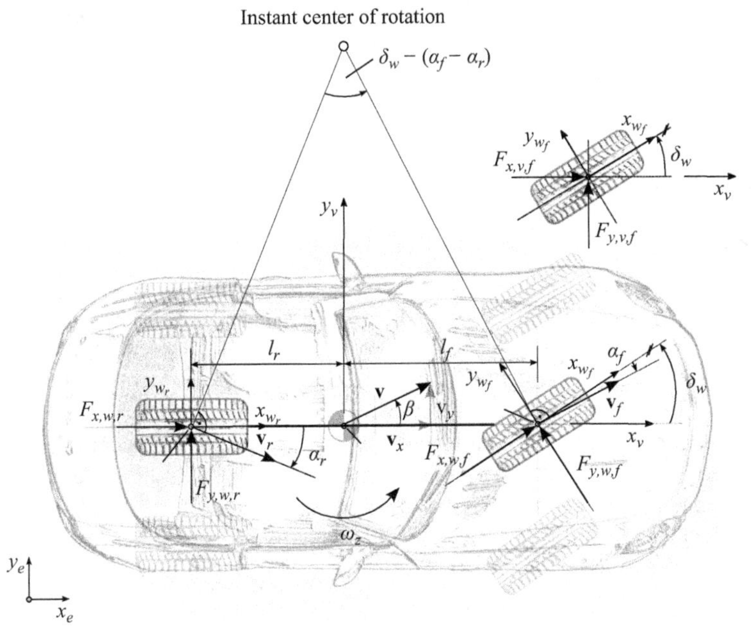

Figure 7.2 Schematics of the single-track model

$F_{x,v,f/r}$, $F_{y,v,f/r}$ are the front/rear axle forces in lateral and longitudinal direction w.r.t. the vehicle-fixed axis system (x_v- and y_v-axes). In contrast, $F_{x,w,f/r}$, $F_{y,w,f/r}$ represent the axle forces w.r.t. to the wheel-axis systems (x_{w_r}-, y_{w_r}- , x_{w_f}- and y_{w_f}-axes). Note that the front wheel axis system is rotated around the z-axis of the vehicle-fixed system by the wheel steering angle δ_w. In general, special care needs to be taken which axis system the tire forces refer to. However, in case of small steering angles δ_w ($<5°$) the trigonometric functions may be linearized and the tire forces related to the vehicle axis and the front wheel axis systems coincide, i.e., $F_{x,v,f} \approx F_{x,w,f}$ and $F_{y,v,f} \approx F_{y,w,f}$. For brevity, the second subscript can be dropped and the forces written as $F_{y,f}$ and $F_{x,f}$. The rear wheel axis system is assumed to coincide with the vehicle-fixed system anyway (front-wheel steered vehicle) and no approximations need to be made. Often the lateral velocity v_y is replaced by the sideslip angle β, which reads as

$$\beta = \text{atan}\left(\frac{v_y}{v_x}\right) \tag{7.2}$$

Again, the trigonometric function is linearized, i.e., the slip angle represents the ratio between lateral and longitudinal velocity. Due to continuous rotation of the vehicle axis system w.r.t. the z-axis the lateral acceleration a_y is given by

$$a_y = \frac{dv_y}{dt} + v_x\omega_z \tag{7.3}$$

Furthermore, under the assumption of an almost constant longitudinal velocity v_x, the lateral acceleration can be approximated by

$$a_y \approx v_x \left(\frac{d\beta}{dt} + \omega_z \right) \tag{7.4}$$

The angles arising between the wheel's velocity vectors $\mathbf{v}_f$, $\mathbf{v}_r$ and the corresponding heading directions are denoted as slip angles α_f, α_r. Linearization of the trigonometric functions yields the definitions:

$$\alpha_f = \delta_w - \beta - \frac{l_f \omega_z}{v_x} \quad \text{and} \quad \alpha_r = -\beta + \frac{l_r \omega_z}{v_x} \tag{7.5}$$

An important fact to be discussed at this place refers to the distances l_f and l_r. From Figure 7.2, it appears that the tire forces act directly at the wheel axis center. In fact this is not accurate, as the force build-up is not equally distributed over the tire contact patch, see e.g., [18] for more details. The displacement of the resulting force between the center and the force contact point is known as pneumatic trail n_p. It is assumed that the distances l_f and l_r already take into account a constant value of n_p and are modified accordingly.

Obviously, the front and rear road–tire forces do have a significant impact on the dynamics of the lateral velocity and yaw rate as shown in (7.1b) and (7.1c). In general, these lateral forces depend on many parameters, such as tire vertical load, lateral/longitudinal velocities, slip angles, camber angles and circumferential slip, to name the most important [17]. Depending on the required complexity and validity range of the tire force model there exist numerous modeling approaches in the literature. The tire models discussed in the following only consider the actual slip angle of the wheel and neglect any effects introduced by the camber angle. Moreover, any changes of vehicle load, tire–road adhesion, etc. are lumped into the model parameters. For small values of lateral acceleration[4] the relation between wheel slip angle and resulting lateral force appears linearly and a model reads as

$$F_y = c_\alpha \alpha \tag{7.6}$$

Therein, c_α denotes the lateral corning stiffness that is defined as

$$c_\alpha := \left. \frac{\partial F_y}{\partial \alpha} \right|_{\alpha=0} \tag{7.7}$$

However, the typical S-shaped characteristics, see Figure 7.3, of the slip–force curve cannot be represented by this linear force model and deviations between the model and real forces quickly become significant as soon as the linear range is left.

A more complex model consisting of at least four parameters is given by the well-known semi-empirical *magic formula* by Pacejka [19]. However, its employment for

[4]Often this refers to $|a_y| < 4\ \mathrm{m{\cdot}s^{-2}}$. It is known as *linear* or *proportional* range when referring to vehicle dynamics related jargon.

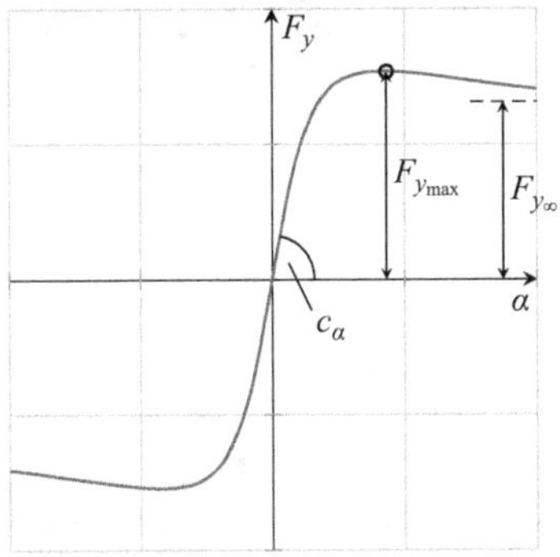

Figure 7.3 Characteristic points of the slip-force curve (TM_Simple [20])

the front and rear wheels requires eight parameters to be identified simultaneously. An alternative relying on three parameters by providing an accurate description is given by the TM_Simple tire model [20]. It combines the ideas of the magic formula and TMeasy [21]. Neglecting any effects due to vertical tire load variations the lateral tire force F_y can be expressed by

$$F_y = D \sin \underbrace{\left[B \left(1 - \mathrm{e}^{\frac{-|\alpha|}{C}} \right) \mathrm{sign}(\alpha) \right]}_{=:f_y(B,C,\alpha)} \tag{7.8}$$

Its parameters D, B and C correspond to characteristic points of the curve as illustrates Figure 7.3. The coherence between parameters and characteristic points (maximum $F_{y\max}$, limit value $F_{y\infty}$, cornering stiffness c_α) is defined by

$$D = F_{y\max} \tag{7.9a}$$

$$B = \pi - \mathrm{asin}\left(\frac{F_{y\infty}}{F_{y\max}}\right) \tag{7.9b}$$

$$C = c_\alpha^{-1} BD \tag{7.9c}$$

The parameters of the TM_Simple model are clearly affected by any vehicle configuration changes (e.g., vehicle load, tire type) as well as environmental influences (e.g., road surface, temperature). Hence, the model validity is increased, if the model parameters are selected from e.g., a look-up table, depending on the actual conditions.

Furthermore, the cornering stiffness of the front axle c_{α_f} accounts not only for the tires' stiffness, but also takes into account any elasticities of the steering system. Hence, it is denoted as *effective* front cornering stiffness $c_{\alpha_{fe}}$.

Merging the discussed findings into (7.1b) and (7.1c) results in a description of the lateral dynamics reading as

$$\frac{d\beta}{dt} = \frac{1}{mv_x}\left[D_f f_y(B_f, C_f, \alpha_f) + D_r f_y(B_r, C_r, \alpha_r)\right] - \omega_z \tag{7.10a}$$

$$\frac{d\omega_z}{dt} = \frac{l_f}{J_z} D_f f_y(B_f, C_f, \alpha_f) - \frac{l_r}{J_z} D_r f_y(B_r, C_r, \alpha_r) \tag{7.10b}$$

with $f_y(B, C, \alpha)$ as in (7.8) and α_f, α_r defined by (7.5).

Up to now, it has been assumed that the tire force build-up is instantaneous and there are no dynamics involved. However, in reality there is some phase lag introduced by the deflection of the tire tread, see e.g., [17]. Often, these dynamics are approximated by a first-order low pass filter with a time constant dependent on lateral tire stiffness,[5] the relaxation length (ratio between lateral cornering stiffness and lateral stiffness) and longitudinal vehicle speed, see e.g., [22].

7.1.2 Chassis roll motion

The rotation of the chassis around the axis pointing into the heading direction of the vehicle is denoted as the roll motion. Due to cornering (or crosswinds, road disturbances and/or obstacles) a roll angle between the vertical axes of the vehicle and its chassis arises. The resulting roll angle φ is an important measure for the handling evaluation as it acts directly on the driver and especially its head. Furthermore, it is also used for safety-related applications, e.g., roll-over detection systems [23].

In literature there exist various roll dynamics modeling approaches that differ in complexity and number of parameters. Given the task to define a model that describes the roll dynamics in a sufficiently accurate manner requiring a low number of parameters,[6] the roll dynamics modeling approach is based on a single degree of freedom assumption. Due to its simplicity and low number of parameters such a model is employed e.g., in [17,22,24–26]. Alternatively, [27] discusses higher-order modeling, but the number of parameters to characterize the model is not suitable for an online identification.

The restricted movement of the chassis can be interpreted as a rigid body rotating around a fixed point, defined as the roll center (RC). Figure 7.4 depicts the chassis' rotation around the longitudinal axis. The proposed simplifications hold if the following assumptions are not asserted:

- The road bank angle is negligible as the experiments are carried out under test track conditions.
- System excitation results from steering actuations rather than any environmental influences, e.g., rough road, obstacles.

[5]Not to be confused with the lateral *cornering* stiffness.

[6]The requirement for a low number of parameters comes from the objective to identify dominant model parameters online from system in- and output signals.

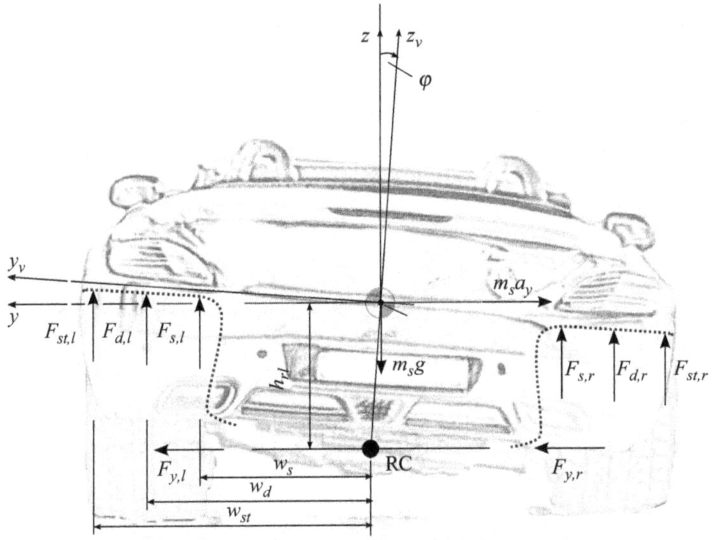

Figure 7.4 Schematics of the chassis roll motion

- Coupling effects between yaw and roll dynamics can be disregarded.
- The track width of the vehicle is constant and not affected by the roll motion.
- The suspension parts (front/rear) show similar characteristics and can be lumped into a single element per side.
- The roll centers (front/rear), i.e., roll axis, are replaced by a single roll center.

Forces acting on the vehicle chassis, illustrated in Figure 7.4, are due to the reactions of the suspension elements to the chassis rotation, the lateral tire forces and lateral acceleration. More precisely, the suspension forces consist of left/right spring forces $F_{s,l/r}$, the damping forces $F_{d,l/r}$ and the forces due to the roll stabilizer $F_{st,l/r}$. The distances between roll center and forces are denoted by w_s (springs), w_d (dampers) and w_{st} (stabilizers). All these distances are assumed constant and symmetric (w.r.t. the roll center). Moreover, h_{rl} defines the distance between CoG and RC. The vehicle mass needs to be separated into sprung and unsprung parts. The latter consists mainly of the axles and tire elements [22] and is abbreviated by m_u. Sprung parts are accumulated into the mass m_s. The moment of inertia of the sprung vehicle parts $J_{x,s}$ (w.r.t. the longitudinal vehicle axis) is adjusted for the roll center location, i.e., $J_{x,c} = J_{x,s} + m_s h_{rl}^2$.

Application of the angular momentum conservation principle w.r.t. RC results in the differential equation describing the roll motion, i.e.:

$$J_{xc}\frac{d\omega_x}{dt} = m_s a_y h_{rl}\cos\varphi + m_s g h_{rl}\sin\varphi + w_s\Delta F_s + w_d\Delta F_d + w_{st}\Delta F_{st} \quad (7.11)$$

The roll angular velocity ω_x, is defined as the time derivative of the roll angle φ, i.e., $\omega_x := \frac{d\varphi}{dt}$. Notice that the lateral acceleration a_y in (7.11) corresponds to the so-called *intermediate*[7] axis system. Its horizontal axis is parallel to the road surface (assuming a flat, non-banked road). However, the vehicle-fixed axis system corresponds to the chassis that is rotated by the angle φ around the longitudinal axis. Now, as the sensing device of the lateral acceleration is attached firmly to the vehicle body its measurements refer to the vehicle-fixed axis coordinates. Hence, its reading is given by

$$a_{y,m} = a_y \cos\varphi + g \sin\varphi \tag{7.12}$$

where the subscript m indicates the measurement. Furthermore, the terms ΔF_s, ΔF_d and ΔF_{st} in (7.11) are defined as differences between left and right suspension forces. For example, ΔF_s is given by

$$\Delta F_s := F_{s,l} - F_{s,r} \tag{7.13}$$

ΔF_d and ΔF_{st} are defined accordingly. For the modeling of the roll motion these forces are approximated by

$$F_s \approx (-1)^i \; c_s(\varphi) \; w_s \varphi \tag{7.14}$$

$$F_d \approx (-1)^i \; d(\omega_x) \; w_d \omega_x \tag{7.15}$$

$$F_{st} \approx (-1)^i \; \frac{c_{st} w_{st}}{2b_{st}^2} \; \varphi \qquad\qquad i = 1, 2 \tag{7.16}$$

The index i defines the direction of the force, i.e., $i = 1$ refers to the left and $i = 2$ to the right side. The spring stiffness is represented by $c_s(\varphi)$ and the damping coefficient by $d(\omega_x)$, respectively, the roll stabilizer lever arm is denoted by b_{st} and the torsional stiffness factor of the passive stabilizer reads as c_{st} [17].

Accumulating the spring and damping force-related coefficients into so-called *effective* stiffness and damping factors $c_\Sigma(\varphi)$ and $d_\Sigma(\omega_x)$ results in

$$c_\Sigma(\varphi) := 2\,w_s^2\, c_s(\varphi) + w_{st}^2 \,\frac{c_{st}}{b_{st}^2} \tag{7.17}$$

$$d_\Sigma(\omega_x) := 2\,w_d^2\, d(\omega_x) \tag{7.18}$$

Inserting these effective coefficients as well as (7.12) into (7.11) leads to

$$J_{xc} \frac{d\omega_x}{dt} = -c_\Sigma(\varphi)\varphi - d_\Sigma(\omega_x)\omega_x + m_s h_{rl} a_{y,m} \tag{7.19}$$

It shall be mentioned at this point that effects due to vertical tire deflection can be lumped into the effective stiffness factor and therefore do not appear explicitly in the model. Moreover, based on the argumentation in [28] the dominant non-linearities of the spring and damping forces do not have to be taken into account due to the opposite suspension deflection (when the chassis rotates around the roll axis). Consequently, the stiffness and damping parameters are modified such, that their dependencies on

[7] For further details the reader is referred to [9].

roll angle and velocity are neglected. In fact, the simplified effective factors $\bar{c}_\Sigma$, $\bar{d}_\Sigma$ result as

$$\bar{c}_\Sigma := 2\,\bar{w}_s^2\,\bar{c}_s + \bar{w}_s^2 \frac{c_{st}}{b_{st}^2} \tag{7.20}$$

$$\bar{d}_\Sigma := 2\,\bar{w}_s^2\,\bar{d} \tag{7.21}$$

with $\bar{w}_s$ being defined as the average distance between RC and the suspension position. The factors $\bar{c}_s$, $\bar{d}$ represent the assumed constant stiffness and damping factors $c_s(\varphi)$, $d(\omega_x)$.

The resulting roll dynamics equation can be formulated as

$$J_{xc}\frac{d\omega_x}{dt} = -\bar{c}_\Sigma\varphi - \bar{d}_\Sigma\omega_x + m_s h_{rl} a_{y,m} \tag{7.22}$$

and contains, in contrast to (7.19), only constant coefficients. This differential equation (or slight variations of it) are often seen within applications dealing with roll angle observation, e.g., [24,25,29].

For brevity, the notation $\bar{c}_\Sigma$ and $\bar{d}_\Sigma$ will be changed from now on to c_Σ and d_Σ.

Coupling effects with the lateral dynamics

Due to the occurrence of a roll angle (and consequently a roll velocity) the front and rear slip angles α_f, α_r, see (7.5), are to be modified. As a result, the effective slip angles α_{f_e} and α_{r_e} are introduced as [28]:

$$\alpha_{f_e} = \delta_w - \beta - \frac{l_f\omega_z}{v_x} - \frac{h_{rl}\omega_x}{v_x} \tag{7.23a}$$

$$\alpha_{r_e} = -\beta + \frac{l_r\omega_z}{v_x} - \frac{h_{rl}\omega_x}{v_x} \tag{7.23b}$$

Clearly, these modifications also need to be integrated into the description of the lateral dynamics as in (7.1b) and (7.1c).

7.1.3 Model parameters and sensor setup

The model-based vehicle handling evaluation approach requires accurate knowledge of both the model structure and its parameters. Otherwise, the simulated driving maneuvers differ significantly from the reality and drawing a conclusion on the vehicle handling is not feasible. Hence, the model parameters that shall be identified need to be defined. The differential equations describing the lateral as well as the roll dynamics, see (7.1b), (7.1c) and (7.22), consist of numerous parameters influencing the model dynamics. A systematic approach based on a sensitivity analysis [30,31] as well as characterization of vehicle parameters likely known from the design phase or measurable a priori (without extensive efforts) is presented in [4]. Herein, only the results are summarized in Table 7.2, i.e., model parameters to be identified online.

Aside the parameter estimation the objective of the proposed mechanisms is to use low-cost measurement equipment. These refer to angular rate and acceleration sensors due to their attractive cost-efficiency compared to angular position and

Table 7.2 Overview of domains, parameters and measurements

Domain	Parameters (known)	Parameters (to be identified)	Measurement
Lateral	l_f, l_r, J_z, m, h_{rl}	$D_{f/r}$, $C_{f/r}$, $B_{f/r}$	ω_z, δ, v_x
Roll	m_s, J_{xc}, h_{rl}	$\bar{c}_\Sigma$, d_Σ	ω_x, $a_{y,m}$

(optical) velocity sensing devices. Supported by a theoretical analysis of the system's observability measures, see [32] for details, the sensors listed in Table 7.2 have been selected to provide crucial information for the state and parameter estimation task.

The presented models of lateral and roll dynamics, parameters selected for the identification task and required in-vehicle measurements represent the basis for the proposed observer design concepts, see Section 7.3. Before discussing any observer designs a rather general overview of the state of the art w.r.t. the field of model-based vehicle handling evaluation and observer-based parameter identification is provided.

7.2 State-of-the-art review

Generally, the kind of problem discussed herein refers to a parameter identification of a *white-box* model. More specifically, this means a model of known structure, but its parameters are not [33]. This is in contrast to *black-box* or *gray-box* models, that assume no or only partial knowledge of the system's structure.

The algorithms for identifying these parameters can be classified into *batch* and *real-time* processing. Whereas *batch* processing, often denoted as an *offline* algorithm, operates on a complete data set, its counterpart the *real-time* processing algorithm only uses actual data samples available from measurements and the complete history of previous data can be dropped. Often real-time processing algorithms are associated with the term *recursive* as the identification result is updated whenever a new measurement value arrives. For an extensive overview and classification of individual identification methods the reader is referred to [33].

Putting the focus of the state-of-the-art review on model-based vehicle handling assessment the number of publications is rather small and mainly found in German-speaking countries. Phenomenological models and their parameter identification by a least-squares algorithm performed offline are discussed in [15]. In contrast, [34] operates on physically motivated model descriptions and identifies parameters based on a least-squares algorithm as well as an optimization procedure operating in the frequency domain. Both approaches are performed offline. An encouraging application of an online identification algorithm based on the Covariance Intersection (CI) method, see [35], is proposed in [16]. This concept does not, as opposed to the standard Kalman filter [36] or Recursive Least Squares [37] approach, assume that estimation errors and noise terms are uncorrelated. In the case of correlated terms this leads to correct estimates using the CI approach, but too optimistic results from the Kalman filter, i.e., underestimation of the covariances. Furthermore, it is claimed that the algorithm does not diverge in case of missing system excitation and it is better suited to operate on non-linear systems compared to standard least-squares-based concepts. However, its implementation is costly w.r.t. computational efforts.

All these proposed approaches have in common that a certain set of sensing equipment, as listed in Table 7.1, is required. But, for the herein considered sensor setup,[8] these approaches are not applicable directly. For example, [16,34] assume that the roll angle is measured in-vehicle, which is not true herein. Therefore, rather than measuring these vehicle responses they can be estimated by state observation schemes. In general, these aim to recover the initial states of the system, given its in- and outputs. Existence of an observer corresponds to the concept of a system's *observability*, see e.g., [38–40], needs to hold. The standard Luenberger observer [41] is well suited for the class of linear time-invariant systems. Extensions to non-linear (deterministic) systems have been proposed by e.g., [42–44]. Extending the scope of systems to the class of stochastic systems the Kalman filter (KF) is of essential importance. This set of very powerful equations aims to minimize the squared Euclidean distance of the estimation error's expected value. The important assumption on the internal process model as well as the measurement values is that these are Gaussian, zero-mean, white and uncorrelated. However, even in the case of non-Gaussian noise the Kalman filter is still *the* optimal linear filter [45]. In terms of practical applications related to the automotive area, the filter is often employed for sensor fusion within the context of positioning systems, e.g., [46] or vehicle state estimation [47]. In reality, the dynamics of the system to be observed often do not obey a linear system description. Therefore, extensions to the standard Kalman filter have been proposed such as the *Extended* KF or *Unscented* KF [48]. The idea of the Extended Kalman filter is to linearize the non-linear system around the filter estimate. It is assumed that the linearized propagation of the stochastic system states, i.e., its mean and covariance, from one time instant to the other is sufficiently close to the non-linear one. However, in the case of "severe" non-linearities the tuning of the Extended KF might be difficult and results in inaccurate[9] estimates [45]. The Unscented KF reduces the linearization errors of the Extended KF as the stochastic system states are not only represented by their means and covariances,[10] but so-called *sigma points*. These describe the probability density function of the stochastic system states and are propagated through the true non-linear system, rather than its linearized approximation. Applications of these two modified filters are manifold within the automotive area, e.g., estimation of tire forces [53], vehicle states [54–56] or load transfer [57].

A rather exhaustive overview of state-of-the-art observer mechanisms applicable to both, linear and non-linear systems, is provided in [58]. In general, the discussed

[8]Angular rate and translational acceleration measurements, rather than angular positions and translational velocities.

[9]For further details on stability properties of the filter consider e.g., [49–52].

[10]The EKF inherently assumes the states as Gaussian random variables, represented uniquely by their means and variances. Note that in general random variables are not represented uniquely by their first and second statistical moments.

state observers require both, good understanding of the system structure and also its parameters. However, the lack of accurate parameter knowledge and unmodeled dynamics paired with the availability of noise-corrupted measurements complicates the task of accurate state recovery significantly [59].

One way to interpret these unwanted effects is to introduce an additional system input that accounts for any parametric and/or modeling deficiencies. But then, standard concepts such as the Luenberger observer or Kalman filter are no longer capable of forcing the estimation error trajectories to the origin, but only some bounded region [60]. Now, the concept of sliding modes introduces not only robustness to perturbations, but invariance to certain classes of disturbances. State estimators based on first- and higher-order sliding modes have been applied successfully within the automotive area, e.g., [60–64], even if disturbances are acting on the system.

Combining these two problems of state estimation and parameter identification is well known as *joint state and parameter estimation* in the literature. The identification of parameters relies on the estimated states and vice versa. Within this work, two paradigms how to transform the problem of state and parameter estimation into a pure state estimation problem will be reviewed.

State augmentation

Given the description of a general system, i.e.:

$$\frac{d\mathbf{x}}{dt} = \mathbf{f}(\mathbf{x}, \boldsymbol{\theta}, \mathbf{u}) \tag{7.24a}$$

$$y = h(\mathbf{x}) \tag{7.24b}$$

with $\mathbf{x}(t) \in \mathbb{R}^n$ being the state vector, $\mathbf{u}(t) \in \mathbb{R}^m$ the input vector, $\boldsymbol{\theta} \in \mathbb{R}^k$ the vector of *constant* parameters to be identified and the function $\mathbf{f} : D_f \subseteq \mathbb{R}^n \times \mathbb{R}^k \times \mathbb{R}^m \to \mathbb{R}^n$. The initial conditions of $\mathbf{x}$ and $\boldsymbol{\theta}$ are defined as $\mathbf{x}_0 := \mathbf{x}(t_0)$ and $\boldsymbol{\theta}_0 := \boldsymbol{\theta}(t_0)$. Then, by definition of a new *augmented* state vector $\mathbf{x}_a(t) \in \mathbb{R}^{n+k}$, i.e.:

$$\mathbf{x}_a := [x_{a_1} \quad \dots \quad x_{a_{n+k}}]^T = [\underbrace{x_1 \quad \dots \quad x_n}_{n\times 1} \vdots \underbrace{\theta_1 \quad \dots \quad \theta_k}_{k\times 1}]^T \tag{7.25}$$

system (7.24) can be rewritten as

$$\frac{d\mathbf{x}_a}{dt} = \underbrace{\begin{bmatrix} f_1(\mathbf{x}_a, \mathbf{u}) \\ f_2(\mathbf{x}_a, \mathbf{u}) \\ \vdots \\ f_n(\mathbf{x}_a, \mathbf{u}) \\ f_{n+1}(\mathbf{x}_a, \mathbf{u}) \\ \vdots \\ f_{n+k}(\mathbf{x}_a, \mathbf{u}) \end{bmatrix}}_{=:\mathbf{f}_a(\mathbf{x}_a, \mathbf{u})} \begin{matrix} \left.\vphantom{\begin{matrix} a \\ b \\ c \\ d \end{matrix}}\right\} n \times 1 \\ \left.\vphantom{\begin{matrix} a \\ b \\ c \end{matrix}}\right\} k \times 1 \end{matrix} \tag{7.26a}$$

$$y = h_a(\mathbf{x}_a) \tag{7.26b}$$

where the last k elements of the function $\mathbf{f}_a(\mathbf{x}_a, \mathbf{u}) : D_f \subseteq \mathbb{R}^{n+k} \times \mathbb{R}^m \to \mathbb{R}^{n+k}$ are equal to zero as the model parameters are assumed constant. Obviously, the increase of the system order from n to $n+k$ necessitates a system observability analysis a priori of any observer design.

For that type of system description common state observation concepts can be applied straightforwardly. Those intrinsically estimate the system states and model parameters due to the modified system description (7.26). A difficulty that may arise due to the size of the parameter vector $\boldsymbol{\theta}$ is the determination of observability. As the system is rendered non-linearly once the state is augmented with unknown parameters the (local) criterion refers to evaluation of the Jacobian matrix of a diffeomorphism, see [59] for details. Note that the regularity of the Jacobian only represents a sufficient condition for local observability!

Provided that system (7.26) is at least locally observable an Extended or Unscented Kalman filter might be employed for the joint estimation of states and parameters. A general discussion on the convergence properties of an Extended KF used for state and parameter estimation is found in [65]. Especially within recent years the Unscented KF gained high attraction due to the better performance and advantages w.r.t. implementation as claimed by [66]. These mechanisms also found great acceptance for combined estimation of states and parameters related to vehicle dynamics, e.g., [54,61,67]. An interesting extension of the Extended Kalman filter is the so-called *Dual* Kalman filter that operates on individual instances of the filter for the state estimation and the parameter identification task [68]. Application to automotive problems is to be found in e.g., [69–71]. A system-immanent difficulty of all Kalman filter variants is the tuning of the process and measurement noise covariance matrices. Even though there exist suggestions for a systematic tuning, e.g., [72,73], it is a time-consuming and challenging process.

Unknown input reconstruction

In contrast to state vector augmentation, the approach of unknown input estimation lumps the uncertain or completely unknown parameters into a virtual input. Again, $\boldsymbol{\theta} \in \mathbb{R}^k$ is a parameter vector holding the uncertain parameters. Exploiting the physical meaning behind the parameters at least a nominal range can be assumed known. That allows separation into a known (nominal) and an unknown part, i.e.:

$$\boldsymbol{\theta} = \bar{\boldsymbol{\theta}} + \Delta\boldsymbol{\theta} \tag{7.27}$$

Here, $\bar{\boldsymbol{\theta}}$ represents the nominal and $\Delta\boldsymbol{\theta}$ the unknown part of vector $\boldsymbol{\theta}$. Now, considering system (7.24) and the separation of the state vector it can be re-written as

$$\frac{d\mathbf{x}}{dt} = \mathbf{f}(\mathbf{x}, \bar{\boldsymbol{\theta}}, \mathbf{u}) + \underbrace{\Delta\mathbf{f}(\mathbf{x}, \Delta\boldsymbol{\theta}, \mathbf{u})}_{=:\xi} \tag{7.28a}$$

$$y = h(\mathbf{x}) \tag{7.28b}$$

The objective of the observer design is now to estimate the system states $\mathbf{x} \in \mathbb{R}^n$ robustly, even if an unknown input $\boldsymbol{\zeta} \in \mathbb{R}^n$ acts on the system.[11] For the considered class of systems so-called *unknown input* observers can be constructed. Ideally, recovery of the input $\boldsymbol{\zeta}$ allows further identification of the parametric uncertainties $\Delta\boldsymbol{\theta}$. However, due to stochastic (measurement noise) and also deterministic (unmodeled system dynamics) noise effects the reconstructed unknown input gets distorted and this results ultimately in inaccurately estimated parameters.

Linear-time invariant systems that are affected by an unknown input require to revisit the notions of observability and detectability. For this system class these concepts are studied in [74]. Moreover, necessary and sufficient existence conditions of an unknown input observer are derived. The introduced notions of *strong* observability and *strong* detectability refer to the location of the system's invariant zeros, i.e., strong observability is given if no invariant zeroes exist and strong detectability in case the zeros are in $\mathbb{C}^-$. These findings are extended in [75] for a more general class of systems. Based on the introduction of *indistinguishable* states[12] the definitions of observability/detectability and the related existence conditions of unknown input observers need to be revised, see [75]. Herein, only the notion of indistinguishable states requires some explanation. Suppose that for system (7.28), $\mathbf{x}_0$ represents an initial state vector and $\mathbf{u}$, $\boldsymbol{\zeta}$ the known and unknown inputs. Then, the following theorem can be formulated as in [75].

Theorem 7.1. *Suppose* $\bar{\mathbf{x}}_0 \neq \mathbf{x}_0$ *and* $\mathbf{y}(t, \mathbf{x}_0, \mathbf{u}, \boldsymbol{\zeta}) = \mathbf{y}(t, \bar{\mathbf{x}}_0, \mathbf{u}, \bar{\boldsymbol{\zeta}})\ \forall t > 0$ *for some, distinct unknown inputs* $\boldsymbol{\zeta}, \bar{\boldsymbol{\zeta}}$, *then* $\bar{\mathbf{x}}_0$ *is said to be a strongly indistinguishable state of* $\mathbf{x}_0$.

For linear time-invariant systems affected by an unknown input, some observation schemes based on conventional sliding modes, i.e., 1-sliding, can be found in [76–78]. For the existence of these observers certain structural criteria need to be fulfilled [79], i.e.,

- The *observer matching condition* needs to hold. This is the analogue to the matching condition of the control scenario, see [80]. In other words, the transfer matrix from unknown input(s) to measured output(s) needs to have a relative degree equal to one.
- For rendering the sliding motion (asymptotically) stable the invariant zeros of the transfer matrix between unknown input(s) and measured output(s) need to be located in $\mathbb{C}^-$.

These conditions follow directly from [74]. The second condition can be weakened by e.g., [81]. Therein additional system outputs are calculated such that a conventional observer for estimation of states and unknown inputs can be formulated. Application

[11] For now, it is only assumed that $\boldsymbol{\zeta}(t)$ is bounded by some constant $\zeta^+ < \infty$. The mathematically more rigorous restrictions on the unknown input are defined in Section 7.3.

[12] This concept has been presented originally for systems without unknown inputs in [39].

of these concepts is often found within fault detection systems, e.g., [82], but also automotive problems, such as optimal braking with unknown road condition estimation [83]. Switching from the conventional sliding mode concepts to higher-order the *super-twisting* algorithm [84] is often employed. In [85] a robust state estimation concept that is based on the second-order super-twisting algorithm is presented. It is extended in [86] by exploiting the concept of equivalent output injection term to provide an estimate of the unknown input. Moreover, from the recovered input parametric uncertainties of the plant are reconstructed employing a time-continuous least-squares algorithm for both, constant and time-varying parameters. These algorithms (with some modifications) are used for vehicle state and parameter estimation in e.g., [60,62,64,87] or without the least-squares algorithm for tire effective radius estimation [88]. Exploiting the concept of equivalent output injection term for the recovery of parameters does have some general drawbacks that reduce the quality of parameter estimates in case of the herein discussed roll dynamics significantly. These are the following:

- The filtration process required to extract the equivalent output injection term from the discontinuous observer correction adds some phase lag on the reconstructed unknown input. In contrast, the states are estimated exactly, i.e., without any time delay compared to their true values. Now, given the case that the unknown input is represented by a linear combination of states and uncertain parameters the phase shift leads to offset-affected estimates.
- Employment of the discontinuous sliding mode observers allows finite time estimation of the states. However, convergence of the least squares-based parameter estimator occurs only asymptotically. Hence, the potential of the overall concept to drive the errors to zero within finite time is asserted.

At the end of this review one property of the observer design related to mechanical systems needs to be emphasized. The standard system setup is regarded as a second-order system, with state variables defined as position and velocity, with perturbations acting on the forces, i.e., derivative of the velocity, and the position is measured. However, for the roll model derived in Section 7.1.2 it is the roll velocity that is measured with perturbations still acting on the time derivative of the latter. Hence, the relative degree of the system output with respect to the unknown input is different from the standard setup. An application of the observer design recipe as in e.g., [85] is not possible!

7.3 Contributions

The main part of the chapter presents the authors' contributions to the specified problem of model-based vehicle handling evaluation. Aside the theoretical discussion of Sections 7.3.1 and 7.3.2, also their evaluation in simulations as well as real-world scenarios is reviewed, see Sections 7.3.3 and 7.3.4.

The main contributions of this work can be summarized as:

- Extension of the standard observer design process, as discussed in e.g., [85], for the application to mechanical systems, where the velocity (rather than position) is measurable.
- Integration of a finite-time parameter estimation algorithm based on higher-order sliding modes into the framework of unknown input observer-based estimation. This addresses directly the discussed issue of the equivalent output injection term recovery.
- Application of these formulated mechanisms to the vehicle dynamics assessment related tasks, such as tire force model and roll model parameter identification.

7.3.1 Robust lateral dynamics parameter estimator

Section 7.1.1 revealed the great importance of the road–tire force characteristics for the modeling of the lateral dynamics. Due to the time-, resource- and cost-intensity of parameter identification schemes based on tire test bench data, e.g., [89], the parameters of TM_Simple are to be estimated in-vehicle.

Clearly, even if the number of parameters is low (compared to more sophisticated tire force models[13]) it is necessary to estimate three values per axle resulting in six parameters that are assumed unknown. It appears reasonable that online identification of the parameters is almost impossible as only one state of system (7.10) is measured, i.e., the vehicle's yaw rate ω_z. See Table 7.2 at the end of Section 7.1 for further details. Recapitulating the definition of the force in (7.8) and also the coherence of the parameters with characteristic points of the curve, see (7.9), an approach is proposed that splits the identification into two subsequent phases.

- Stage one (S1) aims to identify the maximum value of the lateral tire forces during an aggressive vehicle excitation. That might either be a slow sinusoidal steering maneuver or a steady-state circular drive (both up to the limits of adhesion).
- Stage two (S2) addresses the identification of lateral cornering stiffness, see (7.7). Due to its representation as the slope of the slip-force characteristics at the origin, i.e., ($\alpha = 0$), the driving maneuver is designed such that the forces are within a narrow band around the origin. The slip angles shall remain very small. Figure 7.5 depicts the two maneuvers and the regions in terms of tire force and slip angle the vehicle has to be operated in.

The separation of the problem allows for design of slim, tailored observation concepts for online identification of the characteristic points and consequently tire model parameters.

[13]See for example Pacejka's magic formula that requires identification of numerous parameters even in its general form, see [90].

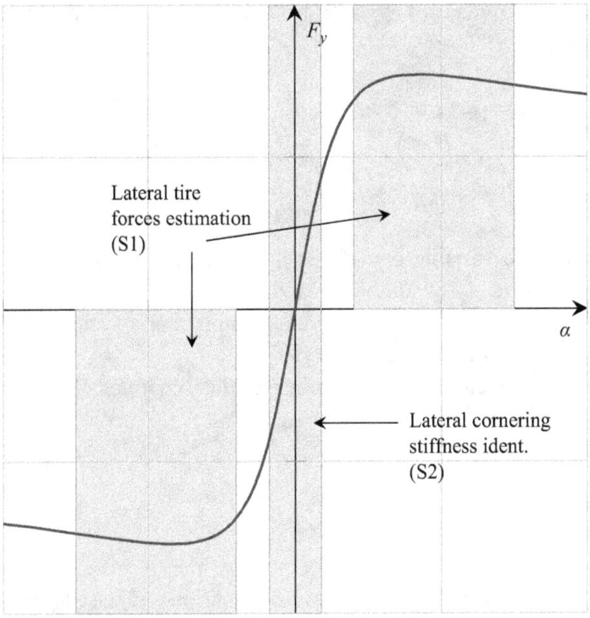

Figure 7.5 Suggested vehicle operational areas for the parameter identification of lateral tire slip-force characteristics

7.3.1.1 Lateral tire force observer

Prior to the design of the sliding mode observer the plant and its state variables, in- and outputs need to be defined. Based on (7.1b), (7.1c), (7.2) and (7.3) the system can be formulated as

$$\frac{dx_1}{dt} = k_1\, u + k_2\, \zeta_u \tag{7.29a}$$

$$y = x_1 \tag{7.29b}$$

with the constants k_1 and k_2 defined by

$$k_1 := \frac{l_f\, m}{J_z} \quad \text{and} \quad k_2 := -\frac{l_f + l_r}{J_z} \tag{7.30}$$

Furthermore, the state variable $x_1 := \omega_z$, input signal $u := a_y$ and uncertainty $\zeta_u := F_{y,r}$ are introduced. Note that the considered class of uncertainties is (a) bounded and (b) has a Lipschitz-continuous time-derivative for $t \geq 0$.

System (7.29) is said to be strongly observable as the output y is differentiated once until the unknown input appears. Hence, the relative degree equals the system order and no internal dynamics exist [91]. Therefore, an unknown input observer of the form [92]:

$$\frac{d\hat{x}_1}{dt} = k_1\, u + \lambda_1 \lfloor e_1 \rceil^{\frac{1}{2}} + \gamma \tag{7.31a}$$

$$\frac{d\gamma}{dt} = \lambda_2 \lfloor e_1 \rceil^0 \tag{7.31b}$$

can be designed. The variable e_1 defines the estimation error that is given by the deviation between true state variable x_1 and its estimate $\hat{x}_1$, i.e., $e_1 := x_1 - \hat{x}_1$. The observer gains λ_1 and λ_2 are both positive constants. For a systematic observer gain tuning approach the reader is referred to [93].

Re-writing the system in error coordinates, the dynamics, i.e.:

$$\frac{de_1}{dt} = \omega - \lambda_1 \lfloor e_1 \rceil^{\frac{1}{2}} \tag{7.32}$$

$$\frac{d\omega}{dt} = k_2 \frac{d\zeta_u}{dt} - \lambda_2 \lfloor e_1 \rceil^0 \tag{7.33}$$

with $\omega := k_2\, \zeta_u - \gamma$, reveal the convergence conditions of the algorithm, i.e., both, the unknown input ζ_u and its time derivative need to be upper bounded by constants ζ_1^+, ζ_2^+, i.e.:

$$\left|\zeta_u(t)\right| < \zeta_1^+ < \infty \tag{7.34a}$$

$$\left|\frac{d}{dt}\zeta_u(t)\right| < \zeta_2^+ < \infty \qquad \forall t > 0 \tag{7.34b}$$

Then, an appropriate choice of the observer gains (e.g., based on the upper bound ζ_2^+) ensures convergence of e_1 and ω to zero within finite time denoted by T_1, i.e.:

$$\frac{de_1(t)}{dt} = 0 \quad \Rightarrow \quad \omega(t) = 0 \quad \Rightarrow \quad k_2\, \zeta_u(t) = \gamma(t) \quad \forall t \geq T_1 \tag{7.35}$$

Hence, from knowledge of γ the unknown input ζ_u can be reconstructed. It shall be noted at this point that explicit filtration of the discontinuous term is <u>not</u> required.

From the estimated ζ_u, that equals the rear tire force and measurement of the lateral acceleration a_y, the front lateral tire force $F_{y,f}$ can be recovered, see also (7.3) and (7.1b).

7.3.1.2 Lateral cornering stiffness observer

The observer design is based on a modified representation of the classical single-track model as presented in (7.10a), (7.10b), (7.5) and (7.6).[14] However, rather than using the standard system input, i.e., steering angle δ_w, (7.10a) and (7.10b) are used to

[14] It should be noted at this point that the influence of the roll dynamics on the slip angle definition, as in (7.23), can be neglected due to the low excitation level of the vehicle being applied.

re-write the system such, that lateral acceleration represents the new system input.[15] Aside the input affine system description an additional advantage is the appearance of the uncertainty at only one (system) channel. The underlying system representation for the observer design reads as

$$\frac{dx_1}{dt} = \tilde{a}_{11}\, c_{\alpha_r}\, u_2^{-1}\, x_1 + \tilde{a}_{12}\, c_{\alpha_r}\, x_2 + b_{11}\, u_1 \tag{7.36a}$$

$$\frac{dx_2}{dt} = -x_1 + u_1 u_2^{-1} \tag{7.36b}$$

$$y = x_1 \tag{7.36c}$$

with the constants $\tilde{a}_{11}$, $\tilde{a}_{12}$ and b_{11} abbreviating the terms:

$$\tilde{a}_{11} := -\frac{(l_f + l_r)\, l_r}{J_z} \qquad \tilde{a}_{12} := \frac{(l_f + l_r)}{J_z} \qquad b_{11} := \frac{l_f\, m}{J_z} \tag{7.37}$$

The state variables x_1 and x_2 are defined as the yaw rate ω_z and the sideslip angle β, i.e., $x_1 := \omega_z$, $x_2 := \beta$. As discussed, u_1 is defined as the lateral acceleration and u_2 represents the actual longitudinal velocity of the vehicle, i.e., $u_1 := a_y$ and $u_2 := v_x$.

Now, formulating the problem of parameter identification as an unknown input estimation task requires the parameter c_{α_r} to be separated into a nominal $\bar{c}_{\alpha_r}$ and an uncertain part Δc_{α_r}, i.e.:

$$c_{\alpha_r} = \bar{c}_{\alpha_r} + \Delta c_{\alpha_r} \tag{7.38}$$

Merging (7.38) with the system description (7.36) results in a system reading as

$$\frac{dx_1}{dt} = \tilde{a}_{11}\, \bar{c}_{\alpha_r}\, u_2^{-1}\, x_1 + \tilde{a}_{12}\, \bar{c}_{\alpha_r}\, x_2 + b_{11}\, u_1 + \zeta(x_1, x_2) \tag{7.39a}$$

$$\frac{dx_2}{dt} = -x_1 + u_1 u_2^{-1} \tag{7.39b}$$

$$y = x_1 \tag{7.39c}$$

Therein, ζ represents an unknown system input that constitutes of uncertain parameters and system states, i.e.:

$$\zeta(x_1, x_2) := \underbrace{[\, \tilde{a}_{11}\, u_2^{-1}\, x_1 + \tilde{a}_{12}\, x_2 \,]}_{=:\Gamma(x_1, x_2)}\, \Delta c_{\alpha_r} \tag{7.40}$$

Assuming the system inputs u_1 and u_2 to be bounded, which is reasonable due to the physical meaning of these values, the state variables are also bounded.[16] Due to these

[15] That idea is based on [94] and leads to an input affine system representation.
[16] Consequently, system (7.39) is Bounded-Input Bounded-Output stable, see [31] for details.

facts and the bounded value of the cornering stiffness, the unknown input $\zeta(x_1,x_2)$ is upper bounded by

$$|\zeta(x_1, x_2)| < \zeta^+ < \infty \tag{7.41}$$

Obviously, the assumption of a bounded unknown input is asserted for a vanishing input u_2, i.e., longitudinal velocity v_x. However, estimation of cornering stiffness for a vehicle (almost at standstill) does not make sense whatsoever and therefore v_x will be definitely above some feasible threshold.

Before discussing an observer design for a robust state estimation of system (7.39), the indistinguishability of the states shall be checked. Applying Theorem 7.1 to (7.39), a second instance of the system needs to be introduced (states z_1, z_2) with a different unknown input denoted by $\bar{\zeta}(z_1,z_2)$. The system reads as

$$\frac{dz_1}{dt} = \tilde{a}_{11}\,\bar{c}_{\alpha_r}\,u_2^{-1}\,z_1 + \tilde{a}_{12}\,\bar{c}_{\alpha_r}\,z_2 + b_{11}\,u_1 + \bar{\zeta}(z_1, z_2) \tag{7.42a}$$

$$\frac{dz_2}{dt} = -z_1 + u_1 u_2^{-1} \tag{7.42b}$$

$$y = z_1 \tag{7.42c}$$

Defining the state errors $e_1 := x_1 - z_1$ and $e_2 := x_2 - z_2$ their dynamics can be formulated by

$$\frac{de_1}{dt} = \tilde{a}_{11}\,\bar{c}_{\alpha_r}\,u_2^{-1}\,e_1 + \tilde{a}_{12}\,\bar{c}_{\alpha_r}\,e_2 + \Delta\zeta(e_1, e_2) \tag{7.43a}$$

$$\frac{de_2}{dt} = -e_1 \tag{7.43b}$$

with $\Delta\zeta(e_1,e_2) := \zeta(x_1,x_2) - \bar{\zeta}(z_1,z_2)$. Now, assuming that the output signals of the two systems are identical as argued in Theorem 7.1 yields $e_1 \equiv 0$.

From (7.43b) it follows that $e_2(t) = C$, where C denotes some constant in general different from zero.[17] Inserting these findings into (7.43a) the relation:

$$-\tilde{a}_{12}\,\bar{c}_{\alpha_r}K = \Delta\zeta(0, e_2) \tag{7.44}$$

with K being some arbitrary constant different from zero, follows directly. In other words, even though the system outputs y are equal the differences in terms of unknown inputs are compensated for by the second system state. Hence, the dynamics of (7.39) are said to be indistinguishable. This leads to the inability to design an observer allowing for recovery of the system's initial states. These findings coincide with the fact that the system is neither strongly observable nor strongly detectable, which is the weaker criterion, see [74] for details. In other words, it is the transfer function from unknown input ζ to system output y that shows a zero at the origin of the s-plane. Hence, the internal dynamics are stable, but not asymptotically!

[17] Considering an initial state $z_2(t_0) \neq 0$.

Now, the challenging question is how to overcome that problem. In fact, the initial condition of state x_2 can be selected almost at choice by the design of experiments. Assuming that there is no lateral motion of the vehicle before conductance of the identification experiments it is feasible to consider the initial condition *before* any lateral excitation is applied.[18] Reconsidering (7.39b) its input $\tilde{u}_1 := u_1 u_2^{-1}$, state x_1 and also its initial condition are known. Hence, numerical integration provides an estimate of x_2, i.e.:

$$\hat{x}_2(t) = \int_0^t -x_1(\tau) + \tilde{u}_1(\tau) d\tau \tag{7.45}$$

The initial condition $x_2(0)$ is assumed zero. Clearly, obtaining estimates from numerical integration is generally not recommended,[19] but due to a short experiment duration, i.e., integration period, the method appears to be suitable. Moreover, its feasibility is evaluated by simulation-based work, see Section 7.3.3, and also in real-world experiments, Section 7.3.4, and the results are encouraging.

Now, reviewing again the problem of uncertain lateral cornering stiffness identification the system to be considered can be reformulated to

$$\frac{dx_1}{dt} = \tilde{a}_{11}\, \bar{c}_{\alpha_r}\, u_2^{-1}\, x_1 + \tilde{u}_a + \zeta(x_1, x_2) \tag{7.46a}$$

$$y = x_1 \tag{7.46b}$$

as the second state variable is now interpreted as a known system input and lumped into $\tilde{u}_a := \tilde{a}_{12}\bar{c}_{\alpha_r}\hat{x}_2 + b_{11}u_1$. This system has definitely distinguishable dynamics, is strongly observable and allows the introduction of a finite time parameter estimator [95] that is based on higher-order sliding modes. In contrast to standard least-squares-based approaches, e.g., [33], showing an asymptotic convergence behavior, it provides parameter estimates in finite time. It is based on a *generalized* super-twisting algorithm (GSTA), see [96]. It includes linear correction terms, that have more influence on the dynamics than their non-linear pendants, given that the trajectories are far away from the origin. However, given a certain set of parametrization, the generalized super-twisting algorithm degenerates to the standard one. The finite time parameter estimator reads as

$$\frac{d\hat{x}_1}{dt} = \tilde{a}_{11}\, \bar{c}_{\alpha_r}\, u_2^{-1}\, \hat{x}_1 + \tilde{u}_a + \Gamma(\hat{x}_1, \hat{x}_2)\,\hat{\theta} - \alpha_1 \phi_1(e_1) \tag{7.47a}$$

$$\frac{d\hat{\theta}}{dt} = -\alpha_2\, \phi_2(e_1)\, \Gamma(\hat{x}_1, \hat{x}_2) \tag{7.47b}$$

[18]It shall be emphasized that the discussed initial condition of the vehicle, does not refer to the very initial state of the vehicle, i.e., at its resting position. It moreover refers to the condition before any lateral excitation of the vehicle is conducted. Hence, it might be driven in a straight line beforehand.

[19]Given the fact that x_2 is based on measured values of x_1 and also $\tilde{u}_1$ their noise might not be perfectly zero-mean. Therefore, the longer the integration interval is, the more influence the mean value of the measurement noise will have.

with the correction terms $\phi_1(e_1)$ and $\phi_2(e_1)$:

$$\phi_1(e_1) := \mu_1 \lfloor e_1 \rceil^{\frac{1}{2}} + \mu_2 e_1 \tag{7.48a}$$

$$\phi_2(e_1) := \frac{1}{2}\mu_1^2 \lfloor e_1 \rceil^0 + \frac{3}{2}\mu_1 \mu_2 \lfloor e_1 \rceil^{\frac{1}{2}} + \mu_2^2 e_1 \tag{7.48b}$$

The error e_1 is defined as the difference between estimated and measured yaw rate, i.e., $e_1 := \hat{x}_1 - x_1$. α_1, α_2 and μ_1 are strictly positive constants and $\mu_2 \geq 0$. The variable $\hat{\theta}$ represents an estimate of the uncertain cornering stiffness, i.e., $\hat{\theta} := \Delta\hat{c}_{\alpha_r}$. The definition of $\Gamma(\hat{x}_1, \hat{x}_2)$ is to be found in (7.40). For a proof of stability and finite time convergence the reader is referred to [95].

Aside its finite time convergence capability the algorithm introduces another major advantage, i.e., it provides bounded estimates $\hat{\theta}$ even if persistence of excitation is not guaranteed.

Comment on the extension to front cornering stiffness estimation

The algorithm presented so far only allows identification of the rear cornering stiffness. However, its knowledge does not reveal any information on the front value. Hence, using (7.5), (7.6), (7.10a) and (7.10b) the system needs to be formulated such, that the front cornering stiffness represents the uncertainty in the first system channel, rather than the rear one. From a structural perspective the model looks like (7.39), but the new unknown input ξ is now a function of the states x_1, x_2 and also the input δ_w. Regardless of the structural (internal) changes of the unknown input the parameter estimation algorithm can be designed identically to the one presented previously.

The conversion of the identified characteristic points of the slip-force curve to tire force model parameters will be discussed further in Section 7.3.3.1.

7.3.2 Robust roll dynamics parameter estimator

The objective of the robust state and parameter estimator is to recover the non-measured roll angle and unknown model parameters of effective roll stiffness c_Σ and roll damping d_Σ from roll velocity measurements. Based on (7.22) a system formulation can be obtained by

$$\frac{dx_1}{dt} = x_2 \tag{7.49a}$$

$$\frac{dx_2}{dt} = -\bar{a}_{21}x_1 - \bar{a}_{22}x_2 + b_2 u + \rho(x_1, x_2) \tag{7.49b}$$

$$y = x_2 \tag{7.49c}$$

with $x_1 := \varphi$, $x_2 := \omega_x$. The general parameters a_{21}, a_{22} and b_2 are defined as

$$a_{21} := \frac{c_\Sigma}{J_{xc}} \qquad a_{22} := \frac{d_\Sigma}{J_{xc}} \quad \text{and} \quad b_2 := \frac{m_s h_{rl}}{J_{xc}} \tag{7.50}$$

The known system input is represented by $u := a_{y,m}$, the *measured* (referring to the vehicle-fixed axis system) lateral acceleration, and the output is $y := x_2$, the roll velocity. Furthermore, the unknown input is defined to be $\rho(x_1, x_2)$. Its structure is not further discussed at the moment. In contrast to the unknowns a_{21} and a_{22}, it is assumed that b_2 is perfectly known and constant.[20] Transforming the parameter estimation problem into an unknown input recovery requires the separation of parameters into nominal and uncertain parts, i.e.:

$$a_{21} := \bar{a}_{21} + \Delta a_{21} \quad \text{and} \quad a_{22} := \bar{a}_{22} + \Delta a_{22} \tag{7.51}$$

Denoted by $\bar{a}_{21/22}$ are the nominal parts as opposed to the uncertainties $\Delta a_{21/22}$. Now, the unknown input can be formulated as a linear combination of uncertain parameters and states reading as

$$\rho(x_1, x_2) := -\Delta a_{21} x_1 - \Delta a_{22} x_2 \tag{7.52}$$

For the design of a robust state observer, e.g., based on conventional sliding modes [79] the observer matching condition, as well as strong observability or at least detectability need to be fulfilled. Whereas the former obviously holds for the roll dynamics estimation problem, the latter fails as the transfer function from unknown input ρ to the system output y suffers from a zero at the origin of the s-plane. Hence, it is neither strongly observable nor strongly detectable as the internal dynamics are only stable. Consequently, a conventional unknown input observer does not even exist, see [74,75].

Interpreting system (7.49) as a mechanical one allows the following interpretation: within standard observation problems the position x_1 is measurable and perturbations act on the forces, i.e., the second system channel or time derivative of x_2. Then, observation concepts as presented in [85,86] are able to robustly recover the system states and furthermore provide estimates of the unknown input. However, for the given system setup it is the velocity, rather than the position, that is measured. And as the unknown input cannot be described by the system output y (and its time derivatives), as would be the case for strongly observable systems, the observer cannot be designed.

Introducing a diffeomorphism $\mathbf{\Phi}$ allows a transformation of the state $\mathbf{x}$ to $\mathbf{z} \in \mathbb{R}^n$. It reads as

$$\mathbf{\Phi}(\mathbf{x}, u) := \begin{bmatrix} z_1 \\ z_2 \end{bmatrix} = \begin{bmatrix} c(\mathbf{x}) \\ L_{\mathbf{a}}c(\mathbf{x}) + L_{\mathbf{b}}c(\mathbf{x})u \end{bmatrix} = \begin{bmatrix} y \\ \frac{dy}{dt} \end{bmatrix} \tag{7.53}$$

with the *Lie*-derivative defined as

$$L_{\mathbf{a}}c(\mathbf{x}) := \frac{\partial c(\mathbf{x})}{\partial \mathbf{x}} \mathbf{a}(\mathbf{x}) \tag{7.54}$$

[20]The identification of the parameters a_{21} and a_{22} introduces the advantage that any deficiencies in the knowledge of the moment of inertia J_{xc} are compensated for by the estimates. Hence, even if the absolute values of c_Σ and d_Σ are incorrect, the dynamics of the real system will be approximated correctly by the model representation.

Then, system (7.49) can be re-written in **z**-coordinates as

$$\frac{dz_1}{dt} = z_2 \tag{7.55a}$$

$$\frac{dz_2}{dt} = \bar{f}(z_1, z_2, v) + \zeta(z_1, z_2) \tag{7.55b}$$

$$y = z_1 \tag{7.55c}$$

with v representing the new input and being defined as $v := \frac{du}{dt}$. The functions $\bar{f}$ and ζ represent the nominal dynamics and the unknown input respectively, i.e.:

$$\bar{f}(z_1, z_2, v) := -\bar{a}_{21}z_1 - \bar{a}_{22}z_2 + b_2 v \tag{7.56a}$$

$$\zeta(z_1, z_2) := -\Delta a_{21}\, z_1 - \Delta a_{22}\, z_2 \tag{7.56b}$$

Both functions are assumed bounded in any compact set of the state space. Accounting for the underlying mechanical character of the system it is clear that these assumptions hold. The upper bound of the uncertain term is introduced explicitly as it is of importance for the observer gain tuning. It is denoted by ζ^+ and reads as

$$|\,\zeta(z_1, z_2)\,| < \zeta^+ < \infty \qquad \forall z_1, z_2 \tag{7.57}$$

Interestingly, that change of coordinates does not affect the model parameters $\bar{a}_{11}$, $\bar{a}_{21}$, b_2 at all. But, the relative degree of the output y w.r.t. the unknown input ζ now equals the system order. Unfortunately, this necessitates the calculation of the time derivative of the input signal. Basically, the state transformation can be interpreted as adding an integrator to the original system input and defining that series connection as the new system. Obviously, the integrator cancels the distracting zero of the original system's transfer function (from unknown input ρ to output y). Even though the diffeomorphism contains uncertain parameters, i.e., a_{21} and a_{22}, this is not a problem as it is only used to transform the system into form (7.55), but the reverse transformation to **x**-coordinates is never used. Now, for system (7.55) a standard higher-order sliding mode observer, e.g., [86], allowing for robust estimation of the states and also recovery of the unknown input (for a subsequent estimation of the uncertain parameters) can be applied. However, for the objective of identifying Δa_{21} and Δa_{22} as well as estimation of x_1 (referring to the chassis' roll angle φ) individual tasks are to be defined for tailored submodules, as depicted by Figure 7.6. First, the time derivative of the system input u needs to be calculated accurately by a robust exact differentiator (RED) [93]. Its output serves as input for system (7.55). Second, a robust state observation concept recovers the system states z_1 and z_2 and might also provide an estimate of the unknown input ζ. Third, a finite time parameter estimation algorithm replaces the standard least-squares approach, as presented in e.g., [86], and provides estimates of the uncertainties Δa_{21} and Δa_{22}. And fourth, an adaptive Luenberger observer is employed to calculate the actual roll angle by taking into account the measured roll velocity and also the uncertain parameter estimates. The latter is necessary, as the recovered system states

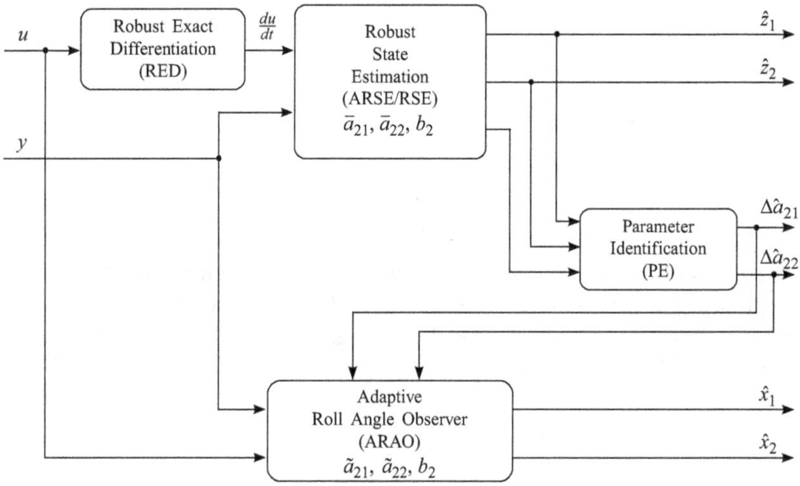

Figure 7.6 Proposed concept schematics of roll model parameter identification

z_1 and z_2 physically refer to roll velocity and acceleration, rather than angle x_1 and velocity x_2.

7.3.2.1 Robust exact differentiator (RED)

Providing exact estimates of the time derivative without introducing any delay to the output due to filtering operations, the robust exact differentiator as presented and discussed in [93,97] is *the* choice for the given task. Generally, it interprets the task of time differentiation as a control problem. An integrator represents an artificial plant whose output $\hat{u}$ ideally tracks some given reference value u. Employing a variable structure concept for the control of the loop provides inherent robustness and also finite time convergence of the control error. The class of input functions u is restricted to bounded functions with Lipschitz-continuous derivatives and some small amount of noise [93]. Introducing the error coordinate:

$$\varepsilon := u - \hat{u} \tag{7.58}$$

it is the objective of the control to force it to zero within finite time T_2, i.e., $\varepsilon \equiv 0 \ \forall t \geq T_2$.

This can be achieved by using a super-twisting controller of the form:

$$v = \bar{v} + \kappa_1 \lfloor \varepsilon \rceil^{\frac{1}{2}} \tag{7.59a}$$

$$\frac{d\bar{v}}{dt} = \kappa_2 \lfloor \varepsilon \rceil^{0} \tag{7.59b}$$

The controller parameters κ_1 and κ_2 are strictly positive constants. For a systematic tuning of the gains an upper bound of the uncertainty needs to be defined. Therefore, the system is re-written in error coordinates and reads as

$$\frac{d\varepsilon}{dt} = \underbrace{\frac{du}{dt} - \bar{\nu}}_{=:\bar{\varepsilon}} - \kappa_1 \lfloor \varepsilon \rceil^{\frac{1}{2}} \tag{7.60a}$$

$$\frac{d\bar{\varepsilon}}{dt} = \frac{d^2 u}{dt^2} - \kappa_2 \lfloor \varepsilon \rceil^0 \tag{7.60b}$$

Obviously, that refers to a standard setup when considering the super-twisting control approach. For proofs of stability and finite time convergence see e.g., [93,98]. Reviewing (7.60b), it becomes clear that the second time derivative of the system input u needs to be upper bounded, i.e.:

$$\left| \frac{d^2}{dt^2} u(t) \right| < u^+ < \infty \qquad \forall t \tag{7.61}$$

Adjustment of the controller gains can be based on the upper bound of the uncertainty and is presented by [84,93]:

$$\kappa_1 = 1.5 \left(u^+\right)^{\frac{1}{2}} \qquad \text{and} \qquad \kappa_2 = 1.1 u^+ \tag{7.62}$$

Moreover, if the input signal u is corrupted by some (bounded) noise term $|n| \leq n^+$, it can be shown that the accuracy of the time differentiation is given by the term:

$$\left| \nu - \frac{du}{dt} \right| < \tilde{\kappa}_1 \sqrt{n^+} \qquad \forall t \geq T_2 \tag{7.63}$$

with $\tilde{\kappa}_1$ being a function of κ_1, κ_2, see [93] for details.

7.3.2.2 Robust state estimation (RSE)

Due to the convenient structure of (7.55) the design of a variable structure observer based on a super-twisting algorithm is straightforward and follows the ideas of [85]. It consists of a plant replica and the correction terms, i.e.:

$$\frac{d\hat{z}_1}{dt} = \hat{z}_2 + \lambda_1 \lfloor e_1 \rceil^{\frac{1}{2}} \tag{7.64a}$$

$$\frac{d\hat{z}_2}{dt} = -\bar{a}_{21}\hat{z}_1 - \bar{a}_{22}\hat{z}_2 + b_2 \nu + \lambda_2 \lfloor e_1 \rceil^0 \tag{7.64b}$$

The estimation errors are introduced as $e_1 := z_1 - \hat{z}_1$ and $e_2 := z_2 - \hat{z}_2$. Then, the error dynamics can be written as

$$\frac{de_1}{dt} = e_2 - \lambda_1 \lfloor e_1 \rceil^{\frac{1}{2}} \tag{7.65a}$$

$$\frac{de_2}{dt} = -\bar{a}_{21} e_1 \underbrace{-\bar{a}_{22}\, e_2 + \zeta(z_1, z_2)}_{=:\xi(z_1, z_2, \hat{z}_2)} - \lambda_2 \lfloor e_1 \rceil^0 \tag{7.65b}$$

Finite time convergence of the estimation errors e_1 and e_2 requires the uncertainty ξ to be upper bounded, i.e.:

$$|\xi(z_1, z_2, \hat{z}_2)| < \xi^+ < \infty \tag{7.66}$$

for any z_1, z_2 and $|\hat{z}_2| \leq 2\,\text{sup}|z_2|$, see [85] for details. System (7.55) is BIBS stable and due to the boundedness of system input v (see Section 7.3.2.1) the existence of ξ^+ is guaranteed. Tuning of the observer gains can be based on the upper bound ξ^+ and (7.62). If the observer gains are tuned correctly, solutions of system (7.65) converge to the origin $(e_1, e_2) = (0, 0)$ within finite time T_3, i.e.:

$$e_1(t) = e_2(t) = 0 \qquad \forall t \geq T_3 \tag{7.67}$$

For a stability proof the reader is referred to e.g., [85,93,98]. Moreover, for the estimation of the finite convergence time T_3 different methods can be found in [85,93,98–100]. These rather conservative estimates can theoretically be used for the activation of the parameter identification algorithm.

7.3.2.3 Adaptive robust state estimation (ARSE)

So far, some knowledge of the uncertainty-related upper bound is assumed known for the tuning of the observer mechanism. Clearly, the better the bound is known, the less noise due to the high switching terms will be introduced into the estimation values. Often, conservatively chosen bounds lead to high gains and degrade the estimation quality. Another scenario might be a time-varying bound of the uncertainty as is the case for the roll model parameter estimation. In terms of experiments design the system will be excited at different frequencies leading to uncertainties whose bounds vary from experiment to experiment. An adaptive gain scheme, e.g., [101–104], that replaces the static gains tuned for a single upper bound of the uncertainty, will be introduced. Now, the observer gains will be denoted by $\tilde{\lambda}_1$, $\tilde{\lambda}_2$ to distinguish them explicitly from their static counterparts λ_1, λ_2, i.e.:

$$\tilde{\lambda}_1 := \tilde{\lambda}_1(e_1, e_2, \ldots, e_n, t) \qquad \text{and} \qquad \tilde{\lambda}_2 := \tilde{\lambda}_2(e_1, e_2, \ldots, e_n, t) \tag{7.68}$$

The algorithm discussed in the following is known as adaptive gain super-twisting algorithm [102] and it varies the observer gains adaptively until a sliding motion is established. The latter is triggered by a detector that determines entering and leaving of a domain $\Omega := |e_1| - \mu$, with μ being some positive constant. Clearly, the definition of the detector is not unique and at the discretion of the designer. For the given application it is chosen as a band around the sliding variable e_1 [105]. Once the trajectories enter the domain Ω the observer gains decrease. In turn, as soon as the trajectories leave the band they are pushed back by increased gains. Mathematically speaking, the algorithm reads as

$$\frac{d\tilde{\lambda}_1}{dt} = \begin{cases} \kappa \lfloor \Omega \rceil^0 & \text{if } \tilde{\lambda}_1 > \tilde{\lambda}^- \\ \eta & \text{else} \end{cases} \tag{7.69a}$$

$$\tilde{\lambda}_2 = \varepsilon \tilde{\lambda}_1 \tag{7.69b}$$

with $\tilde{\lambda}^-$ representing a lower bound for the gain $\tilde{\lambda}_1$ and κ, η, ε being defined as positive constants. Furthermore, the initial value $\tilde{\lambda}_1(t_0)$ needs to be chosen such, that $\tilde{\lambda}_1(t_0) > \tilde{\lambda}^-$ holds. Finally, $\eta \geq \mu$ needs to be ensured.

The rate of observer gain change is driven by the gain κ. Moreover, η is some small positive constant that is employed if the gain $\tilde{\lambda}_1$ falls below the threshold $\tilde{\lambda}^-$. Scaling between $\tilde{\lambda}_1$ and $\tilde{\lambda}_2$ is accomplished by a factor ε. For a stability proof of the algorithm consider [102].

Both super-twisting algorithms (with static or adaptive gains) are able to bring the estimation errors to zero in finite time.[21] Reconsidering (7.65b) and the fact the $e_2 \equiv 0$ after time T_3 one can write:

$$\frac{de_2}{dt} \equiv 0 \quad \Rightarrow \quad 0 \equiv \zeta(z_1, z_2) - \nu_{eq} \qquad t \geq T_3 \tag{7.70}$$

Here, the term ν_{eq} is denoted as the equivalent output injection term that can be approximated by low-pass filtering of the discontinuous term,[22] i.e.:

$$\hat{\nu}_{eq} := \left(\tilde{\lambda}_2 \lfloor e_1 \rceil^0\right)_{\text{lpf}} \tag{7.71}$$

The higher the switching frequency and the smaller the filter constant the more tends $\hat{\nu}_{eq}$ to ν_{eq} [106]. From the approximated equivalent term $\hat{\nu}_{eq}$ an estimate of the unknown input $\hat{\zeta}$ can be recovered. Exploiting that quantity along with the definition of the unknown input, see (7.56b), a least-squares algorithm might be employed to identify the uncertain parameters Δa_{21} and Δa_{22}, see e.g., [86,87].

Herein, an alternative approach is chosen that is based on the finite time parameter estimation concept already presented in Section 7.3.1.2.

7.3.2.4 Parameter identification (PE)

The estimation of uncertain roll dynamics model parameters using the equivalent output injection concept suffers from several disadvantages that will be discussed in the following.

- Assume convergence of the robust state estimator, i.e., $(e_1, e_2) = (0, 0)$. Then, the state estimates z_1 and z_2 are accurate or in other words, there is no time delay between true and estimated values. However, the uncertainty estimation (based on the equivalent output injection term) suffers from a time delay due to the low-pass filtration. That mismatch in terms of timely correlation leads to distorted parameter estimates not converging to the true values.
- Even though the robust exact differentiator and also the robust state estimator converge in finite time the parameter estimation algorithm (e.g., least-squares approach) asserts the finite time convergence property.

[21] Please note that in contrast to the static gain algorithm the adaptive algorithm leads to a *real* sliding mode [102].

[22] Alternatively, to the use of the adaptive gains, this approach also works with static gains.

- In case the vehicle is not excited persistently the activity of a standard least-squares-based estimator shall be stopped as the estimates may diverge otherwise. For the proposed scheme this is not the case as its estimates remain bounded even in the case of non-persistent excitation.

The parameter estimator based on the generalized super-twisting algorithm reads as [95]:

$$\frac{d\tilde{z}_2}{dt} = -\bar{a}_{21}z_1 - \bar{a}_{22}\tilde{z}_2 + b_2 v - \alpha_1 \left(\mu_1 \lfloor \tilde{e}_2 \rceil^{\frac{1}{2}} + \mu_2 \tilde{e}_2 \right) + \mathbf{\Gamma}(\hat{z}_1, \tilde{z}_2)\hat{\boldsymbol{\theta}} \quad (7.72a)$$

$$\frac{d\hat{\boldsymbol{\theta}}}{dt} = -\alpha_2 \left(\frac{\mu_1^2}{2} \lfloor \tilde{e}_2 \rceil^0 + \frac{3}{2}\mu_1\mu_2 \lfloor \tilde{e}_2 \rceil^{\frac{1}{2}} + \mu_2^2 \tilde{e}_2 \right) \mathbf{\Gamma}(\hat{z}_1, \tilde{z}_2)^T \quad (7.72b)$$

with the error $\tilde{e}_2$ defined as $\tilde{e}_2 := \tilde{z}_2 - \hat{z}_2$. The value $\hat{z}_2$ represents the estimate of state z_2 from the robust state estimator and $\tilde{z}_2$ denotes the estimate of $\hat{z}_2$ computed by the parameter estimator algorithm. The gains α_1, α_2 and μ_1 are strictly positive and $\mu_2 \geq 0$. The vectors $\mathbf{\Gamma}(\hat{z}_1, \tilde{z}_2)$ and $\hat{\boldsymbol{\theta}}$ read as

$$\mathbf{\Gamma}(\hat{z}_1, \tilde{z}_2) := -\begin{bmatrix} \hat{z}_1 & \tilde{z}_2 \end{bmatrix} \quad \text{and} \quad \boldsymbol{\theta} := \begin{bmatrix} \Delta\hat{a}_{21} & \Delta\hat{a}_{22} \end{bmatrix}^T \quad (7.73)$$

It shall be noted here again that the employment of the finite time parameter estimator reduces the tasks of the robust state estimator to (as the name suggests) state estimation. However, it does not require unknown input recovery as for the equivalent output injection term approach. Refer to [95] for a stability proof and convergence time estimation.

7.3.2.5 Adaptive roll angle observation (ARAO)

The last issue that still remains open, is the estimation of the roll angle. So far, knowledge of the roll velocity and roll acceleration, as well as effective stiffness and damping parameters is obtained. Now, in order to gain an estimate of the actual roll angle φ a standard Luenberger observer can be designed for (7.49) reading as

$$\frac{d\hat{x}_1}{dt} = \hat{x}_2 + l_1 (y - \hat{x}_2) \quad (7.74a)$$

$$\frac{d\hat{x}_2}{dt} = -\tilde{a}_{21}\hat{x}_1 - \tilde{a}_{22}\hat{x}_2 + b_2 u + l_2 (y - \hat{x}_2) \quad (7.74b)$$

where the expressions $\tilde{a}_{21}$ and $\tilde{a}_{22}$ are time-variant, i.e., $\tilde{a}_{21} = a_{21} + \Delta a_{21}^*(t)$ and $\tilde{a}_{22} = a_{22} + \Delta a_{22}^*(t)$. The terms $\Delta a_{21}^*(t)$ and $\Delta a_{22}^*(t)$ are interpreted as a time-varying difference between the real parameters a_{21}, a_{22} and the estimated ones $\tilde{a}_{21}$, $\tilde{a}_{22}$. Writing the system in error coordinates, with the errors defined as $e_1 := x_1 - \hat{x}_1$ and $e_2 := x_2 - \hat{x}_2$, results in:

$$\frac{de_1}{dt} = (1 - l_1)\, e_2 \quad (7.75a)$$

$$\frac{de_2}{dt} = -a_{21}e_1 - (a_{22} + l_2)\, e_2 + \sigma(\hat{x}_1, \hat{x}_2) \quad (7.75b)$$

with the input σ defined by

$$\sigma(\hat{x}_1, \hat{x}_2) := -\Delta a_{21}^*(t)\hat{x}_1 - \Delta a_{22}^*(t)\hat{x}_2 \tag{7.76}$$

Ideally, if the parameter estimations converge to the true values the differences $\Delta a_{21}^*(t)$, $\Delta a_{22}^*(t)$ vanish and so does the input σ. Due to the eigenvalue placement the error dynamics captured by system (7.75) are asymptotically stable. However, due to imperfections of the parameter estimation algorithms and parasitic dynamics there remains some deviation between the real and the estimated parameters. Hence, analysis of any effects w.r.t. $\sigma(\hat{x}_1, \hat{x}_2)$ on the estimation quality of the roll angle would be beneficial. Calculating the transfer functions of the input σ to the outputs e_1 and e_2 leads to

$$G_1(s) = \frac{\bar{e}_1(s)}{\bar{\sigma}(s)} = \frac{1 - l_1}{s^2 + (a_{22} + l_2)s + a_{21}(1 - l_1)} \tag{7.77a}$$

$$G_2(s) = \frac{\bar{e}_2(s)}{\bar{\sigma}(s)} = \frac{s}{s^2 + (a_{22} + l_2)s + a_{21}(1 - l_1)} \tag{7.77b}$$

Assuming that all components, i.e., robust exact differentiation (7.65), (adaptive) robust state estimation (7.65) and parameter identification (7.72) are tuned correctly, then the corresponding errors converge to zero. That yields bounded uncertainties $\Delta a_{21}^*(t)$ and $\Delta a_{22}^*(t)$ for all times $t \geq T_\Sigma$, with T_Σ denoting an overall estimate of the converging times. These upper bounds can be defined by

$$|\Delta a_{21}^*(t)| \leq \Delta a_{21}^+ \qquad \text{and} \qquad |\Delta a_{22}^*(t)| \leq \Delta a_{22}^+ \qquad \forall t \geq T_\Sigma \tag{7.78}$$

Furthermore, defining the system matrix of (7.75) as

$$\mathbf{A}_e := \begin{bmatrix} 0 & (1 - l_1) \\ -a_{21} & -(a_{22} + l_2) \end{bmatrix} \tag{7.79}$$

it is clear that by choosing the vector $\mathbf{l} = [l_1 \quad l_2]^T$ appropriately, the matrix $\mathbf{A}_e$ is Hurwitz.[23] Then following the assumption of a bounded input σ also the outputs and furthermore the states remain bounded. Hence, definition of the state boundaries $\hat{x}_1^+$ and $\hat{x}_2^+$ is feasible. A conservative approximation of the input signal can be calculated as

$$\begin{aligned} \left|\sigma(\hat{x}_1, \hat{x}_2)\right| &= \left|-\Delta a_{21}^*(t)\hat{x}_1 - \Delta a_{22}^*(t)\hat{x}_2\right| \\ &\leq \left|-\Delta a_{21}^*(t)\hat{x}_1\right| + \left|-\Delta a_{22}^*(t)\hat{x}_2\right| \\ &< \Delta a_{21}^+ \hat{x}_1^+ + \Delta a_{22}^+ \hat{x}_2^+ \end{aligned} \tag{7.80}$$

[23]That means the eigenvalues of $\mathbf{A}_e$ do have negative real parts.

At the end, assuming a harmonic system input σ with frequency ω the bounds of the estimation errors $e_1(t)$ and $e_2(t)$ can be approximated conservatively by

$$e_1(t) \approx |G_1(j\omega)| \left(\Delta a_{21}^{+} \hat{x}_1^{+} + \Delta a_{22}^{+} \hat{x}_2^{+}\right) \sin\left(\omega t + \angle G_1(j\omega)\right) \tag{7.81a}$$

$$e_2(t) \approx |G_2(j\omega)| \left(\Delta a_{21}^{+} \hat{x}_1^{+} + \Delta a_{22}^{+} \hat{x}_2^{+}\right) \sin\left(\omega t + \angle G_2(j\omega)\right) \tag{7.81b}$$

7.3.3 Simulation-based concept evaluation

The evaluation whether the proposed concepts meet certain performance specifications requires extensive testing within a realistic simulation environment. This should sensitize the reader to emerging difficulties and required modifications of the algorithms when being applied under (almost) real-world conditions. Results presented in the sequel serve a threefold strategy. First, the application of the proposed concepts to realistic simulation data provides some insight on the state and parameter estimation capabilities. Second, within the simulation environment every possible measurement can be gathered whereas in the experimental vehicle the number of measurement is up to the instruments installed. Third, the obtained results of the parameter identification tasks can be further used within a model-based vehicle handling evaluation, based on the identified parameter sets.

The multi-body vehicle dynamics simulation software package IPG CarMaker® generates the reference data of the vehicle motion. In general, an embedded 17-DoF (degree of freedom) vehicle model allows accurate simulation of realistic driving behavior and is highly flexible with respect to vehicle configurations. To name a few, it allows user-specific suspension setup, chassis type and geometry selection, engine type and performance adjustments as well as vehicle tires, load configuration, etc. A sophisticated driver model renders the software capable of simulating highly dynamical driving maneuvers even in closed-loop operation. For realistic scenarios of model parameter identification and handling evaluation the simulation outputs of CarMaker® are artificially augmented with measurement noise.[24] Implementation of the observation concepts is performed within the MATLAB®/Simulink® environment. The most important virtual vehicle parameters relevant to the observer design are listed in Table 7.3.

The reference data obtained from IPG CarMaker® is not only used for the validation of the joint parameter and state observation concepts, but also to compare effectively performed standardized (open-loop) handling maneuvers with simulated ones. In fact, the latter take into account the identified vehicle parameters of its lateral and roll dynamics.

Before discussing results of the parameter identification process some comments on the steering system modeling shall be given. An essential number of concepts uses measurements of the steering system as input. A standard modeling approach transforms the steering wheel angle δ_h to the wheel angle δ_w by a static transmission

[24]The noise levels are extracted from a sensor signal analysis modeling the disturbances as Gaussian processes, see [4].

Table 7.3 Summary of the virtual vehicle model parameters (IPG CarMaker®)

Parameter	Description	Value	Unit
l_{f_e}	Distance front axle to CoG (incl. pneumatic trail)	1.058	m
l_{r_e}	Distance rear axle to CoG (incl. pneumatic trail)	1.47	m
J_{xc}	Moment of inertia (chassis) w.r.t. x-axis	360	kg·m^2
J_z	Moment of inertia w.r.t. z-axis	2,152.3	kg·m^2
m	(Total) Vehicle mass	1,463	kg
m_s	Vehicle chassis mass (only sprung parts)	1,301	kg
h_{rl}	Distance between CoG and roll center	0.49	m
c_y	Lateral tire stiffness (front and rear)	200,000	N·m^{-1}

ratio, see e.g., [17]. In fact, elasticities of the steering system and lateral tire forces perturb the resulting wheel angle dynamically. For the purpose of simulation-based evaluation the wheel angle is available as a measurement and used as such. In contrast, the in-vehicle evaluation (Section 7.3.4) requires the use of a static transmission ratio as the wheel angle can neither be measured, nor the influence of the lateral forces estimated as the steering stiffness is commonly not known.

7.3.3.1 Parameter identification results

Lateral dynamics

The two-stage process of tire force model parameter recovery requires the employment of the mechanisms to identify the lateral cornering stiffness, as in Section 7.3.1.2, and the maximum value of the arising lateral tire force, see Section 7.3.1.1. The design of experiments for both tailored solutions is based on ideas illustrated in Figure 7.5. In summary, within the first experiment the vehicle is operated at its limits of adhesion such that the maximum tire force can be recovered. Then, in a second maneuver the lateral excitation of the vehicle is kept low in order to stay within the region of linear characteristics between slip angle α and lateral tire force F_y.

Starting with the recovery of the maximum front/rear lateral tire forces the latter directly result in the model parameters D_f and D_r, see (7.9). Theoretically, the observer can be implemented as proposed in (7.31), but the in-vehicle measured lateral acceleration corresponds to $a_{y,m}$, see (7.12), rather than a_y. A general overview of available measurements is provided in Table 7.4. Therefore, it is recommended to perform the parameter identification of the roll model first. Then the effects of the roll motion on the lateral dynamics can be compensated for by open-loop model-based predictions. Clearly, good attenuation of the roll motion effects can only be achieved if the road bank angle is constant and almost zero.

In terms of maneuver design there exist two different approaches. The first refers to a slowly performed step input with increasing steering wheel angle (until the driver feels the vehicle slips significantly). This maneuver could also be interpreted as a steady-state circular drive (with decreasing curvature radius). Alternatively, the maneuver refers to a sinusoidal excitation, again with increasing steering amplitude.

Table 7.4 Vehicle measurement channels used for simulations-based as well as experimental vehicle-based concept evaluation work

Domain	**Signal**	**Simulations**	**Real-world**
Longitudinal	Velocity v_x	x	x
	(Measured) Acceleration $a_{x,m}$	x	x
Lateral	(Measured) Acceleration $a_{y,m}$	x	x
	Lateral acceleration a_y	x	
	Yaw rate ω_z	x	x
	Vehicle sideslip angle* β	x	x
Tires	Lateral (front) forces $F_{y,f}$	x	
	Lateral (rear) forces $F_{y,r}$	x	
	Front slip angle α_{f_e}	x	
	Rear slip angle α_r	x	
Roll	Roll angle φ	x	
	Roll rate ω_z	x	x
	Roll acceleration α_x	x	

*... Acquisition for gathering ground truth data only. Observer schemes do not use it.

However, due to the arising high transients the dynamic build-up of the lateral tire forces gains more influence on the results compared to the slowly-increasing forces of the moderate step input. Therefore, it is suggested to conduct the former described maneuver. The simulated vehicle responses as well as estimation results in terms of yaw rate $\hat{\omega}_z$ and lateral forces $\hat{F}_{y,f}$, $\hat{F}_{y,r}$ are illustrated in Figure 7.7. Extraction of the maximum values is performed by a simple algorithm that checks in real-time whether the actual value exceeds the currently saved maximum. If so it is set as new maximum, otherwise dropped. Once the estimates of the maximum values settle, the values of D_f and D_r can be read off directly.

The estimation of the remaining parameters C_f, C_r, B_f and B_r necessitates the identification of the front and rear cornering stiffness values $c_{\alpha_{fe}}$ and c_{α_r}, exploiting (7.47) and (7.48).

In terms of maneuver design the vehicle is driven at constant velocity and a sinusoidal steering wheel excitation performed with an amplitude of $\delta_h = \pm 15°$ and frequency $f_{ex} = 0.2$ Hz. The yaw rate of the vehicle is measured, the sideslip angle β and the cornering stiffness values estimated. Figure 7.8 shows the results of the observer concept designed for the rear axle, i.e., c_{α_r}. Identification of the front stiffness value $c_{\alpha_{fe}}$ requires an additional observer concept that is based on the ideas of (7.36), but the system re-formulated such that the only unknown parameter refers to the front cornering stiffness value. Even though theoretically the parameter estimation would remain bounded in case of missing excitation, the overall performance is superior when using a sophisticated activation scheme.[25]

[25]Details are to be found in [4]. Activation is annotated by t_{on}, deactivation by t_{off} within corresponding figures.

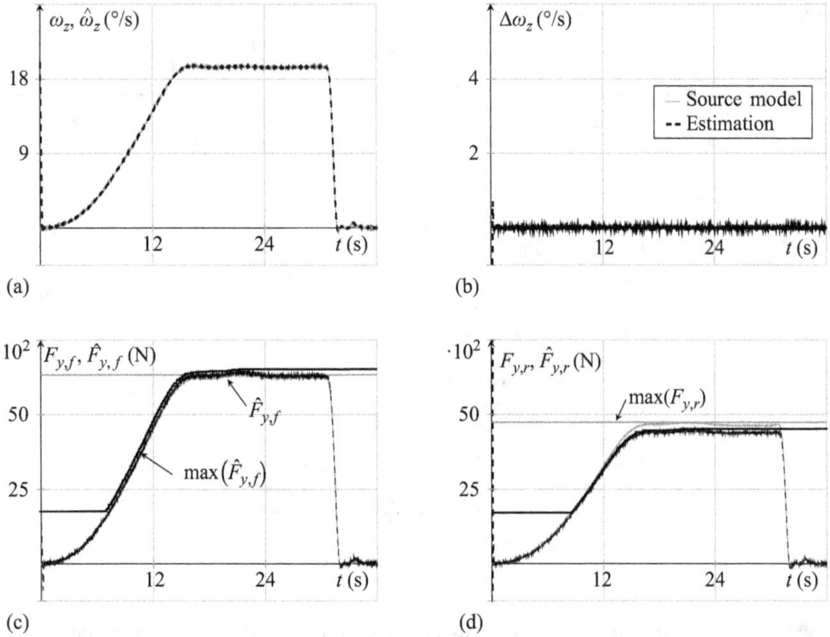

Figure 7.7 Simulation-based estimation results of the lateral tire forces estimation. Maneuver: slow step steering with $\delta_{h,max} = 120°$, $\frac{d\delta_h}{dt} \approx 4°/s$, $v_x = 80 \pm 3$ km·h^{-1}. The recovered tire forces of the front and rear axles are depicted in (c) and (d): (a) Estimation $\hat{\omega}_z$ of ω_z, (b) estimation error $\Delta\omega_z = \omega_z - \hat{\omega}_z$, (c) estimation $\hat{F}_{y,f}$ of $F_{y,f}$, and (d) estimation $\hat{F}_{y,r}$ of $F_{y,r}$.

From Figure 7.8(e), it can be seen that the estimates converge nicely to almost constant values. The fluctuations of the parameters are from their variations to compensate parasitic dynamics. Once the estimates have settled and are within a certain band averaging leads to single values of the front and rear stiffnesses. Now, reconsidering the definition of the tire model (7.8) and coherence with characteristic points (7.9), it remains still an open issue how to calculate three unknown model parameters from the slope of the curve at the origin and its maximum. The idea is based on the influence of parameter B on the slip-force characteristics, i.e., it determines the ratio between end value $F_{y\infty}$ and $F_{y_{max}}$. Simulations have shown that setting the end value to approx. 90% of the maximum value results in slightly declining characteristics of the curve. Using that fixed ratio it is straightforward to identify the parameters C_f and C_r from the cornering stiffness values.

Finally, Figure 7.9 compares the real slip-force characteristics with its recovered pendants. It shall be emphasized that the identified set of parameters is only valid for the actual vehicle setup and road surface characteristics. Changing any of these requires a re-iteration of the parameter identification!

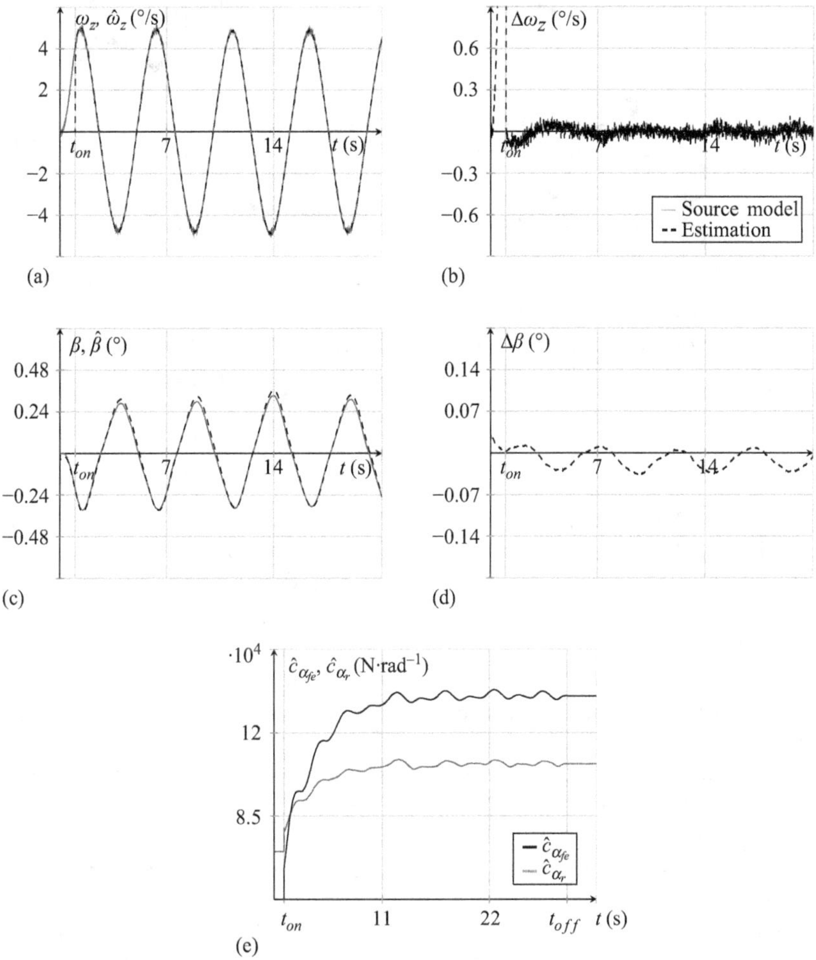

Figure 7.8 Simulation-based estimation results related to the (effective) lateral cornering stiffness identification. Maneuver: sinusoidal steering with $\delta_h \pm 15°$, $f_{ex} = 0.2$ *Hz,* $v_x = 80 \pm 2$ $km{\cdot}h^{-1}$*: (a) Estimation* $\hat{\omega}_z$ *of* ω_z*, (b) estimation error* $\Delta\omega_z = \omega_z - \hat{\omega}_z$*, (c) estimation* $\hat{\beta}$ *of* β*, (d) estimation error* $\Delta\beta = \beta - \hat{\beta}$*, and (e) estimations* $\hat{c}_{\alpha_{fe}}$*,* $\hat{c}_{\alpha_r}$ *of* $c_{\alpha_{fe}}$*,* c_{α_r}*.*

Roll dynamics

Before discussing the evaluation of the individual mechanisms that are involved within the roll dynamics model parameter identification the overall objective shall be recapitulated. Basically, it aims to recover the dominating model parameters, namely effective roll stiffness c_Σ and damping d_Σ as well as roll angle, by measurement of the chassis roll velocity.

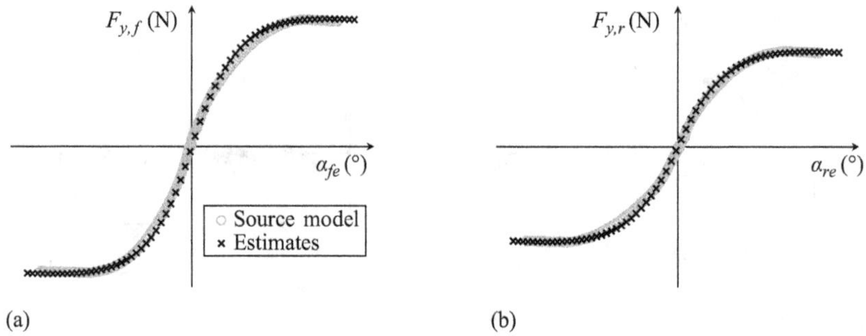

Figure 7.9 Comparison of the real and estimated lateral slip–force characteristic curves: (a) front lateral tire force $F_{y,f}$ vs. slip angle α_{f_e} and (b) rear lateral tire force $F_{y,r}$ vs. slip angle α_{r_e}

Design of experiments goes back to the ideas formulated in [34]. Therein continuous excitation of the vehicle with varying frequency ($f_{ex} \approx 0.2$–3.0 Hz) is suggested. Ideally, the excitation frequencies are chosen equidistantly and once steady-state operation is reached the identification mechanism can be started. During the overall procedure the applied steering wheel angle is to be kept constant. In practice, however, this time-consuming maneuver is often replaced by a sinusoidal chirp function with steadily increasing frequency. Herein, as the reference data is generated from simulation work and time-efficiency is not critical, it is the former maneuver that is executed. Therefore, the following operating points in terms of frequency are selected: $0.1, 0.2, 0.5, 1.0, 1.5$ and 2.0 Hz. To this extent it is assumed that higher input frequencies are realistic only for test bench experiments. A set of characteristic model parameters is identified for every operating point and the parameter variation w.r.t. to excitation frequency obtained.

Tuning of the estimator concept depicted in Figure 7.6 requires adherence to a certain tuning procedure. In fact, the gain adjustment starts with the RED structure until it shows good estimates of the system input's time derivative. Inherently, any high-frequency terms of the signal, e.g., measurement noise, will be amplified yielding an estimation signal with reduced signal-to-noise ratio. Consequently, a low-pass filter is applied to the output of the robust exact differentiator [93]. Once the tuning of the RED concept is finished, it is the robust state estimator that needs to be provided with observer gain values. Referring to the static gain concept, see Section 7.3.2.2, its tuning is based on the upper bound of the uncertainty. However, due to (7.56b) it is obvious that the uncertainty magnitude changes with the upper bounds of the states z_1 and z_2 (assuming that Δa_{21}, Δa_{22} are constants). Hence, the adaptive gain algorithm is superior especially in terms of tuning effort when compared to the static gain concept. For the finite time parameter estimator it is the objective to find a good balance between the linear and non-linear correction term. This ensures fast convergence of the algorithm with low estimation fluctuations once the system is in steady-state. Finally, the adaptive roll dynamics observer requires specification of error dynamics eigenvalues. As its system parameters vary during the identification process these are selected for the nominal case. Obviously, this affects the specified convergence dynamics, but updating the Luenberger gains at every time instant is assumed unnecessary.

Figure 7.10 shows the evaluation of the observer concepts for the case of an excitation frequency $f_{ex} = 1.0$ Hz. Plots (a) and (b) depict the measured lateral acceleration and jerk as well as their estimates (and filtered signals) obtained from the robust exact differentiator.[26] The filtering of the estimates becomes required due to the measurement noise on the signal that affects the time-derivative estimate.

The roll velocity and acceleration estimations and errors are plotted in (c)–(f). Furthermore, the adaptive character of the observer gains can be extracted from (d). In addition to the estimates of the adaptive robust state observation (ARSE) mechanism subplots (e)–(h) illustrate the results of the parameter estimation algorithm. The subplots (e) and (f) show the state estimate and its errors. The uncertainties of the effective stiffness and damping parameters are given by (g) and (h).

Now, these parameter estimates are forwarded to the adaptive roll angle observer resulting in Figure 7.11(a) and (b).

From the encouraging results shown in Figures 7.10 and 7.11, the focus is put on the consistency of the parameter estimates with respect to excitation frequency. It is well known that the roll dynamics model shall represent the real-world system accurately for a frequency range of 0.2–3.0 Hz. Figure 7.12(a) and (b) depicts the effective roll stiffness and damping, respectively, for numerous different excitation frequencies. Interestingly, both the effective stiffness as well as the damping characteristics vary with frequency. For comparison [4] proposes an alternative parameter estimation concept based on an Extended Kalman Filter concept that shows consistent results w.r.t. the sliding mode-based approach. Investigating the underlying model of the roll dynamics more closely an interpretation of the damping factor increase is based on the typically non-linear characteristics of damping forces and deflection velocity. For higher frequencies parasitic dynamics (arising due to the simplifications assumed during the modeling phase, Section 7.1.2) become more evident than for the lower ones. For the subsequent generic vehicle handling evaluation a weighted average of the values[27] is obtained and employed.

7.3.3.2 Handling evaluation results

Following the idea of model-based vehicle handling evaluation, i.e., simulation of standardized driving maneuvers exploiting identified vehicle model parameters, see Table 7.5, this paradigm will be further analyzed within this section. The results obtained for the tire force model as well as roll dynamics parameters are integrated to a single-track model based on (7.1b), (7.8), (7.22) and (7.23).

[26]Exactly speaking, due to the filtering process the capability of ideal derivative calculation is lost, see [93] for details.

[27]The weighting is performed such that the estimates of the extremal values (in terms of frequency) are weighted lower than the remainders.

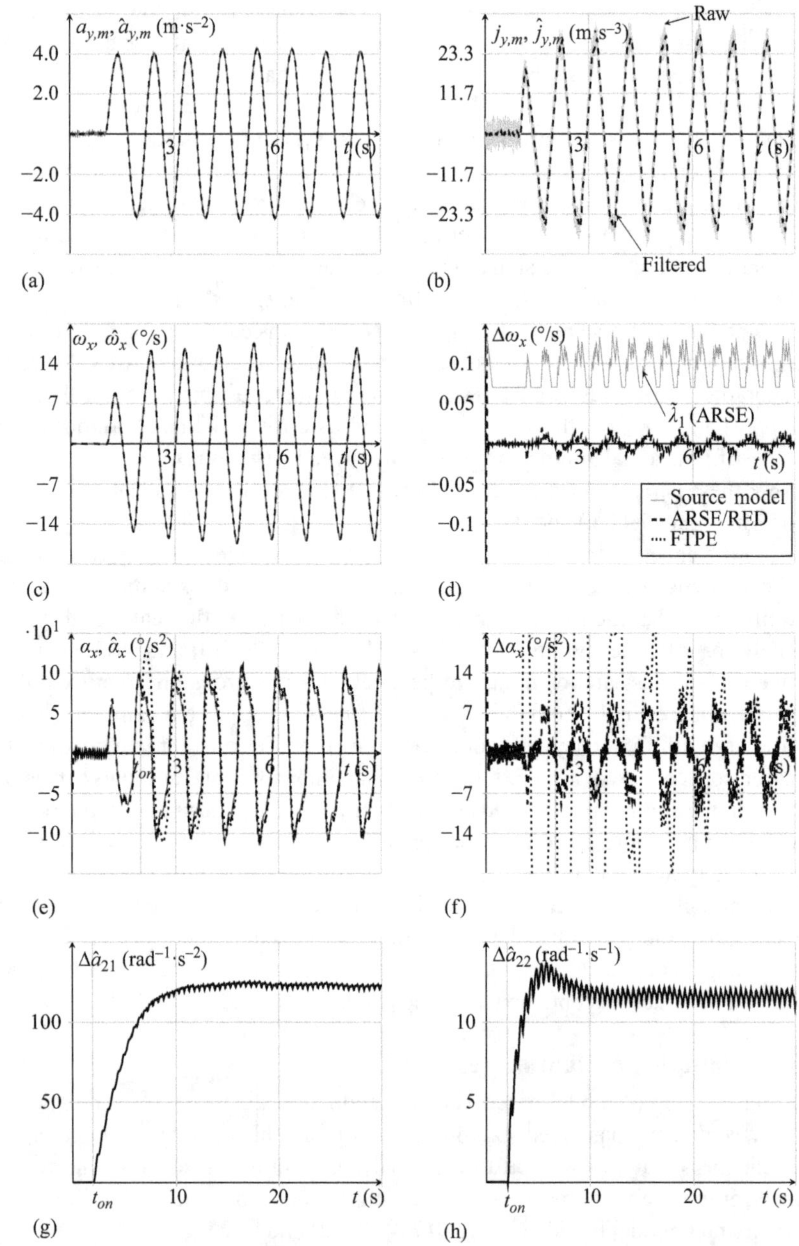

Figure 7.10 Simulation-based estimation results of the uncertain parameters Δa_{21} and Δa_{22} as well as states (roll velocity ω_x and acceleration α_x). Maneuver: sinusoidal steering with $\delta_h = \pm 45°$, $f_{ex} = 1.0$ Hz, $v_x = 80 \pm 2$ km·h^{-1}: (a) Estimation $\hat{a}_{y,m}$ of $a_{y,m}$ (RED), (b) estimation $\hat{j}_{y,m}$ of $j_{y,m}$ (RED), (c) estimation $\hat{\omega}_x$ of ω_x (ARSE), (d) estimation error $\Delta\omega_x = \omega_x - \hat{\omega}_x$, (e) estimation $\hat{\alpha}_x$ of α_x (ARSE), (f) estimation error $\Delta\alpha_x = \alpha_x - \hat{\alpha}_x$, (g) estimation $\Delta\hat{a}_{21}$ of Δa_{21} (PE), and (h) estimation $\Delta\hat{a}_{22}$ of Δa_{22}.

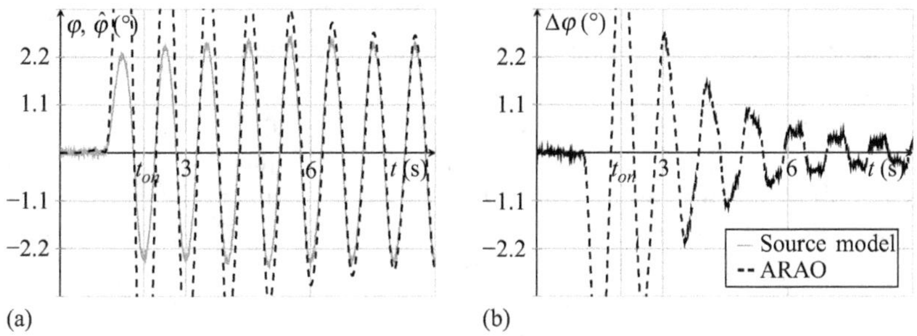

Figure 7.11 Roll angle φ estimation using the adaptive observer (ARAO). Maneuver as in Figure 7.10. (a) Estimation $\hat{\varphi}$ of φ (ARAO) and (b) estimation error $\Delta\varphi = \varphi - \hat{\varphi}$.

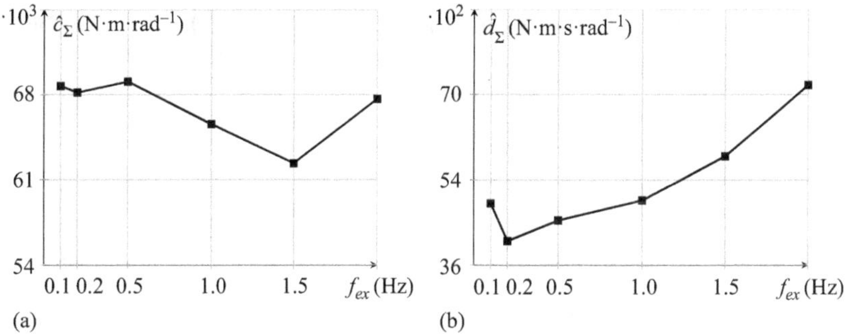

Figure 7.12 Simulation-based estimation results of the roll model parameters c_Σ and d_Σ for an excitation frequency range of $f_{ex} = 0.1$–2.0 Hz. The results reveal the effects of unmodeled dynamics on the estimations of the parameters. (a) Effective roll stiffness $\hat{c}_\Sigma$ vs. f_{ex} and (b) effective roll damping $\hat{d}_\Sigma$ vs. f_{ex}.

Table 7.5 Summary of identified model parameters used for the handling evaluation

Parameter	Description	Value	Unit
c_Σ	Effective roll stiffness	66,292	N·m·rad^{-1}
d_Σ	Effective roll damping	5,193	N·m·s·rad^{-1}
D_f	TM_Simple parameter (front)	6,355	N
B_f	TM_Simple parameter (front)	1.888	–
C_f	TM_Simple parameter (front)	8.38×10^{-2}	rad
D_r	TM_Simple parameter (rear)	4,736	N
B_r	TM_Simple parameter (rear)	1.974	–
C_r	TM_Simple parameter (rear)	8.36×10^{-2}	rad

It is the objective of this evaluation to find out whether the identified vehicle model replicates the realistic vehicle model (represented by IPG CarMaker®) accurately enough to use the simple model (rather than the complex one) for the simulation of standardized handling maneuvers. In fact, it would be ideal to perform this comparison using real-world experimental vehicle data, rather than a complex simulation environment. But, unfortunately the available experimental data from vehicle measurements, analyzed further in Section 7.3.4, is not sufficient to perform the comparison in a feasible manner.

The simulated open-loop maneuvers are selected from the set of maneuvers as suggested by standards such as ISO7401 [9] or ISO4138 [8]. These refer to a steady-state circular drive (ISO4138) and a step steering input (ISO7401). The two maneuvers reveal the stationary and transient driving characteristics of the vehicle. Referring to the steady-state circular drive it is the task of the driver to follow a circular trajectory (curvature radius $R = 42$ m) for an increasing longitudinal velocity. More specifically, it is changed step-wise and then kept constant for some time in order to gather the measurements. Figure 7.13 shows the results of the source model (IPG CarMaker®)

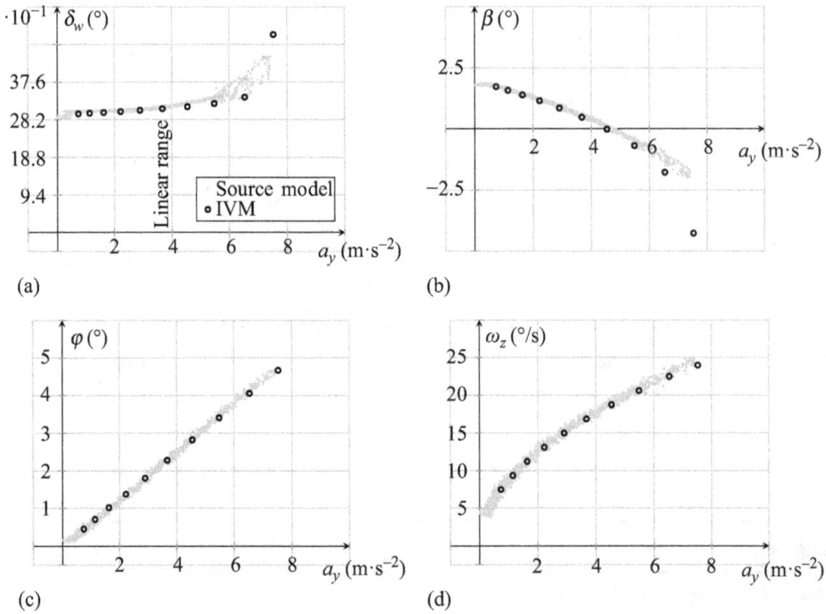

Figure 7.13 Result comparison between source model (IPG CarMaker®) and identified vehicle model. Maneuver: steady-state circular drive with curvature radius $R = 42$ m, $v_x = 0$–70 km·h^{-1}. The identified vehicle model does not provide any estimates of the steering torque T_h (as suggested in [8]), therefore it is replaced by the yaw rate ω_z in (d). (a) Lateral acceleration a_y vs. Wheel angle δ_w, (b) lateral acceleration a_y vs. sideslip angle β, (c) lateral acceleration a_y vs. roll angle φ, and (d) lateral acceleration a_y vs. yaw rate ω_z.

and the predictions using the identified vehicle model (IVM). In contrast to the suggestions of [9], i.e., to plot steering torque versus lateral acceleration level there is no mechanism implemented yet to provide simulation values of the steering torque. Hence, to complete the series of plots it is the yaw rate that takes over the place of steering torque.

In general, the fitting between source model and IVM results is encouraging, despite the deviations (especially w.r.t. the sideslip angle β) that appear for higher levels of lateral acceleration (i.e., beyond the "linear" range). However, with regards to the models' simplicity and the fact that the objective metrics, see [4] for details, are extracted for maximum levels of $|a_y| \leq 6\ \mathrm{m{\cdot}s^{-2}}$ this is acceptable.

The step steering input maneuver is performed for three different levels of accelerations, see [9]. In terms of maneuver execution parameters the step shall be performed such, that the steering velocity is between 200°/s and 500°/s. An interesting aspect of the simulation results illustrated by Figure 7.14 is the correlation between the identified vehicle and source model even for higher lateral acceleration levels, e.g., $a_y = 6\ \mathrm{m{\cdot}s^{-2}}$, that hardly appeared during the identification process. For a quantitative evaluation of the result matching the permissible sensor accuracy limits, as listed in Table 7.1, envelope the simulation results, see the gray shades in Figure 7.14. Ideally,

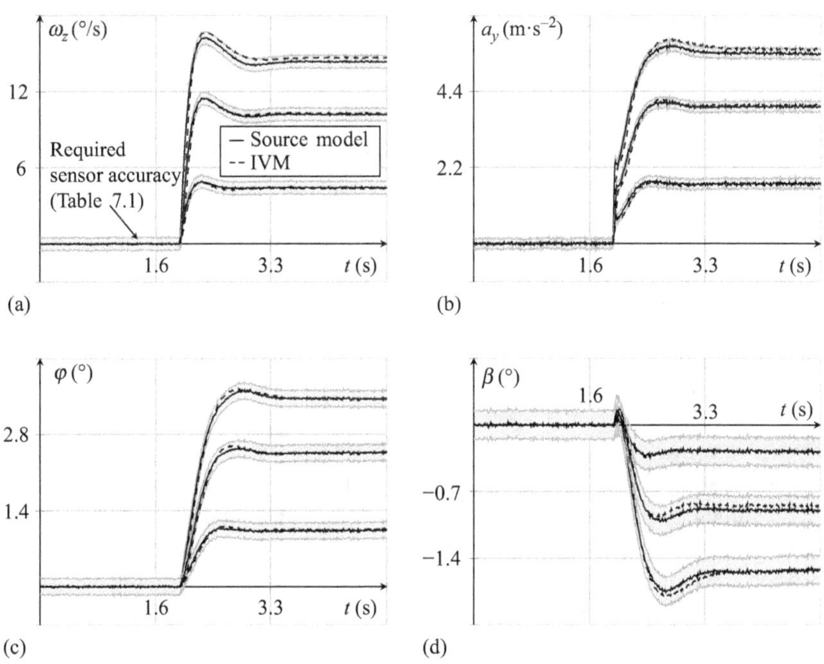

Figure 7.14 Result comparison between source and identified vehicle model. Maneuver: step steering with $\delta_h = 14/30/42°$, $\frac{d\delta_h}{dt} \approx 500°/s$, $v_x = 80 \pm 2\ km{\cdot}h^{-1}$. The required sensor accuracy envelops the references. (a) Yaw rate ω_z, (b) lateral acceleration a_y, (c) roll angle φ, and (d) vehicle sideslip angle β.

the results of IVM stay within those limits. Then, the accuracy of the model forecasts has the potential to support the handling evaluation process substantially.

7.3.4 Experimental results

The last step of the concept evaluation process is its deployment to real-world systems. In general, this serves two purposes: first, the observation concepts can be tested under real conditions and second, the identified parameters can be used again to compare model predictions and effectively performed, standardized handling maneuver. Either of the two cases requires an experimental vehicle instrumented with measurement devices as listed in Table 7.1. Unfortunately, the lack of a highly instrumented vehicle only allows a "soft" validation of the concepts, that will be presented in the following.

Generally, in comparison to the rather complex simulation environment consisting of e.g., IPG CarMaker®, it is expected for real-world application that effects due to measurement noise, time discretization and parasitic dynamics play an even more prominent role.

Table 7.4 lists the sensor signals acquired for the estimation process. The observer sampling time is given by $T_s = 0.001$ s. Even though the standard CAN bus is normally operated by $T_s = 0.01$ s this is not a restriction for the actual problem due to the following reason. The observation results are not used for any real-time control application and hence an artificial time lag can always be introduced without any side effects. In other words, if the actual measurement values (gathered at $T_s = 0.01$ s) of all required sensors are written to a buffer and then up-sampled to $T_s = 0.001$ s the vehicle state estimates as well as the parameters will be delayed by some time. However, as this lag is known exactly it can be easily compensated for *a posteriori* of any calculations.

In the sequel of this chapter the main results of the experimental evaluation will be scaled due to data confidentiality.[28] The notation:

$$x^*(t) := \frac{x(t)}{x_{scal}} \tag{7.82}$$

defines a scaled variable $x^*(t)$, with x_{scal} being a constant and $x(t)$ the original measurement or estimation value.

Lateral dynamics

The identification of the tire slip-force characteristics refers to the estimation of the model parameters in (7.8) discussed in Section 7.3.1. In contrast to the simulations-based evaluation the quantity of steering angle is not available in-vehicle. Hence, a static (non-linear) steering transmission model is identified before conductance of any measurements on a test track using wheel alignment turn plates. Clearly, the model's simplicity is at the expense of inaccuracies that arise due to the complete neglect of steering elasticities.

[28]To some extent the presented measured vehicle data is part of industrial collaborations and requires protection of the data. Moreover, any vehicle related parameters as well as observer tunings are omitted on purpose.

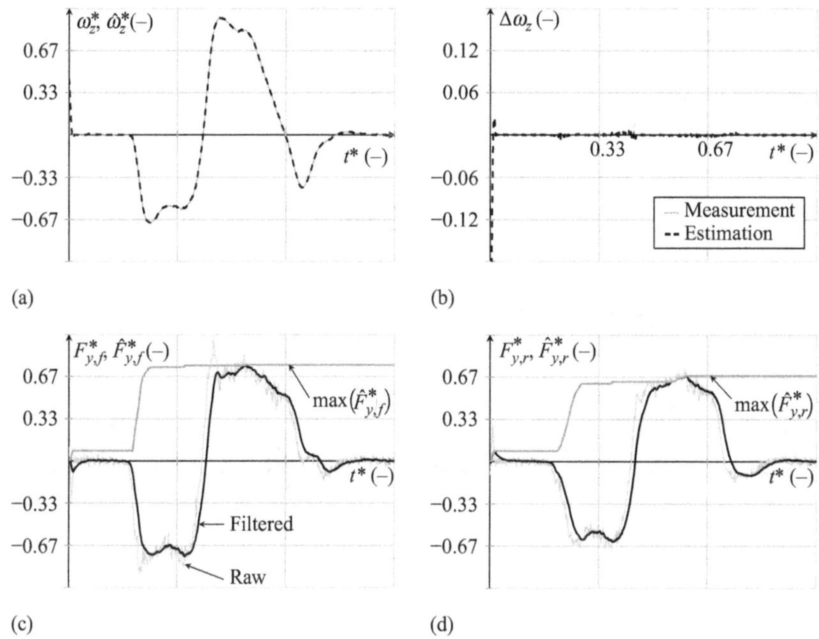

Figure 7.15 Concept evaluation for estimation of maximum lateral tire forces (front and rear wheels). Maneuver: highly dynamic lane change driven at approximately constant longitudinal velocity. (a) Estimation $\hat{\omega}_z^*$ of ω_z^*, (b) estimation error $\Delta\omega_z^* = \omega_z^* - \hat{\omega}_z^*$, (c) estimation $\hat{F}_{y,f}^*$ of $F_{y,f}^*$, and (d) estimation $\hat{F}_{y,r}^*$ of $F_{y,r}^*$.

Starting with the estimation mechanism to reveal the maximum lateral tire forces the vehicle needs to be excited significantly, as depicted in Figure 7.5. Herein, the conducted handling maneuver refers to a dynamically driven lane change as shows Figure 7.15. The subplots (a) and (b) illustrate the estimated yaw rate and its measured reference, whereas the tire forces of front and rear axle are shown in (c) and (d). In order to achieve good transients of the estimation values the observer is tuned quite aggressively and it is recommended to apply a low-pass filter for smooth force estimates. Even if not illustrated by Figure 7.15 it shall be mentioned that the in-vehicle measured lateral acceleration $a_{y,m}$ needs to be compensated for any effects of the chassis' roll motion. Therefore, it is suggested to apply the roll dynamics parameter estimator before the actual one. This allows roll angle prediction for further calculation of a compensated lateral acceleration a_y.

Knowledge extraction of the front and rear cornering stiffness values requires the vehicle to be operated within the so-called *linear* range, see Figure 7.5. The gentle excitation and its vehicle responses are plotted in Figure 7.16. Activation of the parameter estimation algorithm is based on a sophisticated threshold-based

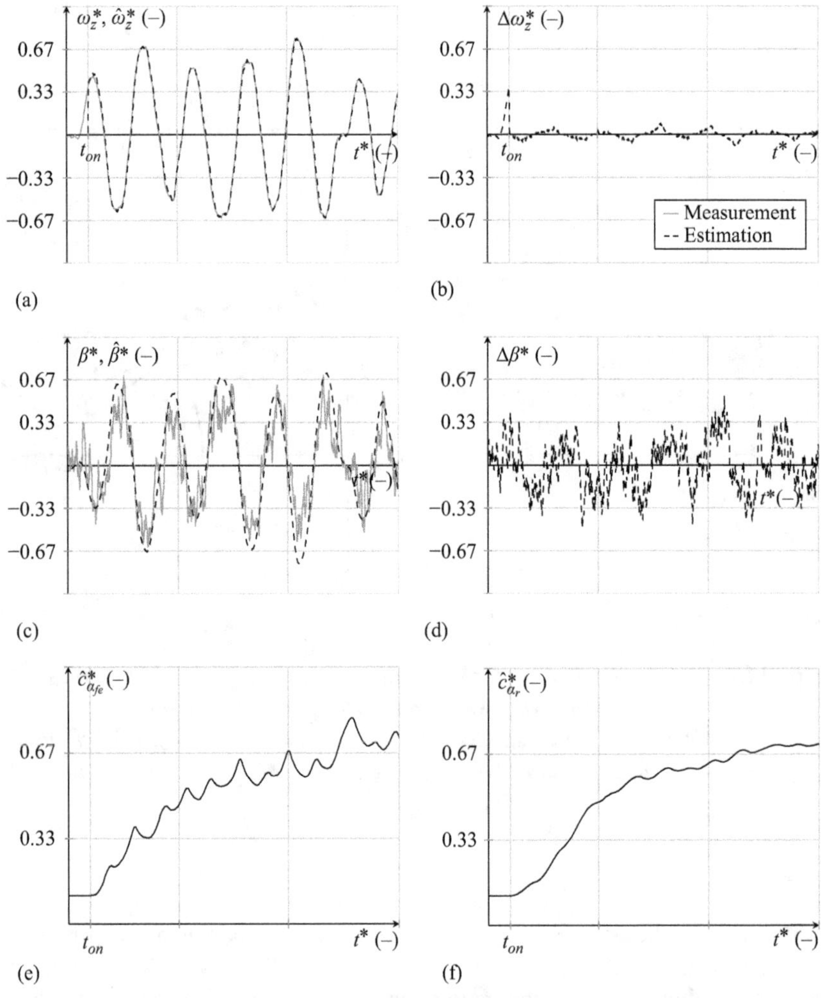

Figure 7.16 Experimental evaluation of the lateral cornering stiffness estimation process. Maneuver: sinusoidal steering with low amplitude in order to stay within the linear range. From (c) it is obvious that the signal of the measured sideslip angle is significantly affected by measurement noise for those low levels of excitation. (a) Estimation $\hat{\omega}_z^*$ of ω_z^*, (b) estimation error $\Delta\omega_z^* = \omega_z^* - \hat{\omega}_z^*$, (c) estimation $\hat{\beta}^*$ of β^*, (d) estimation error $\Delta\beta^* = \beta^* - \hat{\beta}^*$, (e) estimation $\hat{c}_{\alpha_{fe}}^*$ of $c_{\alpha_{fe}}^*$, and (f) estimation $\hat{c}_{\alpha_r}^*$ of $c_{\alpha_r}^*$.

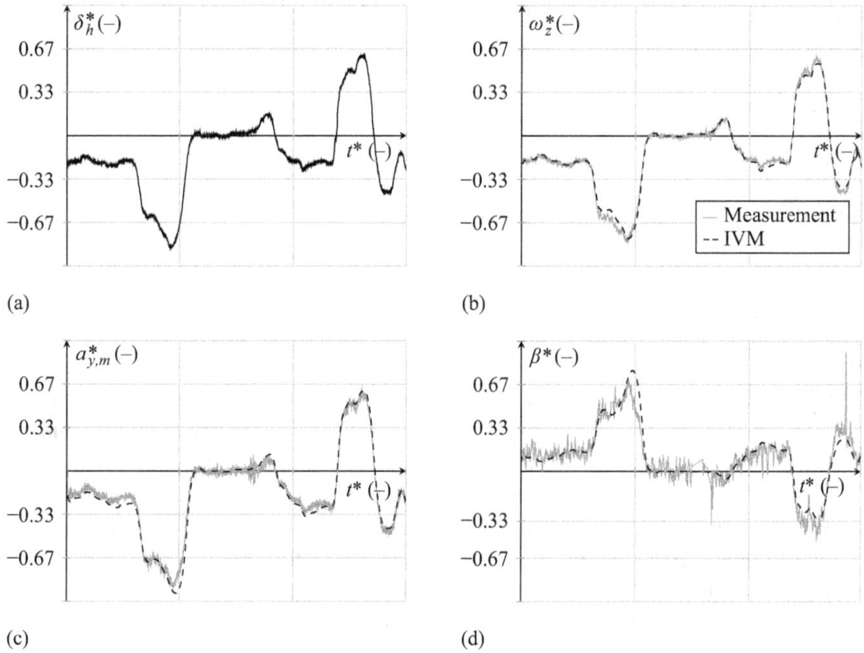

Figure 7.17 Experimental validation of the lateral dynamics parameter estimation results. Maneuver: dynamic cornering resulting in lateral accelerations $|a_{y,m}| > 4\ m{\cdot}s^{-2}$. Here, the non-linear characteristics between force and slip gain importance. (a) Steering wheel angle δ_h^*, (b) yaw rate ω_z^*, (c) measured lateral acceleration $a_{y,m}^*$, and (d) vehicle sideslip angle β^*.

mechanism further discussed in [4], but not necessarily required. Repeating the findings from Section 7.3.1.2, it is known that even for non-persistent excitation the estimates remain bounded. The most interesting results are depicted in plots (e) and (f). These represent the estimates of front and rear cornering stiffness values. As suggested in the simulations-based evaluation single values can be extracted from averaging the estimation time series once the values have settled within some user-definable band.

Now, from the identified characteristic points the model parameters can be obtained as suggested in Section 7.3.3.1. Unfortunately, for the experimental vehicle no ground truth data of the tire force model parameters is available. Hence, the quality of the identified parameters can only be evaluated by a comparison of vehicle response measurements gathered during a driving maneuver and open-loop model predictions (of vehicle responses) exploiting the parameter estimates.

Figure 7.17 shows the results for a maneuver that covers the linear and non-linear operational range (in terms of slip-force characteristics). The matching between measurements and predictions is impressive taking into account the simplicity of the model and identification process. Clearly, and this should be emphasized explicitly, the identified model parameters are only valid for the actual vehicle configuration and road condition. Any changes of these require a re-iteration of the identification process.

Roll dynamics

Application of the joint state and parameter estimation scheme of Section 7.3.2 requires the experimental vehicle to be equipped with a lateral acceleration and roll velocity sensor, see also Table 7.3. Ideally, for generating ground truth data the vehicle is also equipped with a roll angle sensor. Unfortunately, this is asserted by the available experimental vehicle and hence the evaluation of the resulting parameter estimates needs to be conducted alternatively. In fact, it refers to a similar procedure as presented beforehand for the lateral dynamics concept. It accommodates the identified parameters into the underlying model of the observer scheme and compares open-loop model predictions with real vehicle measurements.

Referring to the design of experiments it is suggested in Section 7.3.3.1 to excite the vehicle either by a sinusoidal chirp function or use numerous sinusoids (with varying frequencies all within the range of interest) and identify the parameters for each operating point. The latter is demonstrated for the simulations-based work, but is a lot more time-consuming compared to the chirp signal.

As a consequence the steering wheel is actuated sinusoidally, with approximately constant steering wheel angle, but increasing frequency ($f_{ex} \approx 0.3$–2.2 Hz). Due to the involvement of several submodules yielding roll dynamics related states and parameters the analysis is quite comprehensive. Figure 7.18 presents the results of observer mechanisms when exciting the vehicle with the sinusoidal chirp signal.

First, a correct time differentiation can only be achieved if the RED algorithm is tuned correctly. Subplots (a) and (b) show the measurements and estimates in terms of lateral acceleration $a_{y,m}$ and its time derivative, i.e., lateral jerk $j_{y,m}$.

Estimation of the states, i.e., roll velocity ω_x and roll acceleration α_x are provided by (c) and (d). The adaptive gain adjustments can be extracted from (c) and the state estimate of the finite time parameter estimation algorithm can also be obtained from (d). Accounting for the aggressive tuning of the parameter estimation algorithm resulting in fast transients, a low-pass filter smooths the nervous estimates, see Figure 7.19. Again, as for the lateral dynamics (i.e., cornering stiffness estimation), activation of the algorithm relies on a sophisticated maneuver detection algorithm. In fact, once the parameter observations have settled sufficiently, taking averages results in single values of the effective stiffness and damping factors. Importantly, the adaptive roll angle observer receives the parameter estimates and takes it into account for the actual observations of roll angle and velocity. Figure 7.20 depicts the measurements and estimates that match nicely after some settling time. Referring to subplot (a) it appears that the roll angle is measured in-vehicle, but this is not true! In fact, the gray, thin line refers to a pseudo-reference, as it is calculated offline on the basis of the roll velocity measurements employing a sophisticated signal processing algorithm. Rather than being used as ground truth data this signal shall only provide an idea of the roll angle estimation quality.

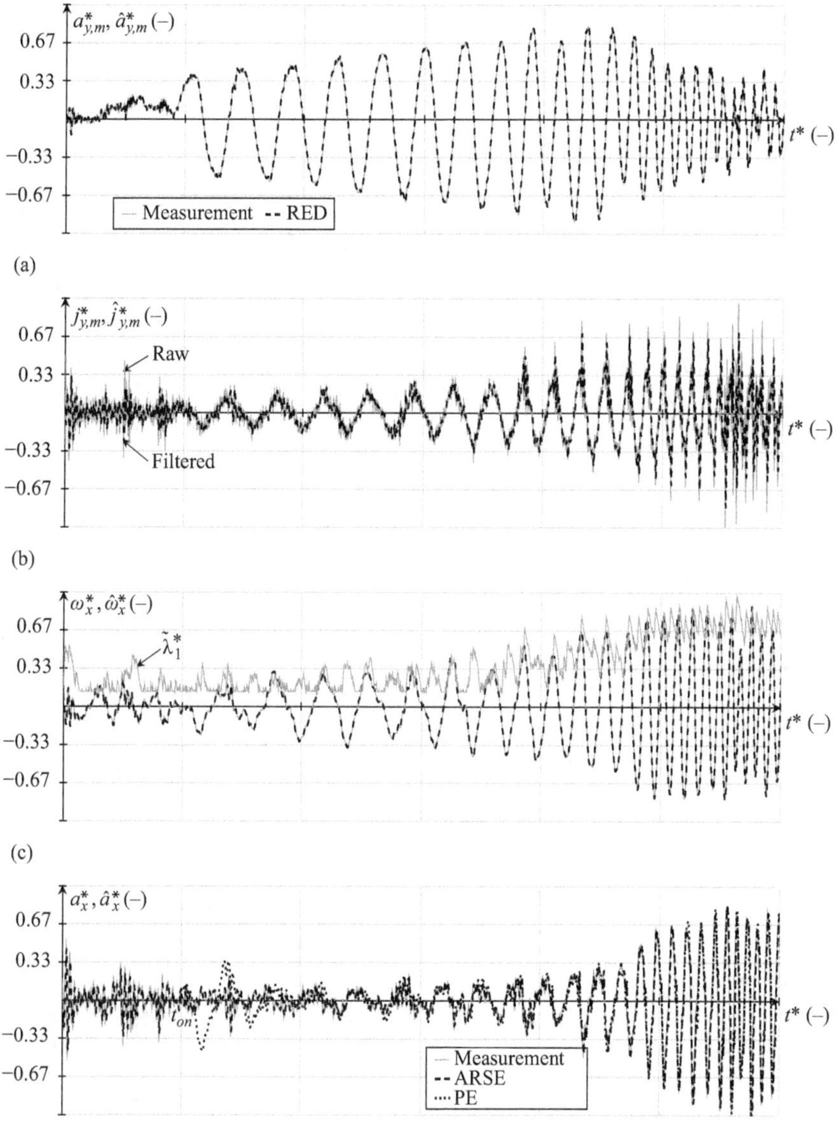

Figure 7.18 Roll dynamics parameter identification – experimental evaluation (RED, ARSE and PE). Maneuver: sinusoidal steering with constant steering angle, but varying frequency (f_{ex} = 0.3–2.2 Hz). The longitudinal velocity is kept constant $v_x \approx 80$ km·h^{-1}. (a) Estimation $\hat{a}^*_{y,m}$ of $a^*_{y,m}$ (RED), (b) estimation $\hat{j}^*_{y,m}$ of $j^*_{y,m}$ (RED), (c) estimation $\hat{\omega}^*_x$ of ω^*_x (ARSE), and (d) estimation $\hat{\alpha}^*_x$ of α^*_x (ARSE, PE).

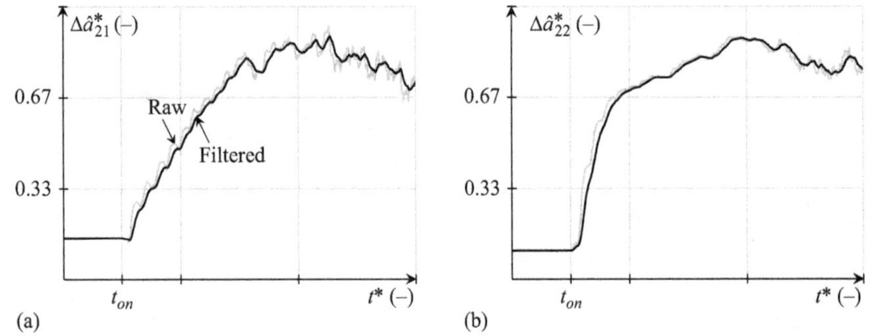

Figure 7.19 *Roll dynamics parameter identification (PE). Maneuver: as in Figure 7.18. The estimated parameters are smoothed by a second-order low-pass filter. (a) Estimation $\Delta\hat{a}^*_{21}$ of Δa^*_{21} and (b) estimation $\Delta\hat{a}^*_{22}$ of Δa^*_{22}.*

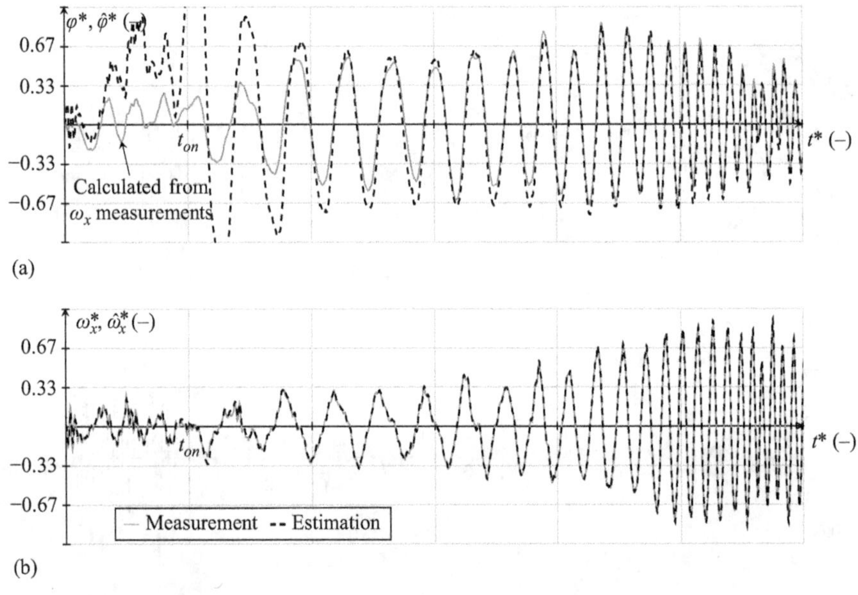

Figure 7.20 *Roll dynamics parameter identification (ARAO). Maneuver: as in Figure 7.18. The roll angle signal in (a) is obtained by sophisticated offline signal processing algorithms based on the measured roll rate ω_x and serves as pseudo-reference. (a) Estimation $\hat{\varphi}^*$ of φ^* (ARAO) and (b) estimation $\hat{\omega}^*_x$ of ω^*_x (ARAO).*

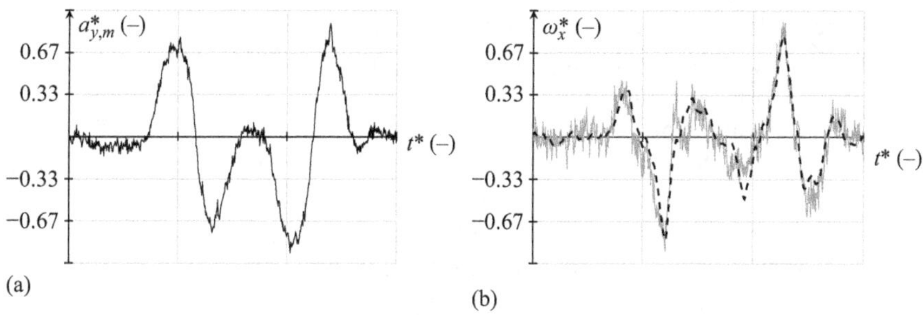

Figure 7.21 Validation of roll model parameter estimation. Maneuver: double lane change (longitudinal velocity $v_x \approx 70$ km·h^{-1}). (a) Measured lateral acceleration $a^*_{y,m}$ (–) and (b) roll rate ω^*_x (–).

Exploiting the identified parameters of effective roll stiffness c_Σ and damping d_Σ these can be fed into the roll dynamics model (7.22). Then, conducting an evaluation handling maneuver, e.g., a double lane change, the resulting lateral acceleration can be forwarded to the roll dynamics model and its predictions as well as the in-vehicle measurements of the roll rate compared.

Figure 7.21 depicts the vehicle excitation signal (a) as well as the predicted and measured response (b). The sound matching between them reveals the potential of the concepts to provide good estimates of model parameters related to the real vehicle dynamics.

7.4 Conclusions

Parameter identification schemes based on unknown input observation have been presented and discussed in this chapter. Exploiting the sliding mode paradigm the proposed concepts ensure robustness to uncertainties and also provide for finite time convergence of the algorithms. Within the context of lateral dynamics a simple, straightforward method for parameter estimation of a non-linear tire force model is proposed employing a finite time parameter identification scheme. For the considered roll dynamics a general observation concept is discussed that allows recovery of unknown system parameters and states related to a mechanical system, if the velocity is measured and not, as in the standard case, the position. Even though the suggested framework involves several submodules and is more complex, the evaluation in simulations and also real-world experiments demonstrates the high potential of the proposed mechanisms. The integration of a finite time parameter identification algorithm ensures on the one-hand convergence within finite time (in contrast to asymptotic behavior of e.g., least-squares-based estimators) and renders the necessity to extract the equivalent output injection term from the discontinuous observer correction term obsolete.

The general advantage to provide the driver a feedback regarding the actual state of the parameter estimation quality ensures that the overall identification process is time- and therefore cost-efficient.

Within the context of vehicle handling evaluation the obtained results from both simulations and experiments are promising and encourage for further investigations, when being applied in-vehicle. Even though a preliminary method validation based on experimental work could already be performed the support of the mechanisms (with regards to model-based handling assessment) on a daily basis needs further work and accuracy evaluation.

Acknowledgment

The authors would like to thank Prof. Fridman, Prof. Moreno and Prof. Ferrara for their interest in the work and inspiring discussions.

References

[1] U. Seiffert and G. Rainer, *Virtuelle Produktentstehung für Fahrzeug und Antrieb im Kfz*, 1st ed. Vieweg+Teubner Verlag, 2008.

[2] B. Heißing and M. Ersoy, *Chassis Handbook – Fundamentals, Driving Dynamics, Components, Mechatronics, Perspectives*, 1st ed. Vieweg + Teubner Verlag | Springer Fachmedien Wiesbaden GmbH, 2011.

[3] A. Zomotor, *Fahrwerktechnik: Fahrverhalten*. Vogel Buchverlag, 1991.

[4] R. Tafner, "Observer-based parameter identification techniques: a framework for vehicle handling assessment," Ph.D. dissertation, Graz University of Technology, 2015.

[5] D. Chen, "Subjective and objective vehicle handling behaviour," Ph.D. dissertation, University of Leeds, School of Mechanical Engineering, 1997.

[6] A. Matsishita, K. Takanami, N. Takeda, and M. Takahashi, "Subjective evaluation and vehicle behavior in lane-change maneuvers," *SAE Technical Paper 800845*, 1980, Society of Automotive Engineers.

[7] C. Kraft, *Gezielte Variation und Analyse des Fahrverhaltens von Kraftfahrzeugen mittels elektrischer Linearaktuatoren im Fahrwerksbereich*, ser. Karlsruher Schriftenreihe Fahrzeugsystemtechnik. KIT Scientific Publ., 2011.

[8] ISO4138, "Passenger cars – steady-state circular driving bahaviour – open-loop test methods," International Organization for Standardization, 2012.

[9] ISO7401, "Road vehicles – lateral transient response test methods," International Organization for Standardization, 1988.

[10] K. Rompe and B. Heißing, *Objektive Testverfahren für die Fahreigenschaften von Kraftfahrzeugen*. Verlag TÜV Rheinland GmbH, Köln, 1984.

[11] ISO/TR 3888:1975, "Road vehicles – test procedure for a severe lane-change manoeuvre," International Organization for Standardization, 1999.

[12] M. Nybacka, X. He, Z. Su, L. Drugge, and E. Bakker, "Links between subjective assessments and objective metrics for steering, and evaluation of driver ratings," *Vehicle System Dynamics*, vol. 52, pp. 31–50, 2014.
[13] H. Totutake, Y. Sugimoto, and T. Shirakata, "Real-time identification method of driver model with steering manipulation," *Vehicle System Dynamics*, vol. 51, pp. 109–121, 2013.
[14] C. Schimmel, "Entwicklung eines fahrerbasierten Werkzeugs zur Objektivierung subjektiver Fahreindrücke," Ph.D. dissertation, Technische Universität München, 2010.
[15] D. Meljnikov, "Entwicklung von Modellen zur Bewertung des Fahrverhaltens von Kraftfahrzeugen," Ph.D. dissertation, Fakultät für Maschinenbau, Universität Stuttgart, 2003.
[16] Z. A. Zomotor, "Online-Identifikation der Fahrdynamik zur Bewertung des Fahrverhaltens von Pkw," Ph.D. dissertation, Institut A für Mechanik der Universität Stuttgart, 2002.
[17] D. Schramm, M. Hiller, and R. Bardini, *Vehicle Dynamics: Modeling and Simulation*. Springer, Berlin, 2014.
[18] T. D. Gillespie, *Fundamentals of Vehicle Dynamics*, 1st ed. Society of Automotive Engineers, Inc., 1992.
[19] H. B. Pacejka and I. J. M. Besselink, "Magic formula tyre model with transient properties," *Vehicle System Dynamics*, vol. 27, sup001, pp. 234–249, 1997.
[20] W. Hirschberg, "TM_simple 4.0: a simple to use tyre model," Graz University of Technology, 2009.
[21] W. Hirschberg, G. Rill, and H. Weinfurter, "Tire model TMeasy," *Vehicle System Dynamics*, vol. 45, sup1, pp. 101–119, 2007.
[22] R. Rajamani, *Vehicle Dynamics and Control*, 2nd ed., Mechanical Engineering Series. Springer, New York, 2012.
[23] H. Dahmani, M. Chadli, A. Rabhi, and A. El Hajjaji, "Road bank angle considerations for detection of impending vehicle rollover," in *Sixth IFAC Symposium Advances in Automotive Control*, 2010, pp. 31–36.
[24] M. Doumiati, A. Victorino, D. Lechner, G. Baffet, and A. Charara, "Observers for vehicle tyre/road forces estimation: experimental validation," *Vehicle System Dynamics*, vol. 48, no. 11, pp. 1345–1378, 2010.
[25] F. Saglam and Y. S. Unlusoy, "Identification of low order vehicle handling models from multibody vehicle dynamics models," in *IEEE International Conference on Mechatronics (ICM)*, 2011, pp. 96–101.
[26] B. C. Hamblin, R. D. Martini, J. T. Cameron, and S. N. Brennan, "Low-order modeling of vehicle roll dynamics," in *Proceedings of American Control Conference*, 2006, pp. 4008–4015.
[27] T. Shim and G. Chinar, "Understanding the limitations of different vehicle models for roll dynamics studies," *Vehicle System Dynamics: International Journal of Vehicle Mechanics and Mobility*, vol. 45, no. 3, pp. 191–216, 2008.
[28] D. Ammon, *Modellbildung und Systementwicklung in der Fahrzeugdynamik*. B.G. Teubner Stuttgart, 1997.

[29] R. Rajamani, D. Piyabongkarn, V. Tsourapas, and J. Y. Lew, "Parameter and state estimation in vehicle roll dynamics," *IEEE Transactions on Intelligent Transportation Systems*, vol. 12, no. 4, pp. 1558–1567, 2011.

[30] A. G. Nalecz, "Application of sensitivity methods to analysis and synthesis of vehicle dynamic systems," *Vehicle System Dynamics*, vol. 18, no. 1–3, pp. 1–44, 1989.

[31] H. K. Khalil, *Nonlinear Systems*, 3rd ed. Upper Saddle River, NJ: Prentice-Hall, 2002.

[32] N. F. Benninger and J. Rivoir, "Ein neues konsistentes Maß zur Beurteilung der Steuerbarkeit in linearen, zeitinvarianten Systemen," *Regelungstechnik*, vol. 34, no. 12, pp. 473 – 479, 1986.

[33] R. Isermann and M. Münchhof, *Identification of Dynamic Systems*. Springer Verlag, Berlin, 2011.

[34] C. Kobetz, "Modellbasierte Fahrdynamikanalyse durch ein an Fahrmanövern parameteridentifiziertes querdynamisches Simulationsmodell," Ph.D. dissertation, Fakultät für Maschinenbau, Technische Universität Wien, 2003.

[35] C. I. W. G. (CIWG), "A culminating advance in the theory and practice of data fusion, filtering, and decentralized estimation," 1997.

[36] R. E. Kalman, "A new approach to linear filtering and prediction problems," *Transactions of the ASME–Journal of Basic Engineering*, vol. 82, no. Series D, pp. 35–45, 1960.

[37] L. Ljung, *System Identification: Theory for the User*, 2nd ed. Upper Saddle River, NJ: Prentice-Hall Ptr, 1999.

[38] R. E. Kalman, "Lectures on controllability and observability," in *Controllability and Observability*, ser. C.I.M.E. Summer Schools, E. Evangelisti, Ed. Springer, Berlin, 2011, vol. 46, pp. 1–149.

[39] R. Hermann and A. J. Krener, "Nonlinear controllability and observability," *IEEE Transactions on Automatic Control*, vol. 22, no. 5, pp. 728–740, 1977.

[40] J. P. Gauthier and G. Bornard, "Observability for any $u(t)$ of a class of nonlinear systems," *IEEE Transactions on Automatic Control*, vol. 26, no. 4, pp. 922–926, 1981.

[41] D. G. Luenberger, "An introduction to observers," *IEEE Transactions on Automatic Control*, vol. 16, no. 6, pp. 596–602, 1971.

[42] M. Zeitz, "The extended Luenberger observer for nonlinear systems," *Systems Control Letters*, vol. 9, no. 2, pp. 149–156, 1987.

[43] J. Birk and M. Zeitz, "Extended Luenberger observer for non-linear multivariable systems," *International Journal of Control*, vol. 47, no. 6, pp. 1823–1836, 1988.

[44] G. Ciccarella, M. Dalla Mora, and A. Germani, "A Luenberger-like observer for nonlinear systems," *International Journal of Control*, vol. 57, no. 3, pp. 537–556, 1993.

[45] D. Simon, *Optimal State Estimation*. NJ, USA: John Wiley & Sons, Inc., 2006.

[46] C. Barrios and Y. Motai, "Improving estimation of vehicle's trajectory using the latest global positioning system with Kalman filtering," *IEEE*

Transactions on Instrumentation and Measurement, vol. 60, no. 12, pp. 3747–3755, 2011.

[47] D. Hodgson, B. C. Mecrow, S. M. Gadoue, H. J. Slater, P. G. Barras, and D. Giaouris, "Effect of vehicle mass changes on the accuracy of Kalman filter estimation of electric vehicle speed," *IET Electrical Systems in Transportation*, vol. 3, no. 3, pp. 67–78, 2013.

[48] S. J. Julier and J. K. Uhlmann, "A new extension of the Kalman filter to nonlinear systems," in *Proceedings of AeroSense: The 11th International Symposium on Aerospace/Defence Sensing, Simulation and Controls*, vol. 3, 1997.

[49] B. La Scala, R. Bitmead, and M. James, "Conditions for stability of the extended Kalman filter and their application to the frequency tracking problem," *Mathematics of Control, Signals and Systems*, vol. 8, no. 1, pp. 1–26, 1995.

[50] K. Reif, S. Gunther, E. Yaz, and R. Unbehauen, "Stochastic stability of the discrete-time extended Kalman filter," *IEEE Transactions on Automatic Control*, vol. 44, no. 4, pp. 714–728, 1999.

[51] K. Rapp, "Nonlinear estimation and control in the iron ore pelletizing process: an application and analysis of the extended Kalman filter," Ph.D. dissertation, Norwegian University of Science and Technology, 2004.

[52] K. K. Tønne, "Stability analysis of EKF-based attitude determination and control," Master's thesis, Department of Engineering Cybernetics, Norwegian University of Science and Technology, 2007.

[53] J. Kim, "Identification of lateral tyre force dynamics using an extended Kalman filter from experimental road test data," *Control Engineering Practice*, vol. 17, no. 3, pp. 357–367, 2009.

[54] M. C. Best, T. J. Gordon, and P. J. Dixon, "An extended adaptive Kalman filter for real-time state estimation of vehicle handling dynamics," *Vehicle System Dynamics*, vol. 34, no. 1, pp. 57–75, 2000.

[55] S. Antonov, A. Fehn, and A. Kugi, "Unscented Kalman filter for vehicle state estimation," *Vehicle System Dynamics*, vol. 49, no. 9, pp. 1497–1520, 2011.

[56] M. Doumiati, A. Victorino, A. Charara, and D. Lechner, "Onboard real-time estimation of vehicle lateral tire-road forces and sideslip angle," *IEEE/ASME Transactions on Mechatronics*, vol. 16, no. 4, pp. 601–614, 2011.

[57] M. Doumiati, A. Victorino, A. Charara, and D. Lechner, "Lateral load transfer and normal forces estimation for vehicle safety: experimental test," *Vehicle System Dynamics*, vol. 47, no. 12, pp. 1511–1533, 2009.

[58] G. Besançon, *Nonlinear Observers and Applications*, Lecture Notes in Control and Information Sciences. Springer, Berlin, 2007, vol. 363.

[59] A. Poznyak, "Deterministic output noise effects in sliding mode observation," *Variable Structure Systems: From Principles to Implementation*, pp. 45–79, London, UK: The Institution of Electrical Engineers, 2004.

[60] H. Imine, L. Fridman, H. Shraim, and M. Djemai, *Sliding Mode based Analysis and Identification of Vehicle Dynamics*, Lecture Notes in Control and Information Sciences. Springer Verlag, 2011, vol. 414.

[61] G. Baffet, A. Charara, and D. Lechner, "Experimental evaluation of a sliding mode observer for tire–road forces and an extended Kalman filter for vehicle sideslip angle," in *46th IEEE Conference on Decision and Control*, 2007, pp. 3877–3882.

[62] H. Shraim, M. Ouladsine, and L. Fridman, "Vehicle parameter and states estimation via sliding mode observers," in *Modern Sliding Mode Control Theory*, ser. Lecture Notes in Control and Information Sciences. Springer, Berlin, 2008, vol. 375, pp. 345–362.

[63] L. Menhour, D. Lechner, and A. Charara, "Embedded unknown input sliding mode observer to estimate the vehicle roll and road bank angles: experimental evaluation," in *Seventh IFAC Symposium on Intelligent Autonomous Vehicles (IAV 2010)*, Italy, 2010, pp. 623–628.

[64] H. Imine, L. Fridman, and T. Madani, "Identification of vehicle parameters and estimation of vertical forces," *International Journal of Systems Science*, vol. 46, no. 16, pp. 2996–3009, 2015.

[65] L. Ljung, "Asymptotic behavior of the extended Kalman filter as a parameter estimator for linear systems," *IEEE Transactions on Automatic Control*, vol. 24, no. 1, pp. 36–50, Feb. 1979.

[66] R. Van der Merwe and E. A. Wan, "The square-root unscented Kalman filter for state and parameter-estimation," in *IEEE International Conference on Acoustics, Speech, and Signal Processing*, vol. 6, 2001, pp. 3461–3464.

[67] C. Cheng and D. Cebon, "Parameter and state estimation for articulated heavy vehicles," *Vehicle System Dynamics*, vol. 49, no. 1–2, pp. 399–418, 2011.

[68] S. Haykin, *Kalman Filtering and Neural Networks*. NJ, USA: John Wiley & Sons, Inc., 2001.

[69] T. A. Wenzel, K. J. Burnham, M. V. Blundell, and R. A. Williams, "Dual extended Kalman filter for vehicle state and parameter estimation," *Vehicle System Dynamics*, vol. 44, no. 2, pp. 153–171, 2006.

[70] S. Lee, K. Nakano, and M. Ohori, "On-board identification of tyre cornering stiffness using dual Kalman filter and GPS," *Vehicle System Dynamics*, vol. 53, no. 4, pp. 437–448, 2015.

[71] S. Hong, C. Lee, F. Borrelli, and J. K. Hedrick, "A novel approach for vehicle inertial parameter identification using a dual Kalman filter," *IEEE Transactions on Intelligent Transportation Systems*, vol. 16, no. 1, pp. 151–161, 2015.

[72] V. A. Bavdekar, A. P. Deshpande, and S. C. Patwardhan, "Identification of process and measurement noise covariance for state and parameter estimation using extended Kalman filter," *Journal of Process Control*, vol. 21, no. 4, pp. 585–601, 2011.

[73] M. Saha, R. Ghosh, and B. Goswami, "Robustness and sensitivity metrics for tuning the extended Kalman filter," *IEEE Transactions on Instrumentation and Measurement*, vol. 63, no. 4, pp. 964–971, 2014.

[74] M. L. J. Hautus, "Strong detectability and observers," *Linear Algebra and Its Applications*, vol. 50, no. 0, pp. 353–368, 1983.

[75] J. A. Moreno and D. Dochain, "Global observability and detectability analysis of uncertain reaction systems and observer design," *International Journal of Control*, vol. 81, no. 7, pp. 1062–1070, 2008.
[76] V. Utkin, *Sliding Modes in Control and Optimization*. Springer-Verlag, Berlin, 1992.
[77] B. Walcott and S. Zak, "State observation of nonlinear uncertain dynamical systems," *IEEE Transactions on Automatic Control*, vol. 32, no. 2, pp. 166–170, 1987.
[78] C. Edwards and S. K. Spurgeon, "On the development of discontinuous observers," *International Journal of Control*, vol. 59, no. 5, pp. 1211–1229, 1994.
[79] S. K. Spurgeon, "Sliding mode observers: a survey," *International Journal of Systems Science*, vol. 39, no. 8, pp. 751–764, 2008.
[80] B. Draženović, "The invariance conditions in variable structure systems," *Automatica*, vol. 5, no. 3, pp. 287–295, 1969.
[81] T. Floquet, C. Edwards, and S. K. Spurgeon, "On sliding mode observers for systems with unknown inputs," in *Ninth International Workshop on Variable Structure Systems (VSS)*, 2006, pp. 214–219.
[82] C. Edwards, S. K. Spurgeon, and R. J. Patton, "Sliding mode observers for fault detection and isolation," *Automatica*, vol. 36, no. 4, pp. 541–553, 2000.
[83] N. Patel, C. Edwards, and S. K. Spurgeon, "Optimal braking and estimation of tyre friction in automotive vehicles using sliding modes," *International Journal of Systems Science*, vol. 38, no. 11, pp. 901–912, 2007.
[84] A. Levant, "Sliding order and sliding accuracy in sliding mode control," *International Journal of Control*, vol. 58, no. 6, pp. 1247–1263, 1993.
[85] J. Davila, L. Fridman, and A. Levant, "Second-order sliding-mode observer for mechanical systems," *IEEE Transactions on Automatic Control*, vol. 50, no. 11, pp. 1785–1789, 2005.
[86] J. Davila, L. Fridman, and A. Poznyak, "Observation and identification of mechanical systems via second order sliding modes," *International Journal of Control*, vol. 79, no. 10, pp. 1251–1262, 2006.
[87] N. K. M'Sirdi, A. Rabhi, L. Fridman, J. Davila, and Y. Delanne, "Second order sliding-mode observer for estimation of vehicle dynamic parameters," *International Journal of Vehicle Design*, vol. 48, no. 3/4, pp. 190–207, 2008.
[88] C. E. Tannoury, F. Plestan, S. Moussaoui, and G. P. Gil, "A variable structure observer for an on-line estimation of a tyre rolling resistance and effective radius," in *Second International Workshop on Variable Structure Systems (VSS)*, 2012, pp. 167–172.
[89] A. Ortiz, J. A. Cabrera, A. J. Guerra, and A. Simon, "An easy procedure to determine Magic Formula parameters," *Vehicle System Dynamics*, vol. 44, no. 9, pp. 689–718, 2006.
[90] H. B. Pacejka, *Tire and Vehicle Dynamics*, 3rd ed. Butterworth-Heinemann, 2012.
[91] Y. Shtessel, C. Edwards, L. Fridman, and A. Levant, *Sliding Mode Control and Observation*. New York: Springer, 2014.

[92] T. Floquet and J.-P. Barbot, "Super twisting algorithm-based step-by-step sliding mode observers for nonlinear systems with unknown inputs," *International Journal of Systems Science*, vol. 38, no. 10, pp. 803–815, 2007.

[93] A. Levant, "Robust exact differentiation via sliding mode technique," *Automatica*, vol. 34, no. 3, pp. 379–384, 1998.

[94] X. Gao, Z. Yu, J. Neubeck, and J. Wiedemann, "Sideslip angle estimation based on input-output linearisation with tire-road friction adaptation," *Vehicle System Dynamics*, vol. 48, no. 2, pp. 217–234, 2010.

[95] J. A. Moreno and E. Guzman, "A new recursive finite-time convergent Parameter Estimation Algorithm," in *Proceedings of the 18th IFAC World Congress*, vol. 18, Milano, Italy, 2011.

[96] J. A. Moreno, "A linear framework for the robust stability analysis of a Generalized Super-Twisting Algorithm," in *Sixth International Conference on Electrical Engineering, Computing Science and Automatic Control, CCE*, 2009, pp. 1–6.

[97] A. Levant, "Higher-order sliding modes, differentiation and output-feedback control," *International Journal of Control*, vol. 76, no. 9–10, pp. 924–941, 2003.

[98] J. A. Moreno and M. Osorio, "Strict Lyapunov functions for the super-twisting algorithm," *IEEE Transactions on Automatic Control*, vol. 57, no. 4, pp. 1035–1040, 2012.

[99] A. Polyakov and A. Poznyak, "Reaching time estimation for super-twisting second order sliding mode controller via Lyapunov function designing," *IEEE Transactions on Automatic Control*, vol. 54, no. 8, pp. 1951–1955, 2009.

[100] V. Utkin, "On convergence time and disturbance rejection of super-twisting control," *IEEE Transactions on Automatic Control*, vol. 58, no. 8, pp. 2013–2017, 2013.

[101] T. Gonzalez, J. A. Moreno, and L. Fridman, "Variable gain super-twisting sliding mode control," *IEEE Transactions on Automatic Control*, vol. 57, no. 8, pp. 2100–2105, 2012.

[102] Y. B. Shtessel, M. Taleb, and F. Plestan, "A novel adaptive-gain supertwisting sliding mode controller: methodology and application," *Automatica*, vol. 48, no. 5, pp. 759–769, 2012.

[103] V. I. Utkin and A. S. Poznyak, "Adaptive sliding mode control with application to super-twist algorithm: equivalent control method," *Automatica*, vol. 49, no. 1, pp. 39–47, 2013.

[104] C. Edwards and Y. Shtessel, "Adaptive dual layer second-order sliding mode control and observation," in *American Control Conference (ACC)*, July 2015, pp. 5853–5858.

[105] R. Tafner, M. Reichhartinger, and M. Horn, "Estimation of tire parameters via second-order sliding mode observers with unknown inputs," in *13th International Workshop on Variable Structure Systems (VSS)*, 2014, pp. 1–6.

[106] V. Utkin, J. Guldner, and J. Shi, *Sliding Mode Control in Electro-Mechanical Systems*, 2nd ed. Taylor & Francis Group, CRC Press, 2009.

Index

absolute rear wheel slip 91
Acceleration Driven Damper Control (ADD) 207
Ackermann angle 129, 130
ACOCAR active suspension 196
active differential 128, 138
active safety devices 37, 41
actuator model 94–6, 100–1, 212–15
adaptive robust state estimation (ARSE) 259–60, 269
adaptive roll angle observation (ARAO) 261–3
advanced control techniques 78
antilock braking systems (ABS) 33, 36–7, 78
  sliding mode low level actuation in 62
    HIL braking system configuration 63–6
    low level control algorithm 67–9
    low level control results 69–71
anti-skid Sure-Brake system 77
AutoSim symbolic multi-body software 108
auxiliary system 55, 103
average speed values 48
axle roll model, validation of 171

bicycle model 140, 142
  and cornering dynamics 126
BikeSim Hypersport motorcycle 108
BikeSim simulation model 111
Bouc–Wen model 213–15, 219–20
brake-based stability control 136–8
brake-based yaw moment control layout 136
Brake-by-Wire (BBW) solution 101
braking control 81–5, 111–14
braking controller design 107–8
braking in the turn test 152–3
braking torque 81
Burckhardt's 41

center of gravity (CoG), of vehicle 122, 234
center of gravity height (COG) 161
chassis roll motion 238–41
chattering 2, 34, 54–5
Chicane test 181–2
closed-loop performance, analysis of 108
  braking control 111–14
  traction control 108–11
Continuous Damping Control (CDC) 195
controlled mechanical system, phase plane of 9–10
controllers assessment 147
  simulation results 148
    braking in the turn test 152–3
    double lane change test 153–4
    sine with dwell test 151–2
    slowly increasing steer test 149–50
    step steer test 150–1
  vehicle specifications and model validation 147–8
control methods overview 205–10
control problem formulation 131–2
control signal of the linear system driven by SMC theory 13
conventional sliding-modes 10–12
Coriolis and centrifugal forces 7

cornering forces 123
  forces linearization 125–6
  tire–road interaction 124–5
cornering stiffness coefficient 125, 137
Covariance Intersection (CI) method 242
cylinder cut-off 80

diffeomorphism 255–6
double lane change test 153–4
Dual Kalman filter 245
Dugoff's models 41

E-4WD vehicles 33–5, 37
effective front cornering stiffness 237
electro-hydraulic brake (EHB) 101
electrohydraulic shock absorber 196
electro-mechanical actuators 78, 197
electro-mechanical brake (EMB) 101
electronic clutch 80
Electronic Control Unit (ECU) 78, 92
Electronic Throttle Body (ETB) 80, 86, 92, 94
equivalent control approach 3, 5–6, 47, 146
Extended Kalman Filter (EKF) 165, 269

Fastest Acceleration/Deceleration Control (FADC) problem 37
feedforward control 142–3
feedforward stabilizing torque 143
Filippov differential inclusion 4
Filippov's method 3–4
first order sliding mode (FOSM) algorithms 144
first order sliding mode (FOSM) controller 54, 58, 68
first-order sliding-modes
  conventional sliding-modes 10–12
  integral sliding mode (ISM) methods 12–14
    for linear systems 14–17
forces linearization 125–6
frequency response function (FRF) 221
full vehicle model 122

Gain-Scheduling technique 58
generalized super-twisting algorithm (GSTA) 253
Groundhook control 208–9

half car model 204–5
heavy vehicles (HVs) 159
  control 171
    sliding mode observer design 171–3
    steering control 174–9
  main risks of accidents related to 160
    jackknifing criteria 160–1
    lane departure criteria 163
    rollover criteria 161–3
  model 167–8
  validation results 179
    lane keeping results 183–7
    rollover avoidance results 181–3
  vehicle model description 166
    model validation 169–71
high-end motorcycles 78
higher order sliding mode control (HOSM) algorithms 34, 55
high-level controller 140–1, 215–18
high order sliding modes (HOSMs) 2, 17–18, 29
historical background 1–2
Hurwitz matrix 11, 15
hybrid control 208–9
hydraulically actuated brakes (HAB) 100–1
hydropulser test bench 213

identified vehicle model (IVM) 269, 272–4
integral sliding manifold 12, 14, 54
integral sliding mode (ISM) methods 12–14, 51–2
  for linear systems 14–17
integral sliding mode approach (ISM) 215

integral sliding mode control design 145–7
integral suboptimal second order sliding mode (ISSOSM) control 54, 57
in-wheel motors (IWM) 147
IPG Carmaker SW 49, 71
ISMC
  control signal of the linear system driven by 17
  linear system state driven by 16
  sliding variables of the linear system driven by 16

jackknifing criteria 160–1
Jacobian matrix 245

Kalman filter (KF) 242–3, 245

Lagrange equations 122, 167
lane departure criteria 163
lane keeping controller diagram 177
lane keeping results 183
lateral cornering stiffness observer 250–4
lateral dynamics 264–6, 274–8
  coupling effects with 241
lateral tire force observer 249–50
lateral vehicle dynamics control via sliding modes generation 121
  controllers assessment 147
    braking in the turn test 152–3
    double lane change test 153–4
    sine with dwell test 151–2
    slowly increasing steer test 149–50
    step steer test 150–1
    vehicle specifications and model validation 147–8
  vehicle modeling and problem formulation 121
    control problem formulation 131–2
    cornering forces 123–6
    dynamic equations 122–3
    vehicle steering properties 128–31
    yaw moment generation 126–8
  vehicle stability control via sliding mode control 139
    feedforward control 142–3
    integral sliding mode control design 145–7
    overall control structure 140–1
    reference model 141–2
    second-order sliding mode (SOSM) control design 143–5
  yaw rate control survey 132
    brake-based stability control 136–8
    torque-based stability control 138–9
    vehicle stability control 132
Lebesgue-measurable noise 26
Linear Parameter Varying (LPV) model 82, 84
linear single track model 122
linear systems 82
  integral sliding modes for, 14–16
linear system state driven by SMC theory 12
linear variable differential transformer (LDTV) 215
Lipschitz bounded uncertainties/disturbances 2
Lipschitz constant 27
load-transfer 80
load transfer ratio (LTR) 163–4, 174, 182
longitudinal force 39
longitudinal friction coefficient 40
longitudinal slip ratio 39
longitudinal tire model 39–44
longitudinal vehicle dynamics control via SMC 33
  modeling 38
    longitudinal tire model 39–44
    vehicle model 44–6
  sliding mode low level actuation in conventional ABS 62–71

tire–road friction and velocity, sliding mode observer for 46–53
traction control (TC) systems 35–6
wheel-slip control, sliding mode applications in 37–8
wheel-slip control, SMC algorithms for 53
higher order sliding mode control 55–6
integral sliding mode control 54–5
integral suboptimal second order sliding mode control 57
suboptimal second order sliding mode control 56
super twisting sliding mode control 57
lower-level controller 215
Luenberger observer 243, 256, 261
Lyapunov equation 21
Lyapunov function 47, 218

"Magic Formula" 41, 236
MF-Tire approach 148
mixed Skyhook-ADD control 207
model-based objective handling evaluation 233–4
MSC BikeSim®simulation environment 108

normalized linear wheel deceleration 97

objective handling evaluation 232–3
observer-based parameter identification for vehicle dynamics assessment 229
chassis roll motion 238–41
model-based objective handling evaluation 233–4
model parameters and sensor setup 241–2
objective handling evaluation 232–3
robust lateral dynamics parameter estimator 248
lateral cornering stiffness observer 250–4
lateral tire force observer 249–50
robust roll dynamics parameter estimator 254
adaptive robust state estimation (ARSE) 259–60
adaptive roll angle observation (ARAO) 261–3
parameter identification (PE) 260–1
robust exact differentiator (RED) 257–8
robust state estimation (RSE) 258–9
simulation-based concept evaluation 263
handling evaluation results 269–74
parameter identification results 264–9
subjective handling evaluation 231–2

Pacejka tire model 165
parameter identification (PE) 41–2, 229–30, 233, 242, 260–1, 263, 281
lateral dynamics 264–7
roll dynamics 267–9
performance-oriented systems 81–2
pneumatic tires 39
Power Driven Damper Control 207
PROSPER software 166, 169–70, 181

quarter car model 36, 87, 201–4

ramp test 182–3, 187
real-time processing algorithm 242
Recursive Least Squares approach 242
ride comfort 191, 202, 204–5, 207–8, 210, 215, 217
robust exact arbitrary order differentiator 26–8
robust exact differentiator (RED) 26, 256, 268

robust lateral dynamics parameter estimator 248
  lateral cornering stiffness observer 250–4
  lateral tire force observer 249–50
robust roll dynamics parameter estimator 254
  adaptive robust state estimation (ARSE) 259–60
  adaptive roll angle observation (ARAO) 261–3
  parameter identification (PE) 260–1
  robust exact differentiator (RED) 257–8
  robust state estimation (RSE) 258–9
roll angular velocity 239
roll center (RC) 238–9
roll dynamics 174, 231, 233, 238–9, 241–2, 255, 260, 267–9, 275, 278–81
rolling resistance 39–40, 45
rollover avoidance results 181
  Chicane test 181–2
  ramp test 182–3
rollover criteria 161
  load transfer ratio (LTR) 163
  static rollover threshold (SRT) 162–3
root mean square (RMS) tracking error 60

safety-oriented systems 81–2
second order sliding mode (SOSM) 2, 17, 19, 25, 34
  control design 143–5
  controllers 78, 91, 102–4
  high order sliding-mode, concept of 17–18
  suboptimal second-order sliding-mode algorithm 24–6
  super-twisting second-order sliding-modes algorithm 21–3
  twisting second-order sliding-modes algorithm 19–21
selective braking 127–8
self-aligning torques 125, 127
semi-active suspension control 193–6
  active suspension system 196–7
  actuator model 212–15
  allocation of the control demand 218–19
  high-level controller 215–18
  low-level controller 219–20
  methods on suspension control 197
    control methods overview 205–10
    half car model 204–5
    quarter car model 201–4
    single degree of freedom (DoF) system 198–201
    vehicle modelling 197–8
  overall control specifications 215
  problem statement 210
  results and discussions 220–5
  vehicle dynamics model 211–12
simulation-based concept evaluation 263
  handling evaluation results 269–74
  parameter identification results 264
    lateral dynamics 264–7
    roll dynamics 267–9
Simulink®-based in-plane motorcycle model 108
sine with dwell test 151–2
single-corner model 44–5, 47, 96
single degree of freedom (DoF) system 198–201
Single-Input Single-Output (SISO) system 3, 5, 13
single-track model 87, 234–5, 250, 269
skid-pad test 148
Skyhook approach 195, 205–10, 221, 223–5
sliding function 21–2
sliding manifold 3, 6–10, 12, 34, 47–8, 53–5, 57–8, 102, 104, 106–7, 143–4, 146, 179
  2-sliding manifold 25
sliding mode control (SMC) 174
sliding mode observer design 171
  states observation 172–3
  vertical forces estimation 173

sliding mode tire–road friction observer 46
- evaluation 49–53
- vehicle velocity and tire–road friction estimation 48–9

sliding motion 1, 3, 6–7, 9, 14, 259
slip–force curve 236–7, 254
slowly increasing steer test 143, 148–50
spark advance dynamics 80
spark-based control 85
sport utility vehicle (SUV) 147, 202, 220
stability control approach, of heavy vehicles 171
- sliding mode observer design 171
  - states observation 172–3
  - vertical forces estimation 173
- steering control 174
  - lane keeping assistance 176–9
  - rollover avoidance 174–6

stability control systems 125, 137–8
static rollover threshold (SRT) 162–3
steering control 174
- lane keeping assistance 176–9
- rollover avoidance 174–6

step steer test 150–1
subjective handling evaluation 231–2
sub-optimal algorithm 2, 107–8
suboptimal second-order sliding-mode (SSOSM) algorithm 24–6, 54, 56–9
super-twisting algorithm 108, 159, 176, 247, 253, 258–61
super-twisting second-order sliding-modes algorithm 21–3
super twisting sliding mode (STSM) control 54, 57, 60
suspension forces 212, 215, 218, 220, 239–40
suspension system 191, 193, 202
- active 196–7
- semi-active 193–6

tire–road friction and velocity, sliding mode observer for 46–53
tire–road interaction 38, 53, 123–5
torque-based stability control 138–9
torque vectoring 128, 138–9
traction and braking, sliding mode control of 77
- braking control 81–5
- braking controller design 107–8
- closed-loop performance, analysis of 108
  - braking control 111–14
  - traction control 108–11
- control-oriented modeling 88
  - complete motorcycle braking dynamics 96–101
  - complete motorcycle traction dynamics 89–96
- SOSM control approach 102–4
- traction control (TC) 85–8
- traction controller design 104–6

traction control (TC) systems 33, 35–6
transmissibility ratio 202–3
transmission model 96
twisting algorithm (TA) 2, 19, 21
twisting second-order sliding-modes algorithm 19–21
2-sliding manifold 25
two-wheeled vehicles, longitudinal dynamics of
- control-oriented modeling 88
  - complete motorcycle braking dynamics 96–101
  - complete motorcycle traction dynamics 89–96

Variable Structure Systems 1, 34
vehicle acceleration estimation 48
Vehicle Control Unit (VCU) 132
vehicle dynamics control (VDC) systems 33, 35, 77, 79, 132
vehicle model 44–6, 166, 197–8
- validation 169–71

vehicle modeling and problem formulation 121

control problem formulation 131–2
cornering forces 123
forces linearization 125–6
tire–road interaction 124–5
dynamic equations 122–3
vehicle steering properties 128–31
yaw moment generation 126
active differential 128
selective braking 127–8
torque vectoring 128
vehicle specifications and model validation 147–8
vehicle stability control (VSC) 127, 132
yaw rate and side slip angle
parallel utilization of, by stability control 135–6
reference and actual values of 134–5
vehicle steering properties 128–31
vehicle velocity estimation 49–51, 86

wheel cornering forces 124
wheel-deceleration control 81
wheel-slip control, SMC algorithms for 53
higher order sliding mode control 55–6
integral sliding mode control 54–5
integral suboptimal second order sliding mode control 57
simulations 58–62
suboptimal second order sliding mode control 56
super twisting sliding mode control 57
wheel-slip thresholds systems 86

yaw amplification factor 130–2
yaw moment generation 126
active differential 128
selective braking 127–8
torque vectoring 128
yaw rate control survey 132
brake-based stability control 136–8
torque-based stability control 138–9
yaw rate step response test 150

Ziegler–Nichols method 58

www.ingramcontent.com/pod-product-compliance
Lightning Source LLC
LaVergne TN
LVHW011916200926
848484LV00002B/644
9781785612091